D1521330

Springer Series in
ADVANCED MICROELECTRONICS 7

Springer
Berlin
Heidelberg
New York
Barcelona
Hong Kong
London
Milan
Paris
Singapore
Tokyo

Springer Series in
ADVANCED MICROELECTRONICS

The Springer Series in Advanced Microelectronics provides systematic information on all the topics relevant for the design, processing, and manufacturing of microelectronic devices. The books, each prepared by leading researchers or engineers in their fields, cover the basic and advanced aspects of topics such as wafer processing, materials, device design, device technologies, circuit design, VLSI implementation, and subsystem technology. The series forms a bridge between physics and engineering and the volumes will appeal to practicing engineers as well as research scientists.

1 **Cellular Neural Networks**
Chaos, Complexity and VLSI Processing
By G. Manganaro, P. Arena, and L. Fortuna

2 **Technology of Integrated Circuits**
By D. Widmann, H. Mader, and H. Friedrich

3 **Ferroelectric Memories**
By J. F. Scott

4 **Microwave Resonators and Filters for Wireless Communication**
Theory, Design and Application
By M. Makimoto and S. Yamashita

5 **VLSI Memory Chip Design**
By K. Itoh

6 **High-Frequency Bipolar Transistors**
Physics, Modelling, Applications
By M. Reisch

7 **Noise in Semiconductor Devices**
Modeling and Simulation
F. Bonani and G. Ghione

Series homepage – http://www.springer.de/phys/books/ssam/

Fabrizio Bonani Giovanni Ghione

Noise in Semiconductor Devices

Modeling and Simulation

With 101 Figures

Springer

Dr. Fabrizio Bonani
Prof. Giovanni Ghione
Dipartimento di Elettronica
Politecnico di Torino
Corso Duca degli Abruzzi, 24
10129 Torino
Italy
bonani@polito.it
ghione@polito.it

Series Editors:

Dr. Kiyoo Itoh
Hitachi Ltd., Central Research Laboratory
1-280 Higashi-Koigakubo
Kokubunji-shi
Tokyo 185-8601
Japan

Professor Takayasu Sakurai
Center for Collaborative Research
University of Tokyo
7-22-1 Roppongi, Minato-ku,
Tokyo 106-8558
Japan

ISSN 1437-0387
ISBN 3-540-66583-8 Springer-Verlag Berlin Heidelberg New York

Library of Congress Cataloging-in-Publication Data applied for.

Die Deutsche Bibliothek - CIP-Einheitsaufnahme
Bonani, Fabrizio: Noise in semiconductor devices : modeling and simulation / Fabrizio Bonani ; Giovanni Ghione. - Berlin ; Heidelberg ; New York ; Barcelona ; Hong Kong ; London ; Milan ; Paris ; Singapore ; Tokyo : Springer, 2001 (Springer series in advanced microelectronics ; 7) (Physics and astronomy online library)
ISBN 3-540-66583-8

Springer-Verlag Berlin Heidelberg New York
a member of BertelsmannSpringer Science+Business Media GmbH

http://www.springer.de

Typesetting by the authors using a Springer TEX macro package

Cover concept by eStudio Calmar Steinen using a background picture from Photo Studio "SONO". Courtesy of Mr. Yukio Sono, 3-18-4 Uchi-Kanda, Chiyoda-ku, Tokyo
Cover design: *design & production* GmbH, Heidelberg

Printed on acid-free paper SPIN: 10676146 57/mf/3141 - 5 4 3 2 1 0

Preface

The design and optimization of electronic systems often requires an appraisal of the electrical noise generated by active devices, and, at a technological level, the ability to properly design active elements in order to minimize, when possible, their noise. Examples of critical applications are, of course, receiver front-ends in RF and optoelectronic transmission systems, but also front-end stages in sensors and, in a completely different context, nonlinear circuits such as oscillators, mixers, and frequency multipliers. The rapid development of silicon RF applications has recently fostered the interest toward low-noise silicon devices for the lower microwave band, such as low-noise MOS transistors; at the same time, the RF and microwave ranges are becoming increasingly important in fast optical communication systems. Thus, high-frequency noise modeling and simulation of both silicon and compound-semiconductor based bipolar and field-effect transistors can be considered as an important and timely topic. This does not exclude, of course, low-frequency noise, which is relevant also in the RF and microwave ranges whenever it is up-converted within a nonlinear system, either autonomous (as an oscillator) or non-autonomous (as a mixer or frequency multiplier).

The aim of the present book is to provide a thorough introduction to the physics-based numerical modeling of semiconductor devices operating both in small-signal and in large-signal conditions. In the latter instance, only the non-autonomous case was considered, and thus the present treatment does not directly extend to oscillators.

As in many other fields, practical, computationally affordable approaches to mathematical modeling appeared many years after the underlying physical foundation was established. In fact, the general framework of the theory of noise modeling of semiconductor devices has been completely defined since the beginning of the 1960s. Since it was recognized that noise is a small-amplitude perturbation of a DC or large-signal AC steady state, amenable to a Langevin-like model, the mathematical framework of noise analysis as a linear, stochastic problem whose solution could be obtained through Green's function techniques has been laid, particularly with the paper by Shockley et al. [1] on the impedance field method (yet another Green's function tecnique) which also had the merit of providing a comparatively simple introduction to the topic at an engineering level.

In the following years (linear) noise device analysis was the object of extensive investigations by several authors; among the main contributors were A. van der Ziel and C. van Vliet. In the meantime, numerical device simulation underwent a dramatic development. Since the end of the 1970s, numerical techniques for the solution of transport equations in a semiconductor were inplemented in computer codes, thus leading to the full development of TCAD (technological computer aided design) tools which are now the basis for device design and optimization in research and production-oriented environments.

Despite this progress, semiconductor device simulation had to wait till the end of the 1980s to see the inclusion of noise analysis in its framework. In 1989 Layman [2] developed a first implementation of a two-dimensional MOSFET noise analysis based on an equivalent network approach, while in the same year a full two-dimensional numerical implementation of the impedance field method within the framework of a monopolar drift-diffusion model was presented by Ghione et al. [3, 4] on the basis of an efficient numerical technique (the so-called adjoint approach) derived from network analysis. Later, the adjoint approach was extended and generalized to the bipolar case by Bonani et al. [5, 6] within the framework of multidimensional models, and a quasi-2D numerical simulator for compound semiconductor field-effect transistors was developed by the Lille University research group [7], on the basis of the active line approach proposed, in its present form, by Cappy and Heinrich in 1989 [8]. Finally, the noise in large-signal, non-autonomous systems has been the object of investigation during the last few years by Bonani et al. [9, 10] and, within a quasi-2D approach, again by Danneville et al. [11].

In the meantime, circuit-oriented noise modeling of active devices continued to be a keystone in low-noise electronic system design. In this context, the development of accurate circuit-oriented models (also called *compact models*) analytically relating, in an exact or approximate way, the device noise performances to geometrical and physical parameters, is becoming increasingly important for field-effect transistors, particularly MOSFETs for RF applications. Whereas noise in bipolar devices is, at least in low-injection conditions, intrinsically shot-like, field-effect transistors exhibit a much more complex behavior, which is a challenge both from the standpoint of modeling and of device optimization. The number of technical publications that appeared during the last few years in the field of compact noise modeling is too large to allow individual contributions to be singled out; a good introduction to the subject can be found in [12].

The book is structured as follows. In the first chapter, an introduction to fluctuation phenomena in semiconductors is provided from the standpoint of materials and devices. The second chapter introduces physics-based device models (and in particular the drift-diffusion model) and gives a detailed treatment of the Langevin approach to noise simulation, including the Green's function solution technique. The last part of the chapter is devoted to analytical applications of the Green's function approach to field-effect and bipolar devices so as to obtain compact device noise models. The third chapter presents a review of the numerical techniques for device simulation in DC and small-signal operation, including the efficient evaluation of the Green's functions required for noise analysis. Some results and applications to several device classes ranging from the simple bulk resistor to transistors are discussed in Chap. 4. Chapter 5 finally deals with the topic of noise simulation in forced, nonlinear, (quasi-) periodic device operation. Appendix A reviews some basic concepts in probability and random process theory relevant to noise analysis.

The authors wish to acknolwedge the contribution of several colleagues who have actively cooperated to the development of research in the field of physics-based numerical noise modeling. Fabio Filicori of the University of Bologna provided the first seminal idea of efficient numerical noise simulation by suggesting the use of adjoint approaches derived from network theory; this led in 1989 to the implementation of the first monopolar two-dimensional simulation code within the framework of the graduation thesis work of Enrico Bellotti, now at Boston University. Mark R. Pinto and R. Kent Smith (of Bell Laboratories) played a fundamental role in the development and implementation of a bipolar noise model into the PADRE multidimensional device simulator; M. Ashraf Alam of Bell Laboratories cooperated in the application of the PADRE noise simulator to MOSFET devices with Simona Donati Guerrieri of Politecnico di Torino, who also developed the large-signal noise simulation code as part of her PhD thesis. Finally, the cooperation of Marco Pirola of Politecnico di Torino to the numerical noise simulation project throughout most of its stages is gratefully acknowledged, and so are helpful discussions with a number of colleagues and researchers involved in device simulation and modeling, among which we can quote Alain Cappy of Lille University, Lino Reggiani of University of Lecce, Luca Varani and Jean-Pierre Nougier of Montpellier University, and Carlo Naldi of Politecnico di Torino. We also acknowledge the valuable help of Claudio Beccari of Politecnico di Torino in overcoming several LaTeX formatting difficulties.

Torino
April 2001

Fabrizio Bonani
Giovanni Ghione

Contents

List of Symbols

a	Equivalent channel thickness (MESFET and JFET)
a_n	Real part of complex factor for the generalized electron diffusion length
a_p	Real part of complex factor for the generalized hole diffusion length
A	Subset of the set of events
$\boldsymbol{A}$	First Fokker–Planck moment
$\boldsymbol{A}$	Interaction term (of the Hamiltonian operator)
A	Matrix representing the discretized system equations for Green's functions
$\overline{A}$	Complement of set A
$A \cup B$	Union of sets A and B
$A \cap B$	Intersection of sets A and B
$\mathsf{A}_{\alpha\beta}$	Matrix representing the linearized and discretized (frequency domain) small-signal system (output in nodal variables α, input in nodal variables β)
b	Junction electron/hole mobility ratio
b	Matrix representing the discretized forcing term for Green's functions
b_n	Imaginary part of complex factor for the generalized electron diffusion length
b_p	Imaginary part of complex factor for the generalized hole diffusion length
B	Subset of the set of events
$B(\cdot)$	Bernoulli function
B	Second Fokker–Planck moment
B_c	Correlation susceptance
B_g	Input generator susceptance
B_o	Optimum input generator susceptance
c	Free carrier density (electron or hole)
C	Conversion matrix
C	FET noise model dimensionless correlation parameter

C Capacitance conversion matrix
C_{ch} Equivalent channel capacitance (per unit surface)
C_{ox} Oxide capacitance (per unit surface)
$\boldsymbol{C}_{\mathrm{D}}^{+}$ Net equivalent ionized donor concentration, all levels
$C_{\mathrm{D}k}^{+}$ Net equivalent ionized donor concentration, level k
$C_{\mathrm{D}ki}^{+}$ Discretized (nodal) net equivalent ionized donor concentration, level k
C_{GS} Gate-source small-signal capacitance
$C_{X,Y}$ Correlation coefficient of random variables/stochastic processes X and Y
$C_{\delta n,\delta p}$ Electron–hole fluctuations correlation coefficient

$\det\{\cdot\}$ Determinant
D Spectral representation of LS time derivative
D^{ss} Spectral representation of SSLS time derivative
D_0 Equilibrium diffusion coefficient (diffusivity)
D_n Electron diffusion coefficient
D_{n} Noise tensor diffusion coefficient
D_{n}' Complex tensor diffusion coefficient
D_p Hole diffusion coefficient
D_{s} Spreading tensor diffusion coefficient

e Base of natural logarithm
e, e_{s} Applied voltage
$\boldsymbol{e}$ Column unit vector
e Constant matrix exploited in the generalized adjoint approach
$e_{\mathrm{s},k}$ Voltage source for contact k
$\exp(\cdot)$ Exponential function
E Energy level
E_{c} Conduction band edge
E_{t} Trap energy level
E_{v} Valence band edge
E_{F} Fermi level
E_{Fi} Intrinsic Fermi level
$\boldsymbol{\mathcal{E}}$ Electric field

f Frequency
f General function (scalar field)
$\overline{f(t)}$ Time average of $f(t)$
$\langle f\rangle$ Ensemble average (expected value) of f
$f(\gamma, t)$ Diagonal part of the reduced density operator
f^{*} Conjugate part of f

$\boldsymbol{f}^{\dagger}$	Hermitian conjugate of $\boldsymbol{f}$
$\dot{f}$	Time derivative of $f(t)$
$\tilde{f}$	Fourier transform of $f(t)$
f_{h}	Corner frequency for $1/f$ noise spectrum
f_i	Discretized (nodal) value of function f
$f_{g_{\mathrm{m}k}}$	Nonlinear transconductace constitutive relationship (branch k)
$f_{\boldsymbol{r}}$	Position probability density
$f_{\boldsymbol{r},\boldsymbol{k}}$	Free electron distribution function
$f_{\boldsymbol{r},\mathrm{s}}$	Position stationary probability density
$f_{\boldsymbol{r},\boldsymbol{v}}$	Position and velocity joint probability density
$f_{\boldsymbol{r},\boldsymbol{v},\mathrm{s}}$	Position and velocity stationary joint probability density
$f_{\boldsymbol{v}}$	Velocity probability density
$f_{\boldsymbol{v},\mathrm{s}}$	Velocity stationary probability density
f_{C_k}	Nonlinear capacitance constitutive relationship (branch k)
f_{L_k}	Nonlinear inductance constitutive relationship (branch k)
$f_{\boldsymbol{N}}$	Electron level populations joint probability density
f_{R_k}	Nonlinear resistance constitutive relationship (branch k)
f_T	FET cutoff frequency
f_X	Probability density of random variable/stochastic process X
$f_{X,\mathrm{s}}$	Stationary probability density of stochastic process X
$f_{X,Y}$	Joint probability density of random variables/stochastic processes X and Y
$f_{X\|Y}$	Conditional probability density of random variables/stochastic processes X and Y
$\boldsymbol{F}$	Formal representation of the PDE model equations
$\boldsymbol{F}$	General vector field
$\boldsymbol{F}(t)$	External applied field
F	Noise figure
$F_{\min}$	Minimum noise figure
$\langle \boldsymbol{F} \cdot \hat{n} \rangle_{ij}$	Discretized average of the normal component of $\boldsymbol{F}$ along side ij
$(\boldsymbol{F} \cdot \hat{n})_{ij}$	Constant value of the discretized average of the normal component of $\boldsymbol{F}$ along side ij
F_n	n-th order Fokker–Planck moment
F_n	Formal representation of the electron continuity equation
F_p	Formal representation of the hole continuity equation
F_X	Distribution function of random variable/stochastic process X
$F_{X,Y}$	Joint distribution function of random variables/stochastic processes X and Y
$F_{X\|Y}$	Conditional distribution function of random variables/stochastic processes X and Y
F_{φ}	Formal representation of Poisson continuity equation

$\mathcal{F}_{1/2}$	Fermi function of order 1/2
$g(\cdot), \boldsymbol{g}(\cdot)$	Deterministic functions
$g(\cdot,\cdot)$	Auxiliary function exploited in the derivation of the Sharfetter–Gummel discretization scheme
g, G	Conductance
g_m	Saturation transconductance
g_n	Noise conductance, series representation
G_k	k-th spectral component of periodic function $g(t)$
G	Channel conductance per unit length
G_ass	Optimum associated gain
G_c	Correlation conductance
$\tilde{\mathsf{G}}_\mathrm{e}$	Matrix discretized contact potential scalar Green's function
$\tilde{\boldsymbol{G}}_{\mathrm{e},n}$	Discretized electron contact potential scalar Green's function
$\tilde{\boldsymbol{G}}_{\mathrm{e},n}^{\mathrm{adj}}$	Discretized adjoint electron contact potential scalar Green's function
$\tilde{\boldsymbol{G}}_{\mathrm{e},p}$	Discretized hole contact potential scalar Green's function
G_g	Input generator conductance
G_k	Internal conductance of real current generator feeding contact k
G_n	Electron generation rate
$\tilde{\boldsymbol{G}}_n$	Discretized scalar Green's function (monopolar model)
G_n	Noise conductance, parallel representation
$\tilde{\boldsymbol{G}}_n^{\mathrm{adj}}$	Discretized adjoint scalar Green's function (monopolar model)
G_n^{dir}	Electron generation rate, band-to-band process
G_{np}	Short-circuit electron current scalar Green's function
$\boldsymbol{G}_{np}$	Short-circuit electron current vector Green's function
G_n^{srh}	Electron generation rate, trap-assisted process
G_o	Optimum input generator conductance
G_p	Hole generation rate
G_p^{dir}	Hole generation rate, band-to-band process
G_{pn}	Short-circuit hole current scalar Green's function
$\boldsymbol{G}_{pn}$	Short-circuit hole current vector Green's function
G_p^{srh}	Hole generation rate, trap-assisted process
G_DS	Drain small-signal conductance
G_DS0	Drain small-signal conductance for $v_\mathrm{DS} = 0$
G_G	Saturation gate small-signal conductance
G_Q	Gate charge scalar Green's function
$\boldsymbol{G}_Q$	Gate charge vector Green's function
$\hat{G}_{X_1,X_2}$	(Bilateral) correlation spectrum of non-stationary stochastic processes $X_1(t)$ and $X_2(t)$

$G_{\alpha,\beta}$	Scalar Green's function (injection in equation β, observation in variable α)
$\boldsymbol{G}_{\alpha,\beta}$	Vector Green's function (injection in equation β, observation in variable α)
$G_{\varphi,k}$	Potential scalar Green's function (injection in level k rate equation)
$\tilde{\mathsf{G}}_{\varphi\mathrm{i}}$	Matrix discretized potential scalar Green's function, internal nodes
$G_{\varphi,n}$	Potential scalar Green's function (injection in electron continuity equation)
$\boldsymbol{G}_{\varphi,n}$	Discretized potential scalar Green's function (injection in electron continuity equation)
$G_{\varphi,p}$	Potential scalar Green's function (injection in hole continuity equation)
$\boldsymbol{G}_{\varphi,p}$	Discretized potential scalar Green's function (injection in hole continuity equation)
$\tilde{\mathsf{G}}_{\varphi\mathrm{x}}$	Matrix discretized potential scalar Green's function, external nodes
$G_{\varphi,\alpha}$	Scalar Green's function (injection in equation α, observation in potential)
$\boldsymbol{G}_{\varphi,\alpha}$	Vector Green's function (injection in equation α, observation in potential)
$\tilde{\boldsymbol{G}}_{\varphi,\alpha\mathrm{i}}$	Discretized scalar Green's function, internal nodes (injection in equation α, observation in potential)
$\tilde{\boldsymbol{G}}^{\mathrm{adj}}_{\varphi,\alpha\mathrm{i}}$	Discretized adjoint scalar Green's function, internal nodes (injection in equation α, observation in potential)
$\tilde{\boldsymbol{G}}_{\varphi,\alpha\mathrm{x}}$	Discretized scalar Green's function, external nodes (injection in equation α, observation in potential)
$\boldsymbol{G}_{\varphi,\varphi}$	Discretized potential scalar Green's function (injection in Poisson equation)
$\hbar$	Reduced Planck constant
$h(t)$	Impulse response of a time-invariant scalar linear system
$\mathsf{h}(t,u)$	Matrix impulse response of a linear system
H	Hamiltonian operator
H_0	Noninteracting electron/phonon Hamiltonian operator
i	Imaginary unit
i	Current
$\tilde{\boldsymbol{i}}^{\mathrm{ss}}_{\mathrm{c}}$	Small-signal contact currents
$i_{\mathrm{c},k}$	Current entering contact k
$i^{\mathrm{ss}}_{\mathrm{c}k}$	Complex SSLS current (branch k)
i_i	Total current entering contact node i
$\tilde{i}_{ij}$	Current generator injected between network nodes i and j
i_k	Current in branch k

i_k^{adj}	Adjoint branch current (branch k)
i_k^{ss}	SSLS current in branch k
i_{l}	Particle DC flow from left to right
i_{r}	Particle DC flow from right to left
i_{s}	Applied current
$i_{\mathrm{s},k}$	Current source for contact k
i_{B}	Base DC current
i_{C}	Collector DC current
i_{D}	Drain DC current
i_{D}	Diode DC total current
$i_{\mathrm{D,s}}$	Saturation drain DC current
i_{E}	Emitter DC current
i_{S}	Diode inverse saturation current
$\boldsymbol{i}_0, \tilde{\boldsymbol{\imath}}_0$	Short-circuit port currents
$\boldsymbol{I}_k$	Spectral components of LS current $i_k(t)$
$\boldsymbol{I}^{\mathrm{ss},+}$	Current upper sideband spectral amplitudes
$\boldsymbol{I}_k^{\mathrm{ss},+}$	Current upper sideband spectral amplitudes (branch k)
I	Identity matrix
$\mathrm{Im}\{\cdot\}$	Imaginary part
I_N	Identity matrix, dimension $N \times N$
$\boldsymbol{j}, \boldsymbol{J}$	Current density
$\boldsymbol{J}_n$	Electron current density
$\boldsymbol{J}_{n\mathrm{t}}$	Total electron current density
$\boldsymbol{J}_p$	Hole current density
$\boldsymbol{J}_{p\mathrm{t}}$	Total hole current density
$\boldsymbol{J}_{\mathrm{t}}$	Total current density
J	Jacobian matrix
k_{B}	Boltzmann constant
K_{f}	Fukui noise factor
$\mathsf{K}_{\boldsymbol{s},\boldsymbol{s}}$	Microscopic local noise source
$K_{X,Y}$	Covariance function of random variables/stochastic processes X and Y
$\mathsf{K}_{\boldsymbol{X},\boldsymbol{Y}}$	Covariance function of random variables/stochastic processes $\boldsymbol{X}$ and $\boldsymbol{Y}$
$K_{\gamma_{\mathrm{D}},\gamma_{\mathrm{D}}}$	Fundamental trap-level local noise source for GR noise
$K_{\gamma_{\mathrm{D}},\gamma_p}$	Fundamental trap level–hole correlation local noise source for GR noise
K_{γ_i,γ_j}	Fundamental local noise source for GR noise (electron populations, levels i and j)
K_{γ_n,γ_n}	Electron fundamental local noise source for GR noise

$\mathsf{K}_{\gamma_n,\gamma_n}$	Electron fundamental local noise source for GR noise in LS operation
$K_{\gamma_n,\gamma_\mathrm{D}}$	Electron–trap level correlation fundamental local noise source for GR noise
K_{γ_n,γ_p}	Electron–hole correlation fundamental local noise source for GR noise
$\mathsf{K}_{\gamma_n,\gamma_p}$	Electron–hole correlation fundamental local noise source for GR noise in LS operation
K_{γ_p,γ_p}	Hole fundamental local noise source for GR noise
$\mathsf{K}_{\gamma_p,\gamma_p}$	Hole fundamental local noise source for GR noise in LS operation
$K_{\delta n,\delta n}$	Electron equivalent local noise source for GR noise
$K_{\delta n,\delta p}$	Electron–hole correlation equivalent local noise source for GR noise
$K_{\delta p,\delta p}$	Hole fundamental local noise source for GR noise
$\mathsf{K}_{\delta \boldsymbol{J}_n,\delta \boldsymbol{J}_n}$	Electron current density (diffusion, approximate equivalent GR and $1/f$ noise) local noise source
$\mathsf{K}_{\delta \boldsymbol{J}_n,\delta \boldsymbol{J}_p}$	Electron–hole correlation current density approximate equivalent local noise source for GR noise
$\mathsf{K}_{\delta \boldsymbol{J}_p,\delta \boldsymbol{J}_p}$	Hole current density (diffusion, approximate equivalent GR and $1/f$ noise) local noise source
l_{ij}	Length of the box side between nodes i and j
L	Length
L	Channel length
L	Inductance conversion matrix
L_n	Electron diffusion length
$\boldsymbol{L}_n$	Rate equations linearized with respect to n
$\tilde{L}_n$	Generalized electron diffusion length
L_p	Hole diffusion length
$\boldsymbol{L}_p$	Rate equations linearized with respect to p
$\tilde{L}_p$	Generalized hole diffusion length
$\boldsymbol{L}_\varphi$	Rate equations linearized with respect to φ
m^*	Effective mass
m_X	Mean of random variable/stochastic process X
M	Phenomenological relaxation matrix
n	Electron density
$\hat{n}$	External normal versor
n_i	Discretized (nodal) electron density
n_i	Intrinsic concentration
$\tilde{\boldsymbol{n}}_\mathrm{i}^\mathrm{ss}$	Small-signal discretized electron density, internal nodes
n_p	Minority electron density

n'_p	Minority electron excess density
n_{pe}	Equilibrium minority electron density
n'_{p0}	DC minority electron excess density
n_{s}	Schottky contact equilibrium electron density
n^{ss}	Small-signal electron density
n_{t}	Trapped electron density
$n_{\mathrm{t}0}$	DC trapped electron density
$\tilde{\boldsymbol{n}}_{\mathrm{x}}^{\mathrm{ss}}$	Small-signal discretized electron density, external nodes
n_0	DC electron density
$\boldsymbol{n}_\zeta$	Electron states occupation number
N	Number of device ports
N	Total particle number
$N(t)$	Number of particles collected at time t
N	Total electron number (population)
N	Total channel electron density per unit surface
$\boldsymbol{N}$	Collection of total electron populations (all levels)
N^+	Net (positive) ionized doping density
N_{c}	Conduction band effective density of states
N_{c}	Number of ungrounded device terminals
N_i	Level i total electron population
N_{i}	Number of internal nodes
$\boldsymbol{N}_{\mathrm{i}}^{\mathrm{ss},+}$	SSLS upper sideband discretized electron density, internal nodes
N_{n}	Number of network nodes
N_{s}	Number of network sides
N_{t}	Trap-level density
N_{v}	Valence band effective density of states
N_{x}	Number of nodes on the external device contacts
$\boldsymbol{N}_{\mathrm{x}}^{\mathrm{ss},+}$	SSLS upper sideband discretized electron density, external nodes
N_{A}	Acceptor doping level
N_{A}^-	Ionized acceptor doping density
N_{AB}	Acceptor doping level for the base region of a *npn* BJT
N_{D}	Donor doping level
N_{D}^+	Ionized donor doping density
N_{DC}	Donor doping level for the collector region of a *npn* BJT
N_{DE}	Donor doping level for the emitter region of a *npn* BJT
N_{F}	Number of input LS tones
N_{L}	Number of LS tones
N_{S}	Number of tones in SSLS analysis
$\boldsymbol{N}_\eta$	Phonon states occupation number
p	Hole density

p_i	Discretized (nodal) hole density
p_{ij}	Electron density transition rate from i to j
p_{iji}	Discretized (nodal) electron density transition rate from i to j
$\tilde{\boldsymbol{p}}_{\mathrm{i}}^{\mathrm{ss}}$	Small-signal discretized hole density, internal nodes
p_n	Minority hole density
p_n'	Minority hole excess density
p_{ne}	Equilibrium minority hole density
p_{n0}'	DC minority hole excess density
p_{s}	Schottky contact equilibrium hole density
p^{ss}	Small-signal hole density
$\tilde{\boldsymbol{p}}_{\mathrm{x}}^{\mathrm{ss}}$	Small-signal discretized hole density, external nodes
p_0	DC hole density
P	FET noise model dimensionless drain noise parameter
P	Total hole number (population)
P_{ij}	Electron number transition rate from i to j
$\boldsymbol{P}_{\mathrm{i}}^{\mathrm{ss},+}$	SSLS upper sideband discretized hole density, internal nodes
P_{n}	Noise power
$\boldsymbol{P}_{\mathrm{x}}^{\mathrm{ss},+}$	SSLS upper sideband discretized hole density, external nodes
$P(A)$	Probability of the subset A of the set of events
$P(A\|B)$	Conditional probability of A conditioned to B
P_{12}, P_{21}	Average noise power
q	Elementary charge
Q	Gate induced charge
$\boldsymbol{Q}_k^{\mathrm{ss},+}$	Charge upper sideband spectral amplitudes (branch k)
Q_{G}	Total gate charge
r, R	Resistance
$\boldsymbol{r}$	Position
$\boldsymbol{r}_i$	Single particle position
r_{n}	Noise resistance, series representation
R	FET noise model dimensionless gate noise parameter
R	Resistance conversion matrix
R_{c}	Correlation resistance
$\mathrm{Re}\{\cdot\}$	Real part
R_{n}	Noise resistance, parallel representation
R_n	Electron recombination rate
R_n^{dir}	Electron recombination rate, band-to-band process
R_n^{srh}	Electron recombination rate, trap-assisted process
R_{o}	Optimum input generator resistance
R_p	Hole recombination rate
R_p^{dir}	Hole recombination rate, band-to-band process

R_p^{srh}	Hole recombination rate, trap-assisted process
R_{D}	Drain parasitic resistance
R_{DS}	Drain small-signal resistance
R_{G}	Gate parasitic resistance
R_{I}	Intrinsic (FET) resistance
R_{S}	Source parasitic resistance
$R_{X,X}^{(n)}(\tau)$	Cyclic autocorrelation function of the cyclostationary stochastic process $X(t)$
$R_{X,Y}$	Correlation function of random variables/stochastic processes X and Y
$\mathsf{R}_{\boldsymbol{X},\boldsymbol{Y}}$	Correlation matrix of random variables/stochastic processes $\boldsymbol{X}$ and $\boldsymbol{Y}$
$\mathsf{R}_{\delta\boldsymbol{v},\delta\boldsymbol{v}}$	Velocity fluctuations correlation function
$R_{\delta\boldsymbol{v},\delta\boldsymbol{v},\mathrm{s}}$	Velocity fluctuations stationary correlation function
s	Time variable
s_i	Discretized (nodal) scalar source term
s_{ij}	Distance between nodes i and j
s	Scalar source term
$\boldsymbol{s}$	Vector source term
$\boldsymbol{s}_{\mathrm{e}}$	Applied external electrical sources
$\boldsymbol{s}_{\mathrm{e}}^{\mathrm{ss}}$	Applied external small-signal electrical sources
$\boldsymbol{s}_{\mathrm{e}0}$	Applied external DC electrical sources
s_k	Noise source term for the k-th level rate equation
s_n	Electron continuity equation noise source term
s_n'	Equivalent electron continuity equation noise source term for trap-assisted GR noise
s_p	Hole continuity equation noise source term
s_p'	Equivalent hole continuity equation noise source term for trap-assisted GR noise
$\boldsymbol{s}_{\mathrm{s}}^{\mathrm{ss}}$	Set of small-signal applied external electrical sources
$\boldsymbol{s}_{\mathrm{D}}$	Noise source terms for the trap-level rate equations
s_φ	Poisson equation noise source term
s_φ'	Equivalent Poisson equation noise source term for trap-assisted GR noise
S	Cross section
S'	Discretized matrix source term for Green's function (monopolar model)
S_{c}	Discretized matrix source term for auxiliary (injection in contact current) Green's function
S_{c}'	Discretized matrix source term for auxiliary (injection in contact current) Green's function (monopolar model)

S_i	Average distance between two boundary nodes connected to node i
$S_{i_{ij},i_{ij}}$	Spectrum of the current generator injected between network nodes i and j
$\boldsymbol{S}_{i,k}^{\mathrm{ss},+}$	SSLS current source (branch k)
S_n	Discretized matrix source term for electron Green's function
S_n'	Discretized matrix source term for electron Green's function (monopolar model)
S_p	Discretized matrix source term for hole Green's function
$\mathsf{S}_{\boldsymbol{s},\boldsymbol{s}}$	Microscopic noise source correlation (unilateral) spectrum
$S_{s_n',s_n'}$	Electron (unilateral) spectrum of the equivalent noise source term for trap-assisted GR noise
$S_{s_n',s_p'}$	Electron–hole (unilateral) correlation spectrum of the equivalent noise source term for trap-assisted GR noise
$S_{s_n',s_\varphi'}$	Electron–potential (unilateral) correlation spectrum of the equivalent noise source term for trap-assisted GR noise
$S_{s_p',s_p'}$	Hole (unilateral) correlation spectrum of the equivalent noise source term for trap-assisted GR noise
$S_{s_p',s_\varphi'}$	Hole–potential (unilateral) correlation spectrum of the equivalent noise source term for trap-assisted GR noise
$S_{s_\varphi',s_\varphi'}$	Potential (unilateral) correlation spectrum of the equivalent noise source term for trap-assisted GR noise
$\boldsymbol{S}_{v,k}^{\mathrm{ss},+}$	SSLS voltage source (branch k)
S_{v_k,v_l}	Correlation (unilateral) spectrum of open-circuit voltages at ports k and l
$\hat{S}_{X,X}^{(n)}$	Cyclic (bilateral) autocorrelation spectrum of the cyclostationary stochastic processe $X(t)$
$S_{X,Y}$	(Unilateral) correlation spectrum of WSS stochastic processes $X(t)$ and $Y(t)$
$\hat{S}_{X,Y}$	(Bilateral) correlation spectrum of WSS stochastic processes $X(t)$ and $Y(t)$
S_{γ_i,γ_j}	Fundamental correlation (unilateral) spectrum of the noise source for GR noise (electron populations, levels i and j)
$S_{\delta e,\delta e}$	One-port open-circuit noise voltage correlation (unilateral) spectrum
$\mathsf{S}_{\delta\boldsymbol{e},\delta\boldsymbol{e}}$	Open-circuit noise voltage correlation (unilateral) spectrum
$S_{\delta e_c,\delta e_c}$	Correlation (unilateral) spectrum of the correlated part of the input noise voltage
$S_{\delta e_\mathrm{u},\delta e_\mathrm{u}}$	Correlation (unilateral) spectrum of the uncorrelated part of the input noise voltage
$S_{\delta e_\mathrm{DS},\delta e_\mathrm{DS}}$	Drain open-circuit noise voltage correlation (unilateral) spectrum
$S_{\delta i,\delta i}$	One-port short-circuit noise current correlation (unilateral) spectrum

$\mathsf{S}_{\delta \boldsymbol{i},\delta \boldsymbol{i}}$	Short-circuit noise current correlation (unilateral) spectrum
$S_{\delta i_c,\delta i_c}$	Correlation (unilateral) spectrum of the correlated part of the input noise current
$S_{\delta i_u,\delta i_u}$	Correlation (unilateral) spectrum of the uncorrelated part of the input noise current
$S_{\delta i_B,\delta i_B}$	Base short-circuit noise current correlation (unilateral) spectrum
$S_{\delta i_B,\delta i_C}$	Base-collector short-circuit noise current correlation (unilateral) spectrum
$S_{\delta i_B,\delta i_E}$	Base-emitter short-circuit noise current correlation (unilateral) spectrum
$S_{\delta i_C,\delta i_B}$	Collector-base short-circuit noise current correlation (unilateral) spectrum
$S_{\delta i_C,\delta i_C}$	Collector short-circuit noise current correlation (unilateral) spectrum
$S_{\delta i_C,\delta i_E}$	Collector-emitter short-circuit noise current correlation (unilateral) spectrum
$S_{\delta i_D,\delta i_D}$	Drain short-circuit noise current correlation (unilateral) spectrum
$S^{(D)}_{\delta i,\delta i}$	Junction short-circuit noise current correlation (unilateral) spectrum, diffusion noise contribution
$S_{\delta i_E,\delta i_B}$	Emitter-base short-circuit noise current correlation (unilateral) spectrum
$S_{\delta i_E,\delta i_C}$	Emitter-collector short-circuit noise current correlation (unilateral) spectrum
$S_{\delta i_E,\delta i_E}$	Emitter short-circuit noise current correlation (unilateral) spectrum
$S_{\delta i_G,\delta i_G}$	Gate short-circuit noise current correlation (unilateral) spectrum
$S_{\delta i_G,\delta i_D}$	Short-circuit gate-drain noise current correlation (unilateral) spectrum
$S_{\delta i,\delta i}$	Junction short-circuit noise current correlation (unilateral) spectrum
$\mathsf{S}_{\delta \boldsymbol{v},\delta \boldsymbol{v}}$	Velocity fluctuation correlation (unilateral) spectrum
$S^{(GR)}_{\delta i,\delta i}$	Junction short-circuit noise current correlation (unilateral) spectrum, GR noise contribution
$\mathsf{S}_{\delta \boldsymbol{j},\delta \boldsymbol{j}}$	Single-carrier current density fluctuation correlation (unilateral) spectrum
$\mathsf{S}_{\delta x,\delta x}$	SCM of cyclostationary process $\delta x(t)$
$\mathsf{S}_{\delta \boldsymbol{J},\delta \boldsymbol{J}}$	Current density fluctuation correlation (unilateral) spectrum
$\mathsf{S}_{\delta \boldsymbol{J}_n,\delta \boldsymbol{J}_n}$	Electron current density fluctuation correlation (unilateral) spectrum
$\mathsf{S}_{\delta \boldsymbol{J}_p,\delta \boldsymbol{J}_p}$	Hole current density fluctuation correlation (unilateral) spectrum
$S_{\delta Q_G,\delta Q_G}$	Gate charge fluctuation correlation (unilateral) spectrum

$S_{\delta R,\delta R}$	Resistance fluctuation correlation (unilateral) spectrum
$S_{\delta\mu,\delta\mu}$	Mobility fluctuation correlation (unilateral) spectrum
$\mathsf{S}_{\boldsymbol{\xi}_r,\boldsymbol{\xi}_r}$	Position microscopic noise source correlation (unilateral) spectrum
$\mathsf{S}_{\boldsymbol{\xi}_r,\boldsymbol{\xi}_v}$	Position–velocity microscopic noise source correlation (unilateral) spectrum
$\mathsf{S}_{\boldsymbol{\xi}_v,\boldsymbol{\xi}_v}$	Velocity microscopic noise source correlation (unilateral) spectrum
$\mathsf{S}_{\boldsymbol{\Gamma},\boldsymbol{\Gamma}}$	Total electron populations noise source correlation (unilateral) spectrum
t	Time variable
t_{ox}	Oxide thickness (MOS)
T	Temperature
T	Time-frequency transformation matrix
T_n	Electron temperature
T_{n}	Noise temperature
$\tilde{\mathsf{T}}_n$	Electron transfer impedance field
$\tilde{\mathsf{T}}_p$	Hole transfer impedance field
T_{D}	Drain noise temperature
T_{G}	Gate noise temperature
U_n	Electron net recombination rate
U_n^{dir}	Electron net recombination rate, band-to-band process
U_{ni}	Discretized (nodal) electron net recombination rate
U_p	Hole net recombination rate
U_p^{dir}	Hole net recombination rate, band-to-band process
U_{pi}	Discretized (nodal) hole net recombination rate
v	Voltage
v	Contact external potential
$\boldsymbol{v}$	Velocity
v_{bi}	Built-in potential
$v_{\mathrm{bi},k}$	Built-in potential for contact k
$\boldsymbol{v}_{\mathrm{c}}^{\mathrm{ss}}$	Small-signal contact potentials
$v_{\mathrm{c}k}^{\mathrm{ss}}$	Complex SSLS voltage (branch k)
$\boldsymbol{v}_i$	Single particle velocity
v_k	Voltage (branch k)
v_k	Potential of contact k
v_k^{adj}	Adjoint branch voltage (branch k)
v_k^{ss}	SSLS voltage (branch k)
$\boldsymbol{v}_n$	Electron velocity
$\boldsymbol{v}_{n0}$	DC electron velocity
$\boldsymbol{v}_p$	Hole velocity
$\boldsymbol{v}_{p0}$	DC hole velocity

v_{sn}	Equivalent electron surface recombination velocity
v_{sp}	Equivalent hole surface recombination velocity
$\tilde{\boldsymbol{v}}_{\mathrm{c}}^{\mathrm{ss}}$	Small-signal contact potentials
v_{th}	Threshold voltage
v_{D}	Diode applied voltage
v_{DS}	Drain-source applied voltage
v_{DSS}	Saturation drain-source voltage
v_{GS}	Gate-source applied voltage
v_{T}	Electrical equivalent of temperature
$\boldsymbol{v}_0, \tilde{\boldsymbol{v}}_0$	Open-circuit port voltages
V	Volume
$\boldsymbol{V}_k$	Spectral components of LS voltage $v_k(t)$
$\boldsymbol{V}^{\mathrm{ss},+}$	Voltage upper sideband spectral amplitudes
$\boldsymbol{V}_k^{\mathrm{ss},+}$	Voltage upper sideband spectral amplitudes (branch k)
w_n	Length of the n side (pn junction)
w_p	Length of the p side (pn junction)
$w_{\zeta\zeta_1}$	Transition rate from electron state ζ to ζ_1
W	Device width
$W_{\boldsymbol{N}\boldsymbol{N}'}$	Transition rate from electron total population state $\boldsymbol{N}$ to $\boldsymbol{N}'$
$W_{\gamma\overline{\gamma}}$	Transition rate from γ to $\overline{\gamma}$
$\hat{x}$	x axis versor
x	Large-signal forcing term
x^{ss}	Small-signal perturbation of LS signal $x(t)$
$x_{\mathrm{c}}^{\mathrm{ss}}$	Complex signal representing x^{ss}
x	Matrix representing the discretized unknown for Green's functions
x_n	Depletion region extension, n side
x_p	Depletion region extension, p side
X	Random variable/stochastic process
$\boldsymbol{X}^{\mathrm{ss},+}$	Collection of (upper) sideband amplitudes of $x^{\mathrm{ss}}(t)$
X_k	Complex amplitude of tone k for LS signal $x(t)$
$X(t)$	Poisson increments process
$\langle X \rangle_{X_0}$	Conditional ensemble average of random variable/stochastic process X
X_{c}	Correlation reactance
X_{g}	Input generator reactance
X_{o}	Optimum input generator reactance
y	Large-signal system response
y, Y	Admittance
y, Y	Admittance matrix
Y	Random variable/stochastic process

Y	Admittance conversion matrix
Y_{c}	Correlation admittance
Y_{g}	Input generator admittance
Y_{i}	Input admittance
$\mathsf{Y}_{i_{\mathrm{c}}v_{\mathrm{c}}}$	Discretized admittance formulation of monopolar drift-diffusion model $((i_{\mathrm{c}}, v_{\mathrm{c}})$ term)
$\mathsf{Y}_{i_{\mathrm{c}}\varphi_{\mathrm{i}}}$	Discretized admittance formulation of monopolar drift-diffusion model $((i_{\mathrm{c}}, \varphi_{\mathrm{i}}$ term)
$\mathsf{Y}_{n_{\mathrm{i}}v_{\mathrm{c}}}$	Discretized admittance formulation of monopolar drift-diffusion model $((n_{\mathrm{i}}, v_{\mathrm{c}})$ term)
$\mathsf{Y}_{n_{\mathrm{i}}\varphi_{\mathrm{i}}}$	Discretized admittance formulation of monopolar drift-diffusion model $((n_{\mathrm{i}}, \varphi_{\mathrm{i}})$ term)
Y_{o}	Optimum input generator admittance
Y_{BB}	Base-base small-signal admittance
Y_{BC}	Base-collector small-signal admittance
Y_{BE}	Base-emitter small-signal admittance
Y_{CB}	Collector-base small-signal admittance
Y_{CC}	Collector-collector small-signal admittance
Y_{CE}	Collector-emitter small-signal admittance
Y_{D}	Diode small-signal admittance
Y_{EB}	Emitter-base small-signal admittance
Y_{EC}	Emitter-collector small-signal admittance
Y_{EE}	Emitter-emitter small-signal admittance
Z	Impedance
Z	Impedance matrix
Z	Impedance conversion matrix
$\mathsf{Z}^{\mathrm{adj}}$	Adjoint impedance matrix
Z_{c}	Correlation impedance
Z_{g}	Input generator impedance
Z_{kl}	Transimpedance between ports k and l
Z_{kl}^{adj}	Transimpedance between ports k and l of the adjoint network
$\tilde{Z}_n$	Scalar electron impedance field
$\tilde{\boldsymbol{Z}}_n$	Nodal values of the discretized scalar electron impedance field
$\tilde{\boldsymbol{Z}}_n$	Vector electron impedance field
Z_{o}	Optimum input generator impedance
$\tilde{Z}_p$	Scalar hole impedance field
$\tilde{\boldsymbol{Z}}_p$	Vector hole impedance field
$\tilde{\boldsymbol{Z}}_p$	Nodal values of the discretized scalar hole impedance field
Z_{D}	Drain small-signal impedance
$\mathbf{0}$	Null vector

$\mathsf{0}_{M,N}$	Null matrix, dimension $M \times N$
α_{H}	Hooge constant
$\alpha_{\mathrm{H}n}$	Electron Hooge constant
$\alpha_{\mathrm{H}p}$	Hole Hooge constant
β	Gate noise factor
β	BJT DC current gain
γ	Flicker noise exponent
γ	Drain noise factor
γ	Junction injection efficiency
$\langle\gamma\|$	Quantum state (bra)
$\|\gamma\rangle$	Quantum state (ket)
γ_i	Scalar Langevin source for i-th level electron density population
γ_n	Scalar Langevin source for electron GR noise
γ_p	Scalar Langevin source for hole GR noise
γ_{D}	Scalar Langevin source for trap-level GR noise
$\mathit{\Gamma}$	Boundary of control volume $\mathit{\Sigma}$
$\boldsymbol{\Gamma}$	Vector Langevin source for total electron populations rate equations
$\mathit{\Gamma}_{\alpha,\beta}$	Coefficients for the equivalent noise sources in the reduced trap-level drift-diffusion noise model
$\delta(\cdot)$	Dirac delta function
$\delta\boldsymbol{e}, \delta\tilde{\boldsymbol{e}}$	Open-circuit voltage fluctuation (noise generator)
$\delta\tilde{e}_{\mathrm{c}}$	Correlated part of the open-circuit voltage noise generator
δe_k	Open-circuit voltage fluctuation at terminal k
$\delta\tilde{e}_{\mathrm{u}}$	Uncorrelated part of the open-circuit voltage noise generator
δe_{DS}	Drain open-circuit noise voltage generator
δi	Current fluctuation
δi	Injected current source
$\delta\boldsymbol{i}, \delta\tilde{\boldsymbol{\imath}}$	Short-circuit current fluctuation (noise generator)
$\delta\tilde{\imath}_{\mathrm{c}}$	Correlated part of the short-circuit current noise generator
$\delta\tilde{\imath}_{\mathrm{c},k}$	Short-circuit current fluctuation at terminal k
δi_{l}	Fluctuation of particle flow from left to right
$\delta\tilde{\imath}_n$	Injected electron current source
$\delta\tilde{\imath}_p$	Injected hole current source
δi_{r}	Fluctuation of particle flow from right to left
$\delta\tilde{\imath}_{\mathrm{u}}$	Uncorrelated part of the short-circuit current noise generator
δi_{D}	Drain short-circuit noise current generator
$\delta\boldsymbol{j}$	Single-carrier current density fluctuation
δn	Electron density fluctuation

$\delta\tilde{\boldsymbol{n}}_\mathrm{i}$	Fluctuation of discretized electron density, internal nodes
$\delta n'_p$	Minority excess electron density fluctuation
δn_t	Trapped electron density fluctuation
$\delta\tilde{\boldsymbol{n}}_\mathrm{x}$	Fluctuation of discretized electron density, external nodes
δp	Hole density fluctuation
$\delta\tilde{\boldsymbol{p}}_\mathrm{i}$	Fluctuation of discretized hole density, internal nodes
$\delta p'_n$	Minority excess hole density fluctuation
$\delta\tilde{\boldsymbol{p}}_\mathrm{x}$	Fluctuation of discretized hole density, external nodes
$\delta\boldsymbol{r}$	Position fluctuation
$\delta\boldsymbol{r}_i$	Single particle position fluctuation
δv	Voltage fluctuation
$\delta\boldsymbol{v}$	Velocity fluctuation
$\delta\tilde{\boldsymbol{v}}_\mathrm{c}$	Contact potentials fluctuation
$\delta\boldsymbol{v}_i$	Single particle velocity fluctuation
$\delta\boldsymbol{C}^+_{\mathrm{D}k}$	Net equivalent ionized donor concentration fluctuation
$\delta\boldsymbol{\mathcal{E}}$	Electric field fluctuation
δN	Electron number fluctuation
$\delta\boldsymbol{N}$	Fluctuation of total electron populations (all levels)
$\delta\boldsymbol{J}$	Current density fluctuation
$\delta\boldsymbol{J}_n$	Electron current density fluctuation
$\delta\boldsymbol{J}_p$	Hole current density fluctuation
δQ	Gate induced charge fluctuation
δQ_G	Total gate charge fluctuation
δR	Resistance fluctuation
$\delta X(t)$	Fluctuation of the Poisson increments process
$\delta\boldsymbol{X}$	Spectral respresentation of cyclostationary process $\delta x(t)$
$\delta_{\alpha,\beta}$	Kronecker symbol
$\delta\mu$	Mobility fluctuation
$\delta\varphi$	Potential fluctuation
$\delta\tilde{\boldsymbol{\varphi}}_\mathrm{i}$	Fluctuation of discretized electrostatic potential, internal nodes
$\delta\tilde{\boldsymbol{\varphi}}_\mathrm{x}$	Fluctuation of discretized electrostatic potential, external nodes
Δ	Equivalent thickness of the two-dimensional electron gas (HEMT)
Δ_{ij}	Normalized potential difference between nodes i and j
ε	Dielectric constant
ε_ox	Oxide dielectric constant (MOS)
η	Mean of a Gaussian random variable
λV	Electron–phonon interaction Hamiltonian operator
Λ_n	Part of the linearized electron continuity equation
Λ'_n	Part of the linearized equivalent electron continuity equation for trap-assisted GR noise

Λ_p	Part of the linearized hole continuity equation
Λ'_p	Part of the linearized equivalent hole continuity equation for trap-assisted GR noise
Λ_φ	Part of the linearized Poisson equation
Λ'_φ	Part of the linearized equivalent Poisson equation for trap-assisted GR noise
μ	Free carrier mobility
μ'	Free carrier small-signal (differential) mobility
μ_n	Electron mobility
μ_p	Hole mobility
$\boldsymbol{\xi_r}$	Vector Langevin source for position fluctuations
$\boldsymbol{\xi_v}$	Vector Langevin source for velocity fluctuations
ρ	Charge density
ρ'	Single-particle charge density
ρ_{D}	Density operator
σ	Electrical conductivity
σ	Standard deviation of a Gaussian random variable
σ_X^2	Variance of random variable/stochastic process X
Σ	Boundary surface of the device volume
Σ	Control volume (finite box)
$\Sigma_{\mathrm{c},k}$	Boundary surface of the device volume pertaining to terminal k
Σ_i	Measure of the control volume pertaining to node i
Σ_{kl}	Area of the control subregion pertaining to node k and element l
τ	Time difference
τ_{c}	Collision time
τ_{d}	Dielectric relaxation time
τ_{eq}	Equivalent lifetime (band-to-band transitions)
τ_n	Electron recombination lifetime
τ_n	Approximate electron lifetime exploited in the linearized band-to-band rate equations
τ_p	Hole recombination lifetime
τ_p	Approximate hole lifetime exploited in the linearized band-to-band rate equations
τ_{r}	Direct recombination lifetime
τ_{t}	Transition time
τ_α	Carrier α recombination lifetime
φ	Electrostatic potential
φ_i	Discretized (nodal) electrostatic potential
φ_0	DC electrostatic potential

φ_{rif}	Reference potential
φ^{ss}	Small-signal electrostatic potential
$\tilde{\boldsymbol{\varphi}}_{\mathrm{i}}^{\mathrm{ss}}$	Small-signal discretized electrostatic potential, internal nodes
$\tilde{\boldsymbol{\varphi}}_{\mathrm{x}}^{\mathrm{ss}}$	Small-signal discretized electrostatic potential, external nodes
$\boldsymbol{\Phi}_{\mathrm{i}}^{\mathrm{ss},+}$	SSLS discretized electrostatic potential, internal nodes
$\boldsymbol{\Phi}_{\mathrm{x}}^{\mathrm{ss},+}$	SSLS discretized electrostatic potential, external nodes
Φ_X	Characteristic function of random variable/stochastic process X
$\boldsymbol{\chi}$	Formal representation of the PDE model boundary conditions
ω	Angular frequency
ω_k^+, ω_k^-	Upper/lower k-th sideband angular frequency
Ω	Set of events
Ω	Device volume
Ω_{kl}	Control subregion pertaining to node k and element l
$\nabla_{\boldsymbol{\alpha}}$	Gradient with respect to $\boldsymbol{\alpha}$
$[\cdot,\cdot]$	Commutator
$\varnothing$	Empty set

1. Noise in Semiconductor Devices

1.1 Fluctuations in Semiconductors

The operation of semiconductor devices is based on charge transport, i.e. on the motion of free carriers in the conduction and valence bands. Under the effect of applied external forces and of the interaction with lattice perturbations or other carriers, electrons and holes undergo a kind of Brownian motion whereby the velocity of each carrier exhibits large fluctuations, while their position slowly drifts according the direction of the applied field or of the concentration gradient. According to a semi-classical approximation, the state of each carrier is described by its trajectory in the phase space $(\boldsymbol{r}_i(t), \boldsymbol{v}_i(t))$. However, owing to the extremely large number of carriers participating in the motion,[1] the individual behavior of each carrier is immaterial, and only average quantities (like the average position $\boldsymbol{r}$ and velocity $\boldsymbol{v}$) are relevant and have an impact on the circuit variables. For instance, the current density (and therefore the total current) in an n-doped semiconductor is proportional to the average electron velocity.

In order to introduce the implications of this collective approach, let us consider N free carriers characterized (in the classical limit) by individual, microscopic position $\boldsymbol{r}_i$ and velocity $\boldsymbol{v}_i$. Solid-state physics suggests that the time scale whereon carriers interact with lattice perturbations is very small (e.g. of the order of less than 1 ps); this implies that $\boldsymbol{r}_i$ and $\boldsymbol{v}_i$ are subject to fast and large fluctuations around a (slowly varying) average value. For instance, consider that the so-called thermal velocity of carriers is of the order of 10^7 cm/s at 300 K in thermodynamic equilibrium, whereas the average velocity is zero. We can therefore conclude that $\boldsymbol{r}_i$ and $\boldsymbol{v}_i$ can be written as:

$$\boldsymbol{r}_i(t) = \overline{\boldsymbol{r}}_i + \delta\boldsymbol{r}_i(t), \tag{1.1}$$

$$\boldsymbol{v}_i(t) = \overline{\boldsymbol{v}}_i + \delta\boldsymbol{v}_i(t), \tag{1.2}$$

where $\overline{\boldsymbol{r}}_i$ and $\overline{\boldsymbol{v}}_i$ are (slowly varying) mean values, and $\delta\boldsymbol{r}_i(t)$ and $\delta\boldsymbol{v}_i(t)$ are fluctuations.

[1] We shall not include here the treatment of so-called *mesoscopic* devices, i.e. devices whose dimension is so small that a coarse graining in space and time cannot be applied. In this case, a more fundamental, quantum-mechanical approach has to be exploited. For an introduction on this topic, see [13] and references therein.

In practice, the large single-particle fluctuations can hardly be detected when considering macroscopic quantities, such as the total current. This occurs because such quantities are related to ensemble averages made on a huge number of particles.[2] In fact, let us collectively describe the particle system in terms of the average position:

$$\boldsymbol{r}(t) = \frac{1}{N}\sum_i \boldsymbol{r}_i(t) \tag{1.3}$$

and of the average velocity:

$$\boldsymbol{v}(t) = \frac{1}{N}\sum_i \boldsymbol{v}_i(t). \tag{1.4}$$

Notice that $\boldsymbol{r}$ and $\boldsymbol{v}$ are defined as ensemble averages on the particle set and are, as a matter of principle, a function of time, which can again be decomposed as the sum of a slowly varying mean and a fluctuation:

$$\boldsymbol{r}(t) = \overline{\boldsymbol{r}} + \delta\boldsymbol{r}(t), \tag{1.5}$$

$$\boldsymbol{v}(t) = \overline{\boldsymbol{v}} + \delta\boldsymbol{v}(t). \tag{1.6}$$

However, since N is extremely large, collective fluctuations are negligibly small with respect to single-particle fluctuations. In fact, assuming particles to be statistically independent, one has

$$\langle \delta\boldsymbol{v}^2(t)\rangle = \frac{1}{N^2}\sum_{i,j}\langle \delta\boldsymbol{v}_i\delta\boldsymbol{v}_j\rangle = \frac{1}{N^2}\sum_i \langle \delta\boldsymbol{v}_i^2\rangle = \frac{1}{N}\langle \delta\boldsymbol{v}_i^2\rangle. \tag{1.7}$$

From a device and circuit standpoint, collective fluctuations, albeit small, propagate to the external terminals of the device thus originating *electrical noise*, i.e. fluctuations in electrical variables such as port voltages and currents. In other words, electrical noise can be interpreted as a macroscopic effect induced by collective fluctuations occurring in the semiconductor device volume (also referred to as *microscopic noise sources*); these are in turn the consequence of single-particle fluctuations due to scattering events. As we shall describe in Chap. 2, the connection between the microscopic noise sources and macroscopic fluctuations is essentially linear, since fluctuations are assumed to be small enough to perturb the noiseless steady state in small-signal operation; moreover, such a connection can be established converting the linearized physical model, which describes device operation under small-signal conditions, into a *Langevin equation* (according to a common definition, e.g. see [14,15], a Langevin equation is any differential equation with a stochastic forcing term, usually termed *Langevin force*) through the addition of stochastic forcing terms (the microscopic noise sources) to the model equations.

[2] Fluctuations in the particle number due to generation–recombination mechanisms are neglected here for the sake of simplicity.

A suitable description of fluctuations taking place in a semi-classical semiconductor must be based on a statistical approach, owing again to the extremely large number of carriers involved, which makes a deterministic picture of the system totally unrealistic. Statistical mechanics techniques allow the single-particle, and therefore the collective fluctuations, to be suitably characterized as random processes whose properties are related to the system parameters. Having established the statistical behavior of the microscopic noise sources, the aim of physics-based device noise modeling is to provide a convenient framework relating those to the fluctuations of the external, or macroscopic, electrical variables. This basic issue will be discussed in detail in Chaps. 2 and 3, while the rest of the present chapter will be devoted to the statistical characterization of microscopic noise sources both in a quasi-classical (Sect. 1.2) and in a quantum (Sect. 1.3) framework, and of the resulting macroscopic, or circuit, electrical noise (Sect. 1.4).

1.2 Microscopic Noise Sources in Semiconductors

Since an extremely high number of carriers is involved in normal semiconductor device operation, proper methods of statistical mechanics have to be exploited in deriving any physics-based device model. Such methods can be outlined as follows. First, the deterministic equations describing the motion of each particle in the system, including the possible scattering events, are written. Due to the huge number of particles participating to the system, a complete solution of these equations, although possible in principle (at least if a classical picture is assumed), is not feasible. Furthermore, this is not even necessary, since the properties of the system are very often effectively described through *macroscopic* quantities which are sufficient to characterize the physical state of the system itself and its interactions with the external world.

Based on these considerations, the system microscopic variables are assumed to be known in a statistical sense, i.e. through the corresponding *probability density,* which can be shown to evolve according to the classical or quantum *Liouville equation,* in the classical and quantum case, respectively. Since, as already discussed from an intuitive viewpoint, macroscopic variables are subject to *fluctuations* around their average values, this procedure clearly results in a loss of information. In order to recover a physical description of fluctuations within the framework of a physics-based device model, two approaches can be followed.

In a fundamental, quantum-mechanical approach (see [16] and references therein, in particular [17–22]), a unified treatment leads to both the physical model and to the statistical characterization of microscopic noise sources in term of suitable (Markov) random processes. Since a complete description of the fundamental approach is beyond the scope of this book, only a brief review will be provided in Sect. 1.3. A second approach, commonly referred to as *mesoscopic*, exploits phenomenological models to obtain a full characterization of the Langevin microscopic noise sources appearing in partial differential equations (PDEs) describing carrier transport [16, 17]. This approach, which is consistent with the fundamental one, will be described in detail in this section. In both cases, the basic mathematical tool exploited for noise analysis is the concept of a Markov stochastic process (see Sect. A.6) obtained as a result of some kind of random forces applied to the physical system under consideration.

As already recalled, two main fluctuations occur in semiconductor materials and, therefore, in devices: fluctuations of carrier *velocity* or *diffusion noise*, associated to the Brownian motion of free carriers in the conduction or valence band, and fluctuations of carrier *number* or *generation–recombination* (GR) *noise*, due to transitions between conduction and valence bands, either direct or trap-assisted. These phenomena are related, respectively, to intraband and interband scattering processes, which have to be described through a quantum-mechanical treatment. In both cases, a full analysis can be carried out, resulting in *white*, or frequency-independent, Langevin microscopic noise sources.

A further noise mechanism has been experimentally observed, whose physical nature is still a matter of discussion: $1/f$ noise, which is particularly important at low frequency. Several explanations have been proposed: the simplest obtains a $1/f$-like noise spectrum as a result of the superposition of several independent GR phenomena, characterized by a proper distribution of carrier lifetimes. Unfortunately, some experimental results suggest that $1/f$ noise is not always related to such a description, and a few competing theories have been proposed. For a review see [23] and [24]. From an application standpoint, $1/f$ noise modeling amounts to introducing microscopic noise sources as a colored (nonwhite) stochastic process, embedding into the source spectrum the $1/f$-like frequency dependence.

1.2.1 Velocity Fluctuations

Within the limits of the mesoscopic approach, free carrier velocity fluctuations are customarily modeled as resulting from Brownian motion of a single particle taking place in the corresponding band (e.g. conduction band for electrons and valence band for holes) [16]. From a statistical standpoint, the modeling of Brownian motion is based on the theory of the Ornstein–Uhlenbeck stochastic process [25], which is thoroughly reviewed in [14]. A complete analysis of Brownian motion can be found in the classical paper by Chandrasekar [26].

In its most general form, Brownian motion analysis takes place in the phase space $(\boldsymbol{r}, \boldsymbol{v})$, and is based on the assumption that such variables are represented by Markov stochastic processes described by a Fokker–Planck equation (A.91) (see the discussion in Sect. A.6). Assuming that the scattering events are characterized by a collision time τ_c, and that the carrier effective mass m^* is isotropic, the first two Fokker–Planck moments are, in thermodynamic equilibrium (i.e. for a free particle) given by [16, 26]:

$$\boldsymbol{A} = \begin{bmatrix} \boldsymbol{v} \\ -\frac{1}{\tau_c}\boldsymbol{v} \end{bmatrix}, \qquad \mathsf{B} = \begin{bmatrix} 0 & 0 \\ 0 & \frac{2}{\tau_c}\frac{k_B T}{m^*}\mathsf{I} \end{bmatrix}, \tag{1.8}$$

while all other moments of order greater than two are identically zero. The corresponding Fokker–Planck equation (A.91), defining an Ornstein–Uhlenbeck process [14], is therefore

$$\frac{\partial f_{\boldsymbol{r},\boldsymbol{v}}}{\partial t} = -\nabla_{\boldsymbol{r}} \cdot (\boldsymbol{v} f_{\boldsymbol{r},\boldsymbol{v}}) + \frac{1}{\tau_c}\nabla_{\boldsymbol{v}} \cdot (\boldsymbol{v} f_{\boldsymbol{r},\boldsymbol{v}}) + \frac{1}{\tau_c}\frac{k_B T}{m^*}\nabla^2_{\boldsymbol{v}} f_{\boldsymbol{r},\boldsymbol{v}}, \tag{1.9}$$

where $f_{\boldsymbol{r},\boldsymbol{v}} = f_{\boldsymbol{r},\boldsymbol{v}}(\boldsymbol{r}, \boldsymbol{v}; t|\boldsymbol{r}_0, \boldsymbol{v}_0)$ is the joint probability density in variables $\boldsymbol{r}$ and $\boldsymbol{v}$. The solution of (1.9), obtained in [26], is quite cumbersome:

$$\begin{aligned} f_{\boldsymbol{r},\boldsymbol{v}}(\boldsymbol{r}, \boldsymbol{v}; t|\boldsymbol{r}_0, \boldsymbol{v}_0) = & \frac{1}{8\pi^3 \left(FG - H^2\right)^{3/2}} \\ & \times \exp\left(-\frac{G|\boldsymbol{R}|^2 - 2H\boldsymbol{R}\cdot\boldsymbol{S} + F|\boldsymbol{S}|^2}{2\left(FG - H^2\right)}\right), \end{aligned} \tag{1.10}$$

where:

$$F = \tau_c^2 \frac{k_B T}{m^*}\left(2\frac{t}{\tau_c} - 3 + 4\mathrm{e}^{-t/\tau_c} - \mathrm{e}^{-2t/\tau_c}\right), \tag{1.11}$$

$$G = \frac{k_B T}{m^*}\left(1 - \mathrm{e}^{-2t/\tau_c}\right), \tag{1.12}$$

$$H = \tau_c \frac{k_B T}{m^*}\left(1 - \mathrm{e}^{-t/\tau_c}\right)^2, \tag{1.13}$$

$$\boldsymbol{R} = \boldsymbol{r} - \boldsymbol{r}_0 - \tau_c \boldsymbol{v}_0 \left(1 - \mathrm{e}^{-t/\tau_c}\right), \tag{1.14}$$

$$\boldsymbol{S} = \boldsymbol{v} - \boldsymbol{v}_0 \mathrm{e}^{-t/\tau_c}. \tag{1.15}$$

Notice that the probability density $f_{\boldsymbol{r},\boldsymbol{v}}$ in (1.10) is a multivariate Gaussian distribution [26], whose mean value clearly shows a dependence on time t; therefore the corresponding stochastic process is non-stationary. Such non-stationarity is, however, eliminated by a coarse graining in time, i.e. assuming the mesoscopic time limit $t/\tau_c \gg 1$, since this procedure results in a stationary probability density $f_{\boldsymbol{r},\boldsymbol{v},\mathrm{s}}(\boldsymbol{r}, \boldsymbol{v}; t|\boldsymbol{r}_0, \boldsymbol{v}_0)$ characterized by:

$$F_s = 2\tau_c \frac{k_B T}{m^*} t, \tag{1.16}$$

$$G_s = \frac{k_B T}{m^*}, \tag{1.17}$$

$$H_s = \tau_c \frac{k_B T}{m^*}, \tag{1.18}$$

$$\boldsymbol{R}_s = \boldsymbol{r} - \boldsymbol{r}_0 - \tau_c \boldsymbol{v}_0, \tag{1.19}$$

$$\boldsymbol{S}_s = \boldsymbol{v}. \tag{1.20}$$

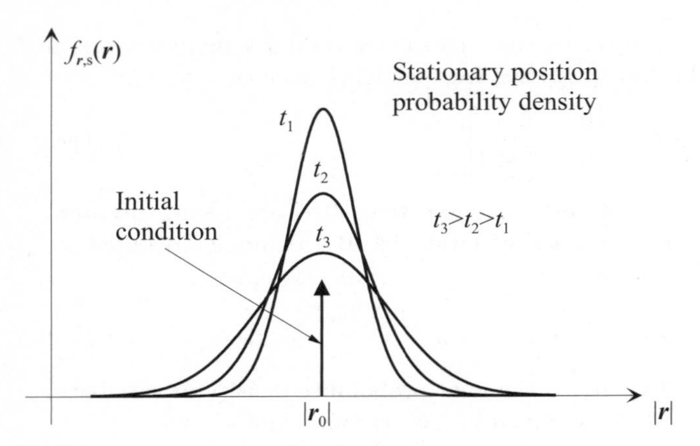

Fig. 1.1. Stationary position distribution obtained in the long-time limit $t/\tau_c \gg 1$

Starting from (1.9), within a coarse-graining limit both in time and space, the stationary spatial probability density$f_{\boldsymbol{r},s}(\boldsymbol{r}; t|\boldsymbol{r}_0, \boldsymbol{v}_0)$ is shown [16, 26] to satisfy the following diffusion equation

$$\frac{\partial f_{\boldsymbol{r},s}}{\partial t} = D_0 \nabla^2_{\boldsymbol{r}} f_{\boldsymbol{r},s}, \tag{1.21}$$

wherein the *diffusion coefficient*, or *diffusivity*, D_0 is given by:

$$D_0 = \frac{k_B T}{m^*} \tau_c. \tag{1.22}$$

The spatial distribution $f_{\boldsymbol{r},s}$ is therefore [16]:

$$f_{\boldsymbol{r},s}(\boldsymbol{r}; t|\boldsymbol{r}_0, \boldsymbol{v}_0) = \frac{1}{8\left(\pi D_0 t\right)^{3/2}} \exp\left(-\frac{|\boldsymbol{r} - \boldsymbol{r}_0|^2}{4D_0 t}\right), \tag{1.23}$$

i.e. the position distribution is a Gaussian with mean equal to the initial condition $\boldsymbol{r}_0$ and standard deviation linearly increasing with time; in other words, starting from a particle with known position $\boldsymbol{r}_0$, the spreading in position, in the long time limit, increases as shown in Fig. 1.1. The mean square value of the position displacement $\delta\boldsymbol{r}(t) = \boldsymbol{r}(t) - \langle \boldsymbol{r}(t) \rangle_{\boldsymbol{r}_0} = \boldsymbol{r}(t) - \boldsymbol{r}_0$, is given by the classical Einstein result:

$$\left\langle \delta \boldsymbol{r}^2(t) \right\rangle_{\boldsymbol{r}_0} = 2D_0 t \mathsf{I}. \tag{1.24}$$

Since the carrier mobility μ is defined as $\mu = \tau_c k_B T/m^*$, (1.22) is the well-known Einstein relation:

$$D_0 = \frac{k_B T}{q} \frac{q \tau_c}{m^*} = v_T \mu. \tag{1.25}$$

Notice that, as pointed out in [25], the long-time limit diffusion equation (1.21) can not be obtained through a limit procedure from the Fokker–Planck equation of the position variable $\boldsymbol{r}$ only, simply because this is not a Markov random process. On the other hand, a better choice is the velocity variable $\boldsymbol{v}$, whose probability density $f_{\boldsymbol{v}}(\boldsymbol{v};t|\boldsymbol{r}_0,\boldsymbol{v}_0)$ satisfies the Fokker–Planck equation[3] [16,26]

$$\frac{\partial f_{\boldsymbol{v}}}{\partial t} = \frac{1}{\tau_{\mathrm{c}}}\nabla_{\boldsymbol{v}} \cdot (\boldsymbol{v} f_{\boldsymbol{v}}) + \frac{1}{\tau_{\mathrm{c}}}\frac{k_{\mathrm{B}}T}{m^*}\nabla^2_{\boldsymbol{v}} f_{\boldsymbol{v}}, \tag{1.26}$$

which yields the non-stationary solution

$$\begin{aligned} f_{\boldsymbol{v}}(\boldsymbol{v};t|\boldsymbol{r}_0,\boldsymbol{v}_0) = &\left[\frac{m^*}{2\pi k_{\mathrm{B}}T\left(1-\mathrm{e}^{-2t/\tau_{\mathrm{c}}}\right)}\right]^{1/2} \\ &\times \exp\left[-\frac{m^*}{2k_{\mathrm{B}}T}\frac{\left|\boldsymbol{v}-\boldsymbol{v}_0\mathrm{e}^{-t/\tau_{\mathrm{c}}}\right|^2}{1-\mathrm{e}^{-2t/\tau_{\mathrm{c}}}}\right] . \end{aligned} \tag{1.27}$$

The corresponding conditional mean of carrier velocity, and square mean value of carrier velocity fluctuation $\delta\boldsymbol{v}(t) = \boldsymbol{v}(t) - \langle\boldsymbol{v}(t)\rangle_{\boldsymbol{v}_0}$ are:

$$\langle\boldsymbol{v}(t)\rangle_{\boldsymbol{v}_0} = \boldsymbol{v}_0\mathrm{e}^{-t/\tau_{\mathrm{c}}}, \tag{1.28}$$

$$\left\langle\delta\boldsymbol{v}^2(t)\right\rangle_{\boldsymbol{v}_0} = \frac{k_{\mathrm{B}}T}{m^*}\left(1-\mathrm{e}^{-2t/\tau_{\mathrm{c}}}\right)\mathsf{I}. \tag{1.29}$$

The stationary limit is obtained for $t/\tau_{\mathrm{c}} \gg 1$:

$$f_{\boldsymbol{v},\mathrm{s}}(\boldsymbol{v};t|\boldsymbol{r}_0,\boldsymbol{v}_0) = \left(\frac{m^*}{2\pi k_{\mathrm{B}}T}\right)^{1/2} \exp\left(-\frac{m^*}{2k_{\mathrm{B}}T}|\boldsymbol{v}|^2\right), \tag{1.30}$$

$$\langle\boldsymbol{v}(t)\rangle_{\boldsymbol{v}_0,\mathrm{s}} = 0, \tag{1.31}$$

$$\left\langle\delta\boldsymbol{v}^2(t)\right\rangle_{\boldsymbol{v}_0,\mathrm{s}} = \frac{k_{\mathrm{B}}T}{m^*}\mathsf{I}, \tag{1.32}$$

i.e. Maxwell's distribution. Notice that the stationary distribution is independent of the initial condition $\boldsymbol{v}_0$ (see Fig. 1.2).

From (1.27), the correlation function of velocity fluctuations can be evaluated. Details of the calculations can be found in [14], yielding:

$$\mathsf{R}_{\delta\boldsymbol{v},\delta\boldsymbol{v}}(t,s) = \langle\delta\boldsymbol{v}(t)\delta\boldsymbol{v}(s)\rangle_{\boldsymbol{v}_0} = \left(-\frac{1}{\tau_{\mathrm{c}}}D_0\mathrm{e}^{-(t+s)/\tau_{\mathrm{c}}} + \frac{2}{\tau_{\mathrm{c}}}D_0\mathrm{e}^{-|t-s|/\tau_{\mathrm{c}}}\right)\mathsf{I}. \tag{1.33}$$

[3] Notice that this equation can be obtained by integrating (1.9) on the whole position space.

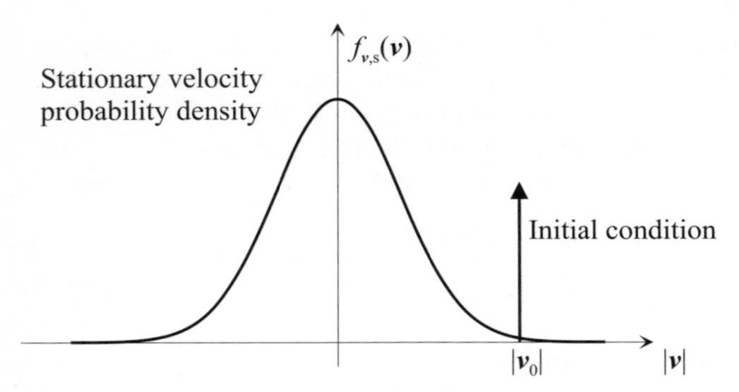

Fig. 1.2. Stationary velocity distribution obtained in the long-time limit $t/\tau_c \gg 1$. The result is independent of the initial condition $\boldsymbol{v}_0$

Again, part of the correlation function is stationary (the term depending on the time difference $(t-s)$) and part can be eliminated within the time coarse-graining procedure, i.e. $t, s \gg \tau_c$ with $t-s=\tau$ held fixed. The corresponding stationary correlation function is therefore:

$$\mathsf{R}_{\delta\boldsymbol{v},\delta\boldsymbol{v},\mathrm{s}}(\tau) = \frac{2}{\tau_c} D_0 \mathrm{e}^{-|\tau|/\tau_c} \mathsf{I}, \tag{1.34}$$

while the related power spectrum is given by Fourier transformation

$$\mathsf{S}_{\delta\boldsymbol{v},\delta\boldsymbol{v}}(\omega) = \frac{4D_0}{1+\omega^2\tau_c^2} \mathsf{I}. \tag{1.35}$$

Notice that the same result could be obtained directly from the Langevin equation (A.94) equivalent to the Fokker–Planck equation (1.9), here written for the variable fluctuations $\delta\boldsymbol{r}$ and $\delta\boldsymbol{v}$:

$$\frac{\mathrm{d}\delta\boldsymbol{r}}{\mathrm{d}t} = \delta\boldsymbol{v} + \boldsymbol{\xi}_{\boldsymbol{r}}, \tag{1.36a}$$

$$\frac{\mathrm{d}\delta\boldsymbol{v}}{\mathrm{d}t} = -\frac{1}{\tau_c}\delta\boldsymbol{v} + \boldsymbol{\xi}_{\boldsymbol{v}}, \tag{1.36b}$$

where the white-noise sources are characterized by their correlation spectrum, related to the second Fokker–Planck moment evaluated in the mean values $\langle\boldsymbol{r}\rangle_{\boldsymbol{r}_0,\boldsymbol{v}_0}$ and $\langle\boldsymbol{v}\rangle_{\boldsymbol{r}_0,\boldsymbol{v}_0}$, see (A.95) and (1.8):

$$\mathsf{S}_{\boldsymbol{\xi},\boldsymbol{\xi}} = \begin{bmatrix} \mathsf{S}_{\boldsymbol{\xi}_r,\boldsymbol{\xi}_r} & \mathsf{S}_{\boldsymbol{\xi}_r,\boldsymbol{\xi}_v} \\ \mathsf{S}_{\boldsymbol{\xi}_v,\boldsymbol{\xi}_r} & \mathsf{S}_{\boldsymbol{\xi}_v,\boldsymbol{\xi}_v} \end{bmatrix} = \begin{bmatrix} 0 & 0 \\ 0 & \frac{4}{\tau_c^2} D_0 \mathsf{I} \end{bmatrix}. \tag{1.37}$$

Equation (1.36) can be solved by means of harmonic analysis (Fourier transformation); although the mathematical background is quite involved (e.g. see [27] for an introduction), the practical application is straightforward. Further, according to the usual properties of harmonic analysis for linear dynamical systems, Fourier transformation yields directly the (harmonic) stationary solution, excluding the transient part of the response. The same consideration applies here, where the "transient" has to be interpreted as the non-stationary (in a statistical sense) part of the stochastic process, explicitly eliminated in the time domain by the time coarse-graining procedure previously described. Harmonic analysis applied to (1.36b) yields:

$$\mathrm{i}\omega\delta\tilde{\boldsymbol{v}} = -\frac{1}{\tau_\mathrm{c}}\delta\tilde{\boldsymbol{v}} + \tilde{\boldsymbol{\xi}}_{\boldsymbol{v}}, \tag{1.38}$$

where $\tilde{x}$ is the Fourier transform of $x(t)$. Equation (1.38) can be explicitly solved as a function of the Langevin source:

$$\delta\tilde{\boldsymbol{v}} = \frac{\tau_\mathrm{c}}{1+\mathrm{i}\omega\tau_\mathrm{c}}\tilde{\boldsymbol{\xi}}_{\boldsymbol{v}}. \tag{1.39}$$

The corresponding fluctuation power spectrum finally is:

$$\mathsf{S}_{\delta\boldsymbol{v},\delta\boldsymbol{v}}(\omega) = \frac{\tau_\mathrm{c}^2}{1+\omega^2\tau_\mathrm{c}^2}\mathsf{S}_{\boldsymbol{\xi}_v,\boldsymbol{\xi}_v} = \frac{4D_0}{1+\omega^2\tau_\mathrm{c}^2}\mathsf{I}. \tag{1.40}$$

In the low-frequency limit $\omega\tau_\mathrm{c} \ll 1$, the velocity-fluctuation spectrum is white, since $\mathsf{S}_{\delta\boldsymbol{v},\delta\boldsymbol{v}} \approx 4D_0\mathsf{I}$. The collision time τ_c is typically of the order of a few ps, therefore the white approximation holds up to frequencies of the order of 100 GHz.

Nyquist Theorem. The equilibrium noise properties of a linear resistor can be derived in a straightforward way from the second-order statistical properties of carrier velocity fluctuations. For the sake of simplicity, let us consider a p-type resistor of length L and cross section S. A single carrier having velocity $\boldsymbol{v}$ yields a current density $\boldsymbol{j} = \rho'\boldsymbol{v}$ where the charge density is $\rho' = q/(LS)$. Therefore, the velocity fluctuations induce current density fluctuations that, for a single carrier, have spectrum $\mathsf{S}_{\delta\boldsymbol{j},\delta\boldsymbol{j}}(\omega) = \rho'^2\mathsf{S}_{\delta\boldsymbol{v},\delta\boldsymbol{v}}(\omega)$. Supposing that different carriers have uncorrelated fluctuations, the current density fluctuations associated with the total current is:

$$\mathsf{S}_{\delta\boldsymbol{J},\delta\boldsymbol{J}}(\omega) = LS\frac{\rho}{q}\mathsf{S}_{\delta\boldsymbol{j},\delta\boldsymbol{j}}(\omega) = \frac{q\rho}{LS}\mathsf{S}_{\delta\boldsymbol{v},\delta\boldsymbol{v}}(\omega) \tag{1.41}$$

where ρ is the total carrier density and $\rho LS/q$ is the total carrier number in the resistor. One therefore obtains:

$$\mathsf{S}_{\delta\boldsymbol{J},\delta\boldsymbol{J}}(\omega) = \frac{q\rho}{LS}\frac{4D_0}{1+\omega^2\tau_\mathrm{c}^2}\mathsf{I}. \tag{1.42}$$

Evaluating the total current fluctuations as the flux of current-density fluctuations on the cross section S one finally obtains:

$$\begin{aligned} S_{\delta i,\delta i}(\omega) &= \frac{Sq\rho}{L}\frac{4D_0}{1+\omega^2\tau_\mathrm{c}^2} = \frac{S}{L}\frac{4q\rho v_\mathrm{T}\mu}{1+\omega^2\tau_\mathrm{c}^2} = \frac{S}{L}\frac{4k_\mathrm{B}T\sigma}{1+\omega^2\tau_\mathrm{c}^2} \\ &= \frac{4k_\mathrm{B}TG}{1+\omega^2\tau_\mathrm{c}^2}, \end{aligned} \tag{1.43}$$

where σ is the material conductivity and G is the low-frequency resistor conductance. At low frequency, i.e. for $\omega\tau_c \ll 1$, the customary Nyquist relationship is obtained:

$$S_{\delta i,\delta i}(\omega) = 4k_B TG. \tag{1.44}$$

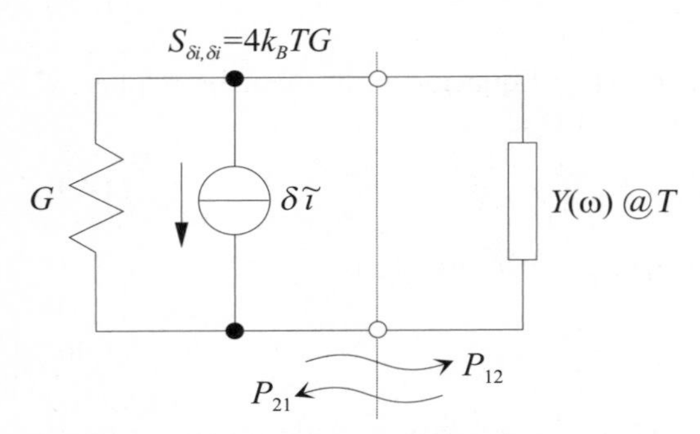

Fig. 1.3. Evaluating the current fluctuation spectrum for an arbitrary noisy one-port in equilibrium

The Nyquist theorem can be directly extended to one-ports operating at or near thermodynamic equilibrium. In such case, the one-port can be characterized in terms of a linear response by its frequency-dependent impedance $Z(\omega)$ or admittance $Y(\omega)$. Let us connect the one-port to a resistor, as shown in Fig. 1.3, and suppose that the system is in equilibrium at temperature T. The average power delivered by the noisy resistor to the two-port is:

$$P_{12} = 4k_B TG\frac{\mathrm{Re}\{Y(\omega)\}}{|G+Y(\omega)|^2}, \tag{1.45}$$

while the average power delivered by the noisy two-port to the resistor is:

$$P_{21} = S_{\delta i,\delta i}(\omega)\frac{G}{|G+Y(\omega)|^2}, \tag{1.46}$$

where $S_{\delta i,\delta i}$ is the power spectrum of the current fluctuations of the noisy one-port. In equilibrium conditions the net power exchange must be zero, i.e. $P_{12} = P_{21}$; this immediately implies:

$$S_{\delta i,\delta i}(\omega) = 4k_B T\,\mathrm{Re}\{Y(\omega)\}, \tag{1.47}$$

where Re indicates the real part. This yields the extension of the Nyquist theorem to frequency-dependent, linear one-ports. An alternative formulation of the Nyquist theorem makes reference to voltage rather than to current fluctuations. As shown in Sect. 1.4.1, the power spectrum of open-circuit voltage fluctuations of a one-port results as (1.140):

$$S_{\delta e,\delta e}(\omega) = 4k_{\mathrm{B}}T\,\mathrm{Re}\,\{Z(\omega)\}\,, \tag{1.48}$$

which becomes, for a resistor:

$$S_{\delta e,\delta e}(\omega) = 4k_{\mathrm{B}}TR. \tag{1.49}$$

Extension to Nonequilibrium. The previous discussion has to be extended to the nonequilibrium case, i.e. for free carriers subject to an external electric field. In such conditions, two diffusion coefficients are introduced, both under the assumption of negligible carrier–carrier scattering [28, 29]:

- the *noise* diffusion coefficient, defined by extending (1.40):

  $$\mathsf{S}_{\delta\boldsymbol{v},\delta\boldsymbol{v}}(\omega) = 4\mathsf{D}_{\mathrm{n}}(\omega); \tag{1.50}$$

  the following complex diffusion coefficient [16, 28] is defined as well:

  $$\mathsf{D}'_{\mathrm{n}}(\omega) = \int_0^{+\infty} \mathsf{R}_{\delta\boldsymbol{v},\delta\boldsymbol{v}}(\tau)\mathrm{e}^{-\mathrm{i}\omega\tau}\,\mathrm{d}\tau. \tag{1.51}$$

  Since the autocorrelation function is a real and even function of time, (1.51) yields:

  $$\mathsf{S}_{\delta\boldsymbol{v},\delta\boldsymbol{v}}(\omega) = 2\left[\mathsf{D}'_{\mathrm{n}}(\omega) + \mathsf{D}'^{\dagger}_{\mathrm{n}}(\omega)\right], \tag{1.52}$$

  so that:

  $$\mathsf{D}_{\mathrm{n}}(\omega) = \frac{1}{2}\left[\mathsf{D}'_{\mathrm{n}}(\omega) + \mathsf{D}'^{\dagger}_{\mathrm{n}}(\omega)\right], \tag{1.53}$$

  where † indicates Hermitian conjugation.
- the *spreading* diffusion coefficient, defined by extending (1.24):

  $$\left\langle \delta\boldsymbol{r}^2(t)\right\rangle = 2\mathsf{D}_{\mathrm{s}}(t)t. \tag{1.54}$$

The connection between these two quantities is provided by the following relation [30, 31]:

$$\lim_{\omega\to 0}\mathsf{D}_{\mathrm{n}}(\omega) = \lim_{t\to +\infty}\mathsf{D}_{\mathrm{s}}(t), \tag{1.55}$$

where the two limits have to be interpreted as $\omega\tau_{\mathrm{c}} \ll 1$ and $t/\tau_{\mathrm{c}} \gg 1$, respectively.

The Einstein relation (1.25), connecting in its most general form the tensor equilibrium diffusivity to the tensor *ohmic* carrier mobility, can be extended as well. Firstly, the *noise temperature* $T_{\mathrm{n}}(\omega)$ is defined [28] by formally introducing a generalized Nyquist relationship between the small-signal impedance $Z(\omega)$ and the open-circuit voltage fluctuation spectrum $S_{\delta e,\delta e}(\omega)$ for a one-port device:

$$S_{\delta e,\delta e}(\omega) = 4k_{\mathrm{B}}T_{\mathrm{n}}(\omega)\,\mathrm{Re}\left\{Z(\omega)\right\}. \tag{1.56}$$

Then, the diagonal components of the small-signal carrier mobility tensor $\mu'_{\alpha\alpha}(\omega)$ are related, at least for homogeneous fields [29], to the noise diffusion coefficient through [28]:

$$D_{\mathrm{n},\alpha\alpha}(\omega) = \frac{k_{\mathrm{B}}T_{\mathrm{n},\alpha}(\omega)}{q}\,\mathrm{Re}\left\{\mu'_{\alpha\alpha}(\omega)\right\}. \tag{1.57}$$

Notice that in implementing this generalized Einstein relation, several difficulties arise:

- for semiconductors with several minima in the conduction band (such as GaAs) the average differential mobility is negative for electric fields in excess of the threshold field for intervalley scattering; in such cases (1.57) must be used with care, i.e. with reference to the population of each equivalent valley;
- the noise temperature is, in general, different from both the lattice and carrier temperature, the latter being defined through carrier energy. Actually, noise and carrier temperature coincide only if the carrier distribution is a displaced Maxwell–Boltzmann distribution function [32], which is hardly the case for a hot-carrier regime (at least at intermediate fields).

To the authors' knowledge, no final results have been established yet for the high-field dependence of the noise diffusivity as a function of the macroscopic variables exploited in physics-based device analysis, e.g. the average energy or the electric field depending on whether non-stationary or drift-diffusion transport models are considered. This leads to serious problems for noise simulations carried out through physics-based models wherein, as in the drift-diffusion approach, both carrier mobility and diffusivity must be introduced as a function of the *local* electric field. Although several (more or less empirical) models relating the carrier mobility to the electric field are available to accurately reproduce the DC and small-signal device behavior, much less information exists on the diffusion coefficient, which directly impacts on the device noise performance. Some high-field diffusivity models are compared in [33].

Equivalent Current Density Fluctuations. Since, in the following, we shall consider the drift-diffusion transport model, which is based on exploiting charge conservation expressed through the carrier continuity equations (see the discussion in Sect. 2.1), the carrier velocity fluctuations have to be expressed in terms of *current density* fluctuations, amenable to be directly inserted into the continuity equations. Let us consider a small semiconductor volume (within the mesoscopic assumption) wherein a single carrier of charge q is moving. Let L be the volume length longitudinal to the carrier motion, and S the corresponding cross section, such that $V = LS$. The current density fluctuation induced by a carrier velocity fluctuation $\delta\boldsymbol{v}$ is then:

$$\delta \boldsymbol{j} = \frac{q}{LS} \delta \boldsymbol{v}, \tag{1.58}$$

so that, according to (1.52):

$$\mathsf{S}_{\delta \boldsymbol{j},\delta \boldsymbol{j}}(\omega) = \frac{q^2}{V^2} \mathsf{S}_{\delta \boldsymbol{v},\delta \boldsymbol{v}}(\omega) = \frac{q^2}{V^2} 2 \left[\mathsf{D}'_{\mathrm{n}}(\omega) + \mathsf{D}'^{\dagger}_{\mathrm{n}}(\omega) \right]. \tag{1.59}$$

Provided $\boldsymbol{r}$ is the position of volume V and $c(\boldsymbol{r})$ is the carrier density (i.e. $c = n$ for electrons and $c = p$ for holes), since carrier interaction is neglected, the corresponding velocity fluctuations are uncorrelated. Therefore the total power spectrum of current density fluctuations is:

$$\mathsf{S}_{\delta \boldsymbol{J},\delta \boldsymbol{J}}(\omega) = \left[c(\boldsymbol{r}) V \right] \mathsf{S}_{\delta \boldsymbol{j},\delta \boldsymbol{j}}(\omega) = \frac{c(\boldsymbol{r}) q^2}{V} 2 \left[\mathsf{D}'_{\mathrm{n}}(\omega) + \mathsf{D}'^{\dagger}_{\mathrm{n}}(\omega) \right]. \tag{1.60}$$

Finally, assuming that current density fluctuations arising in different spatial positions are uncorrelated, one obtains for $V \to 0$:

$$\mathsf{S}_{\delta \boldsymbol{J},\delta \boldsymbol{J}}(\boldsymbol{r}, \boldsymbol{r}'; \omega) = c(\boldsymbol{r}) q^2 2 \left[\mathsf{D}'_{\mathrm{n}}(\omega) + \mathsf{D}'^{\dagger}_{\mathrm{n}}(\omega) \right] \delta(\boldsymbol{r} - \boldsymbol{r}'). \tag{1.61}$$

This relationship can be written for both electrons and holes, thereby yielding:

$$\mathsf{S}_{\delta \boldsymbol{J}_n,\delta \boldsymbol{J}_n}(\boldsymbol{r}, \boldsymbol{r}'; \omega) = \mathsf{K}_{\delta \boldsymbol{J}_n,\delta \boldsymbol{J}_n}(\boldsymbol{r}; \omega) \delta(\boldsymbol{r} - \boldsymbol{r}'), \tag{1.62a}$$

$$\mathsf{S}_{\delta \boldsymbol{J}_p,\delta \boldsymbol{J}_p}(\boldsymbol{r}, \boldsymbol{r}'; \omega) = \mathsf{K}_{\delta \boldsymbol{J}_p,\delta \boldsymbol{J}_p}(\boldsymbol{r}; \omega) \delta(\boldsymbol{r} - \boldsymbol{r}'), \tag{1.62b}$$

where the *local noise source* K for diffusion noise is defined as [28]:

$$\mathsf{K}_{\delta \boldsymbol{J}_n,\delta \boldsymbol{J}_n}(\boldsymbol{r}; \omega) = 2 q^2 n(\boldsymbol{r}) \left[\mathsf{D}'_{\mathrm{n},n}(\omega) + \mathsf{D}'^{\dagger}_{\mathrm{n},n}(\omega) \right], \tag{1.63a}$$

$$\mathsf{K}_{\delta \boldsymbol{J}_p,\delta \boldsymbol{J}_p}(\boldsymbol{r}; \omega) = 2 q^2 p(\boldsymbol{r}) \left[\mathsf{D}'_{\mathrm{n},p}(\omega) + \mathsf{D}'^{\dagger}_{\mathrm{n},p}(\omega) \right]. \tag{1.63b}$$

For scalar noise diffusivity, the local noise sources simplify to:

$$\mathsf{K}_{\delta \boldsymbol{J}_n,\delta \boldsymbol{J}_n}(\boldsymbol{r}; \omega) = 4 q^2 n(\boldsymbol{r}) D'_{\mathrm{n},n}(\omega) \mathsf{I}, \tag{1.64a}$$

$$\mathsf{K}_{\delta \boldsymbol{J}_p,\delta \boldsymbol{J}_p}(\boldsymbol{r}; \omega) = 4 q^2 p(\boldsymbol{r}) D'_{\mathrm{n},p}(\omega) \mathsf{I}. \tag{1.64b}$$

1.2.2 Population Fluctuations

To introduce the analysis of population fluctuations in a semiconductor, let us first consider a homogeneous material, i.e. a semiconductor sample wherein no macroscopic variable depends on position, even outside thermodynamic equilibrium. In other words, transport phenomena are neglected. Let us further assume that in such a material electrons can occupy $s+1$ distinct energy levels E_i ($i = 1, \ldots, s+1$), each characterized by the electron population (occupation number) N_i. According to this terminology, one energy level, e.g.

$E_1 = E_c$, represents the conduction band, another, e.g. $E_{s+1} = E_v$, is the valence band (therefore $N_{s+1} = -P$, the hole occupation number); the other $s-1$ available energies could represent trap levels (see Sect. 2.1.1 for a discussion). Due to charge conservation, the $s+1$ populations are not independent:

$$\sum_{i=1}^{s+1} N_i = \sum_{i=1}^{s} N_i - P = 0; \tag{1.65}$$

however, for symmetry we shall include in the model all the $s+1$ populations.

The transitions between the energy levels i and j are characterized in terms of the corresponding transition rate $P_{ij}(\boldsymbol{N})$, $\boldsymbol{N} = [N_1, \ldots, N_{s+1}]^{\mathrm{T}}$, i.e. the number of electrons jumping from level i to level j per unit time. Taking into account single electron transitions only, the transition rate from state $\boldsymbol{N}$ to $\boldsymbol{N}'$ is given by [34]

$$W_{\boldsymbol{N}\boldsymbol{N}'} = \begin{cases} P_{ij}(\boldsymbol{N}) & \text{if} \begin{cases} \boldsymbol{N} = [N_1, \ldots, N_i, \ldots, N_j, \ldots, N_{s+1}] \\ \boldsymbol{N}' = [N_1, \ldots, N_i - 1, \ldots, N_j + 1, \ldots, N_{s+1}] \end{cases} \\ 0 & \text{in all other cases.} \end{cases} \tag{1.66}$$

According to this description, populations $N_i(t)$ are discrete Markov processes; the dynamics of the related probability density is defined by the Chapman–Kolmogorov equation (see Sect. A.6) for jump processes [14], i.e. the master equation

$$\frac{\partial}{\partial t} f_{\boldsymbol{N}}(\boldsymbol{N}, t|\boldsymbol{N}_0) = \sum_{\boldsymbol{N}'} \left[W_{\boldsymbol{N}'\boldsymbol{N}} f_{\boldsymbol{N}}(\boldsymbol{N}', t|\boldsymbol{N}_0) - W_{\boldsymbol{N}\boldsymbol{N}'} f_{\boldsymbol{N}}(\boldsymbol{N}, t|\boldsymbol{N}_0) \right], \tag{1.67}$$

which can be approximated by a continuous diffusion Markov process [14], i.e. by a Fokker–Planck equation obtained through the Kramers–Moyal expansion (A.92) truncated to second order. The first two Fokker–Planck moments are [16, 34]:

$$A_i(\boldsymbol{N}) = \sum_{\substack{j=1 \\ j \neq i}}^{s+1} \left[P_{ji}(\boldsymbol{N}) - P_{ij}(\boldsymbol{N}) \right], \tag{1.68}$$

$$B_{ii}(\boldsymbol{N}) = \sum_{\substack{j=1 \\ j \neq i}}^{s+1} \left[P_{ji}(\boldsymbol{N}) + P_{ij}(\boldsymbol{N}) \right], \tag{1.69}$$

$$B_{ij}(\boldsymbol{N}) = -P_{ij}(\boldsymbol{N}) - P_{ji}(\boldsymbol{N}) \qquad i \neq j, \tag{1.70}$$

therefore the approximate Fokker–Planck equation reads:

$$
\begin{aligned}
\frac{\partial}{\partial t} f_{\boldsymbol{N}}(\boldsymbol{N}, t|\boldsymbol{N}_0) = &-\sum_{i=1}^{s+1} \frac{\partial}{\partial N_i} \left[A_i(\boldsymbol{N}) f_{\boldsymbol{N}}(\boldsymbol{N}, t|\boldsymbol{N}_0)\right] \\
&+ \frac{1}{2} \sum_{i,j=1}^{s+1} \frac{\partial^2}{\partial N_i \partial N_j} \left[B_{i,j}(\boldsymbol{N}) f_{\boldsymbol{N}}(\boldsymbol{N}, t|\boldsymbol{N}_0)\right],
\end{aligned} \tag{1.71}
$$

which is equivalent to the (stochastic) Langevin equation (A.94)

$$
\frac{\partial \boldsymbol{N}}{\partial t} = -\boldsymbol{A}(\boldsymbol{N}) + \boldsymbol{\Gamma}, \tag{1.72}
$$

provided the Langevin sources $\boldsymbol{\Gamma}$ are white, zero-average stochastic processes characterized by the correlation spectrum (A.95):

$$
\mathsf{S}_{\boldsymbol{\Gamma},\boldsymbol{\Gamma}} = 2\mathsf{B}\left(\langle \boldsymbol{N} \rangle_{\boldsymbol{N}_0}\right), \tag{1.73}
$$

so that:

$$
S_{\Gamma_i,\Gamma_i} = 2 \sum_{\substack{j=1 \\ j \neq i}}^{s+1} \left[P_{ji}\left(\langle \boldsymbol{N} \rangle_{\boldsymbol{N}_0}\right) + P_{ij}\left(\langle \boldsymbol{N} \rangle_{\boldsymbol{N}_0}\right)\right], \tag{1.74}
$$

$$
S_{\Gamma_i,\Gamma_j} = -2 \left[P_{ij}\left(\langle \boldsymbol{N} \rangle_{\boldsymbol{N}_0}\right) + P_{ji}\left(\langle \boldsymbol{N} \rangle_{\boldsymbol{N}_0}\right)\right] \qquad i \neq j. \tag{1.75}
$$

The evolution equation for the average population is derived by taking the conditional average of (1.72):

$$
\frac{\partial \langle \boldsymbol{N} \rangle_{\boldsymbol{N}_0}}{\partial t} = -\langle \boldsymbol{A}(\boldsymbol{N}) \rangle_{\boldsymbol{N}_0}. \tag{1.76}
$$

The zero-average population fluctuations $\delta \boldsymbol{N}$ are then defined as $\delta \boldsymbol{N} = \boldsymbol{N} - \langle \boldsymbol{N} \rangle_{\boldsymbol{N}_0}$, which, under the assumption of small amplitude, yields:

$$
\boldsymbol{A}(\boldsymbol{N}) = \boldsymbol{A}\left(\langle \boldsymbol{N} \rangle_{\boldsymbol{N}_0}\right) - \mathsf{M}\left(\langle \boldsymbol{N} \rangle_{\boldsymbol{N}_0}\right) \cdot \delta \boldsymbol{N} + O\left(\delta \boldsymbol{N}^2\right), \tag{1.77}
$$

where M is the *phenomenological relaxation matrix* [34]

$$
\mathsf{M}\left(\langle \boldsymbol{N} \rangle_{\boldsymbol{N}_0}\right) = - \left.\frac{\partial \boldsymbol{A}}{\partial \boldsymbol{N}}\right|_{\langle \boldsymbol{N} \rangle_{\boldsymbol{N}_0}}. \tag{1.78}
$$

Taking the conditional average of (1.77), one has

$$
\begin{aligned}
\langle \boldsymbol{A}(\boldsymbol{N}) \rangle_{\boldsymbol{N}_0} - \left\langle \boldsymbol{A}\left(\langle \boldsymbol{N} \rangle_{\boldsymbol{N}_0}\right) \right\rangle_{\boldsymbol{N}_0} &= \langle \boldsymbol{A}(\boldsymbol{N}) \rangle_{\boldsymbol{N}_0} - \boldsymbol{A}\left(\langle \boldsymbol{N} \rangle_{\boldsymbol{N}_0}\right) \\
&= O\left(\left\langle \delta \boldsymbol{N}^2 \right\rangle_{\boldsymbol{N}_0}\right),
\end{aligned} \tag{1.79}
$$

so that, to first order in $\delta \boldsymbol{N}$, one can assume:

$$
\langle \boldsymbol{A}(\boldsymbol{N}) \rangle_{\boldsymbol{N}_0} \approx \boldsymbol{A}\left(\langle \boldsymbol{N} \rangle_{\boldsymbol{N}_0}\right). \tag{1.80}
$$

From (1.72), (1.77) and (1.80), the following linearized equation is obtained:

$$\frac{\partial \delta \boldsymbol{N}}{\partial t} = -\mathsf{M}\left(\langle \boldsymbol{N} \rangle_{\boldsymbol{N}_0}\right) \cdot \delta \boldsymbol{N} + \boldsymbol{\Gamma}. \tag{1.81}$$

Notice that for a stationary solution, $\langle \boldsymbol{N} \rangle_{\boldsymbol{N}_0}$ does not depend on time. Therefore (1.76) and (1.80) yield

$$\sum_{\substack{j=1 \\ j \neq i}}^{s+1} P_{ji}(\langle \boldsymbol{N} \rangle_{\boldsymbol{N}_0}) = \sum_{\substack{j=1 \\ j \neq i}}^{s+1} P_{ij}(\langle \boldsymbol{N} \rangle_{\boldsymbol{N}_0}) \tag{1.82}$$

and (1.74) simplifies to

$$S_{\Gamma_i,\Gamma_i} = 4 \sum_{\substack{j=1 \\ j \neq i}}^{s+1} P_{ij}\left(\langle \boldsymbol{N} \rangle_{\boldsymbol{N}_0}\right). \tag{1.83}$$

Since homogeneous conditions are not common in practice, this analysis must be properly extended. First of all, occupation numbers have to be replaced by carrier densities n_i, i.e. the number of electrons per unit volume. Correspondingly, the transition rates p_{ij} are expressed as the number of electrons jumping from level i to level j per unit volume and unit time. The main difference is that the phenomenological equations (1.72) contain an additional term including spatial derivatives, and the corresponding rate equation is converted into a transport equation (the carrier continuity equation, see Sect. 2.1.1) including a stochastic forcing term $\boldsymbol{\gamma}$, whose correlation spectrum is, under the assumption of spatially uncorrelated fluctuations [34]:

$$S_{\gamma_i,\gamma_i}(\boldsymbol{r}, \boldsymbol{r}') = K_{\gamma_i,\gamma_i} \delta(\boldsymbol{r} - \boldsymbol{r}'), \tag{1.84a}$$

$$S_{\gamma_i,\gamma_j}(\boldsymbol{r}, \boldsymbol{r}') = K_{\gamma_i,\gamma_j} \delta(\boldsymbol{r} - \boldsymbol{r}') \qquad i \neq j, \tag{1.84b}$$

where again the spectra K are termed the *local noise sources* for GR noise:

$$K_{\gamma_i,\gamma_i}(\boldsymbol{r}, \boldsymbol{r}') = 2 \sum_{\substack{j=1 \\ j \neq i}}^{s+1} \left[p_{ji}\left(\langle \boldsymbol{n} \rangle_{\boldsymbol{n}_0}\right) + p_{ij}\left(\langle \boldsymbol{n} \rangle_{\boldsymbol{n}_0}\right)\right], \tag{1.85a}$$

$$K_{\gamma_i,\gamma_j}(\boldsymbol{r}, \boldsymbol{r}') = -2\left[p_{ij}\left(\langle \boldsymbol{n} \rangle_{\boldsymbol{n}_0}\right) + p_{ji}\left(\langle \boldsymbol{n} \rangle_{\boldsymbol{n}_0}\right)\right] \qquad i \neq j. \tag{1.85b}$$

For the sake of simplicity, we shall consider the two simplest cases of interband transitions: band-to-band and single-level trap assisted processes. In both cases, the electron and hole recombination (generation) rates[4] will be denoted R_n and R_p (G_n and G_p), respectively.

[4] The recombination rate R_α is the number of carriers of type α recombined per unit volume and time, the generation rate G_α is the number of carriers of type α generated per unit volume and time.

Band-to-band transitions. For band-to-band, or *direct*, transitions, conduction and valence bands are the only energy levels involved. The jump processes taking place in the material are depicted in Fig. 1.4, where arrows denote electron jumps. Since electrons and holes are created or destroyed in pairs, one has:

$$R_n^{\mathrm{dir}} = R_p^{\mathrm{dir}} \qquad G_n^{\mathrm{dir}} = G_p^{\mathrm{dir}}. \tag{1.86}$$

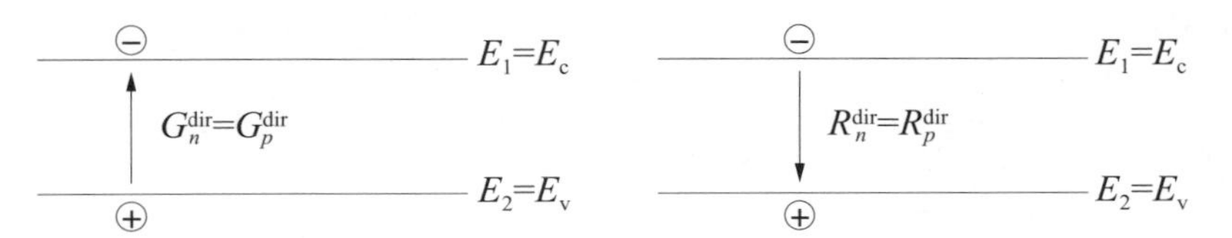

Fig. 1.4. Jump processes taking place for direct transitions: generation (*left*) and recombination (*right*). The *arrows* indicate electron movement

The recombination rate is proportional to the product of free electron and hole densities [35]

$$R_n^{\mathrm{dir}} = R_p^{\mathrm{dir}} = c_{\mathrm{r,dir}} np, \tag{1.87}$$

whilst the generation rate is constant:

$$G_n^{\mathrm{dir}} = G_p^{\mathrm{dir}} = c_{\mathrm{g,dir}}. \tag{1.88}$$

Since in thermodynamic equilibrium electron and hole populations are separately in equilibrium (detailed balance principle)

$$R_n^{\mathrm{dir}}\Big|_{\mathrm{eq}} = G_n^{\mathrm{dir}}\Big|_{\mathrm{eq}} \qquad R_p^{\mathrm{dir}}\Big|_{\mathrm{eq}} = G_p^{\mathrm{dir}}\Big|_{\mathrm{eq}}, \tag{1.89}$$

$c_{\mathrm{r,dir}}$ and $c_{\mathrm{g,dir}}$ are related by

$$c_{\mathrm{r,dir}} \left(np\right)_{\mathrm{eq}} = c_{\mathrm{r,dir}}. \tag{1.90}$$

Therefore, defining the *direct recombination lifetime* [36] $\tau_{\mathrm{r}} = \left(c_{\mathrm{r,dir}} N^+\right)^{-1}$ where N^+ is the net ionized doping density, band-to-band processes are characterized by;

$$R_n^{\mathrm{dir}} = R_p^{\mathrm{dir}} = \frac{np}{\tau_{\mathrm{r}} N^+}, \qquad G_n^{\mathrm{dir}} = G_p^{\mathrm{dir}} = \frac{(np)_{\mathrm{eq}}}{\tau_{\mathrm{r}} N^+}. \tag{1.91}$$

In terms of the transition rates previously exploited, one has

$$p_{12} = R_{n0}^{\mathrm{dir}} = R_{p0}^{\mathrm{dir}}, \qquad p_{21} = G_{n0}^{\mathrm{dir}} = G_{p0}^{\mathrm{dir}}, \tag{1.92}$$

where the generation and recombination rates are evaluated in the noiseless average values $n_0 = \langle n \rangle$, $p_0 = \langle p \rangle$. The correlation spectra of the Langevin sources are then defined by (1.85):

$$K_{\gamma_1,\gamma_1} = 2(p_{12} + p_{21}) = 2\left(R_{n0}^{\text{dir}} + G_{n0}^{\text{dir}}\right), \tag{1.93a}$$

$$K_{\gamma_2,\gamma_2} = 2(p_{12} + p_{21}) = 2\left(R_{p0}^{\text{dir}} + G_{p0}^{\text{dir}}\right), \tag{1.93b}$$

$$K_{\gamma_1,\gamma_2} = -2(p_{12} + p_{21}) = -2\left(R_{n0}^{\text{dir}} + G_{n0}^{\text{dir}}\right). \tag{1.93c}$$

Notice that in the drift-diffusion physical model, transport equations involve electron and hole carrier densities (see the discussion in Sect. 2.1.1), therefore, since holes are positive charges, the Langevin sources are $\gamma_n = \gamma_1$, for the electron continuity equation, and $\gamma_p = -\gamma_2$ for the hole continuity equation; one finally obtains for the local noise source:

$$K_{\gamma_n,\gamma_n} = 2\left(R_{n0}^{\text{dir}} + G_{n0}^{\text{dir}}\right), \tag{1.94a}$$

$$K_{\gamma_p,\gamma_p} = 2\left(R_{p0}^{\text{dir}} + G_{p0}^{\text{dir}}\right), \tag{1.94b}$$

$$K_{\gamma_n,\gamma_p} = 2\left(R_{n0}^{\text{dir}} + G_{n0}^{\text{dir}}\right). \tag{1.94c}$$

Trap-assisted transitions. Among the various possibilities of trap-assisted transitions, we shall consider here only the case of a single trap level within the semiconductor forbidden band which can be ionized by absorbing or emitting a single electron. This process is usually described according to the model proposed by Shockley, Read and Hall [35, 37, 38]. Let N_{t} be the density of trap levels included in the material, n_{t} the density of trapped electrons, τ_n and τ_p the recombination lifetimes for electrons and holes, respectively. Four different processes are assumed to take place:

- electron capture (see Fig. 1.5a), corresponding to the recombination rate [35]:

$$R_n^{\text{srh}} = \frac{n}{\tau_n}\left(1 - \frac{n_{\text{t}}}{N_{\text{t}}}\right); \tag{1.95}$$

- hole capture (Fig. 1.5b), corresponding to the recombination rate:

$$R_p^{\text{srh}} = \frac{p}{\tau_p}\frac{n_{\text{t}}}{N_{\text{t}}}; \tag{1.96}$$

- hole emission (Fig. 1.5c), corresponding to the generation rate:

$$G_p^{\text{srh}} = \frac{p_1}{\tau_p}\left(1 - \frac{n_{\text{t}}}{N_{\text{t}}}\right); \tag{1.97}$$

- electron emission (Fig. 1.5d), corresponding to the generation rate:

$$G_n^{\text{srh}} = \frac{n_1}{\tau_n}\frac{n_{\text{t}}}{N_{\text{t}}}. \tag{1.98}$$

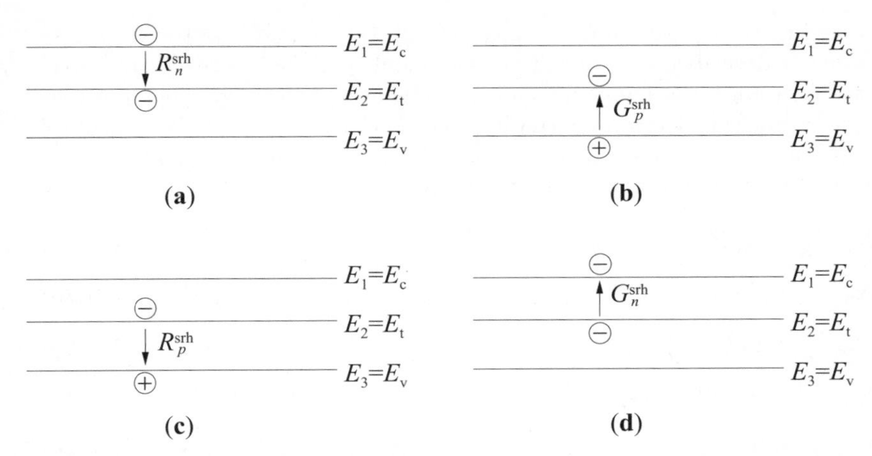

Fig. 1.5. Jump processes taking place for single level trap-assisted transitions: (**a**) electron capture; (**b**) hole capture; (**c**) hole emission; (**d**) electron emission

In (1.97) and (1.98) the following constants are defined:

$$n_1 = N_c \mathcal{F}_{1/2}\left(\frac{E_t - E_c}{k_B T}\right) \qquad p_1 = N_v \mathcal{F}_{1/2}\left(\frac{E_v - E_t}{k_B T}\right), \tag{1.99}$$

where N_c (N_v) is the conduction (valence) band effective density of states, $E_c = E_1$ is the conduction band edge, $E_t = E_2$ is the trap level, $E_v = E_3$ is the valence band edge, and $\mathcal{F}_{1/2}$ is the Fermi function of order 1/2. The transition rates among the three energy levels are therefore:

$$\begin{aligned} p_{12} &= R_{n0}^{srh}, & p_{21} &= G_{n0}^{srh}, \\ p_{13} &= 0, & p_{31} &= 0, \\ p_{23} &= G_{p0}^{srh}, & p_{32} &= R_{p0}^{srh}, \end{aligned} \tag{1.100}$$

where again the generation and recombination rates are evaluated in the noiseless average values $n_0 = \langle n \rangle$, $n_{t0} = \langle n_t \rangle$, $p_0 = \langle p \rangle$. Equation (1.85) yields the correlation spectra of the Langevin sources:

$$K_{\gamma_1,\gamma_1} = 2(p_{12} + p_{21} + p_{13} + p_{31}) = 2\left(R_{n0}^{srh} + G_{n0}^{srh}\right), \tag{1.101a}$$

$$\begin{aligned} K_{\gamma_2,\gamma_2} &= 2(p_{12} + p_{21} + p_{23} + p_{32}) \\ &= 2\left(R_{n0}^{srh} + G_{n0}^{srh} + R_{p0}^{srh} + G_{p0}^{srh}\right), \end{aligned} \tag{1.101b}$$

$$K_{\gamma_3,\gamma_3} = 2(p_{31} + p_{13} + p_{32} + p_{23}) = 2\left(R_{p0}^{srh} + G_{p0}^{srh}\right), \tag{1.101c}$$

$$K_{\gamma_1,\gamma_2} = -2(p_{12} + p_{21}) = -2\left(R_{n0}^{srh} + G_{n0}^{srh}\right), \tag{1.101d}$$

$$K_{\gamma_1,\gamma_3} = -2(p_{13} + p_{31}) = 0, \tag{1.101e}$$

$$K_{\gamma_2,\gamma_3} = -2(p_{23} + p_{32}) = -2\left(R_{p0}^{srh} + G_{p0}^{srh}\right). \tag{1.101f}$$

According to the drift-diffusion model (2.2), the dynamics of the trap-level occupancy is described in terms of an equivalent, positively ionized trap-level density. Therefore, the Langevin sources to be added to the model equations are $\gamma_n = \gamma_1$ for the electron continuity equation, $\gamma_\mathrm{D} = -\gamma_2$ for the trap-level rate equation (2.2d), and $\gamma_p = -\gamma_3$ for the hole-continuity equation. The corresponding local noise sources finally are:

$$K_{\gamma_n,\gamma_n} = 2\left(R_{n0}^{\mathrm{srh}} + G_{n0}^{\mathrm{srh}}\right), \tag{1.102a}$$

$$K_{\gamma_\mathrm{D},\gamma_\mathrm{D}} = 2\left(R_{n0}^{\mathrm{srh}} + G_{n0}^{\mathrm{srh}} + R_{p0}^{\mathrm{srh}} + G_{p0}^{\mathrm{srh}}\right), \tag{1.102b}$$

$$K_{\gamma_p,\gamma_p} = 2\left(R_{p0}^{\mathrm{srh}} + G_{p0}^{\mathrm{srh}}\right), \tag{1.102c}$$

$$K_{\gamma_n,\gamma_\mathrm{D}} = 2\left(R_{n0}^{\mathrm{srh}} + G_{n0}^{\mathrm{srh}}\right), \tag{1.102d}$$

$$K_{\gamma_n,\gamma_p} = 0, \tag{1.102e}$$

$$K_{\gamma_\mathrm{D},\gamma_p} = -2\left(R_{p0}^{\mathrm{srh}} + G_{p0}^{\mathrm{srh}}\right). \tag{1.102f}$$

Approximate equivalent sources for GR noise. Although the microscopic noise sources (1.84), which we shall hereafter denote as *fundamental*, allows to correctly evaluate GR noise in any semiconductor device, a different, albeit to a certain degree equivalent, approach is often found in the literature [28, 39]. This amounts to expressing the GR microscopic noise sources as *current density fluctuations*. In principle, the two approaches can be made fully equivalent if the current density fluctuations are evaluated as a result of the potential and carrier density fluctuations induced in the device by the fundamental sources, but this clearly amounts to solving exactly the fundamental scheme. Therefore, the following simplification is often assumed to derive *approximate* equivalent current density noise sources: these are evaluated according to the relationship

$$\delta\tilde{\boldsymbol{J}}_n = q\boldsymbol{v}_{n0}\delta\tilde{n} \qquad \delta\tilde{\boldsymbol{J}}_p = q\boldsymbol{v}_{p0}\delta\tilde{p}, \tag{1.103}$$

based on the assumption of homogeneous, space-independent conditions, which in turn correspond to approximating the linearized current continuity equations exploited in the physical model into a form akin to (1.81). Such fluctuations are then *assumed* to hold in inhomogeneous conditions as well, yielding [39]:

$$\mathsf{K}_{\delta\boldsymbol{J}_\alpha,\delta\boldsymbol{J}_\beta}(\boldsymbol{r};\omega) = q^2\boldsymbol{v}_{\alpha0}(\boldsymbol{r})\boldsymbol{v}_{\beta0}(\boldsymbol{r})K_{\delta\alpha,\delta\beta}(\boldsymbol{r};\omega) \qquad \alpha,\beta = n,p. \tag{1.104}$$

A further simplification results in the equivalent *monopolar* model, which is based on neglecting minority carrier fluctuations, and their correlation with majority carrier fluctuations. Analytically, this amounts to setting $\mathsf{K}_{\delta\boldsymbol{J}_\alpha,\delta\boldsymbol{J}_\beta} = 0$ unless $\alpha = \beta$, where α denotes the majority carrier.

The detailed expression of the equivalent approximate sources depends on the microscopic GR mechanism considered. For instance, band-to-band transitions yield, in homogeneous conditions:

$$\frac{\partial \delta n}{\partial t} = -\frac{\delta n}{\tau_n} - \frac{\delta p}{\tau_p} + \gamma_n, \tag{1.105}$$

$$\frac{\partial \delta p}{\partial t} = -\frac{\delta n}{\tau_n} - \frac{\delta p}{\tau_p} + \gamma_p, \tag{1.106}$$

where $1/\tau_\alpha = \partial U_\alpha^{\mathrm{dir}}/\partial \alpha\big|_0$ ($U_\alpha^{\mathrm{dir}} = R_\alpha^{\mathrm{dir}} - G_\alpha^{\mathrm{dir}}$, $\alpha = n, p$). Provided the noiseless average values n_0, p_0 are stationary (i.e. time independent), Fourier transformation yields

$$\mathrm{i}\omega\delta\tilde{n} = -\frac{\delta\tilde{n}}{\tau_n} - \frac{\delta\tilde{p}}{\tau_p} + \tilde{\gamma}_n, \tag{1.107}$$

$$\mathrm{i}\omega\delta\tilde{p} = -\frac{\delta\tilde{n}}{\tau_n} - \frac{\delta\tilde{p}}{\tau_p} + \tilde{\gamma}_p. \tag{1.108}$$

This algebraic system can be easily inverted, obtaining:

$$\delta\tilde{n} = \frac{\tau_n}{\mathrm{i}\omega(\tau_n + \tau_p)(1 + \mathrm{i}\omega\tau_{\mathrm{eq}})}\left[(1 + \mathrm{i}\tau_p)\tilde{\gamma}_n - \tilde{\gamma}_p\right], \tag{1.109}$$

$$\delta\tilde{p} = \frac{\tau_p}{\mathrm{i}\omega(\tau_n + \tau_p)(1 + \mathrm{i}\omega\tau_{\mathrm{eq}})}\left[-\tilde{\gamma}_n + (1 + \mathrm{i}\tau_n)\tilde{\gamma}_p\right], \tag{1.110}$$

where $\tau_{\mathrm{eq}} = \tau_n\tau_p/(\tau_n + \tau_p)$. According to (1.94), the corresponding local noise source in terms of carrier density fluctuations (to be converted into current density fluctuations through (1.104)) is

$$K_{\delta n,\delta n} = K_{\delta p,\delta p} = K_{\delta n,\delta p} = \frac{2\tau_{\mathrm{eq}}^2}{1 + \omega^2\tau_{\mathrm{eq}}^2}\left(R_{n0}^{\mathrm{dir}} + G_{n0}^{\mathrm{dir}}\right), \tag{1.111}$$

i.e. the microscopic noise source is no longer white, but exhibits a Lorentzian frequency dependence. More details on this approximation, and a discussion on its validity, can be found in Sect. 4.1.3.

For single-level trap-assisted transitions, the linearized and frequency-transformed rate equations in homogeneous conditions are:

$$(\mathrm{i}\omega\mathsf{I} + \mathsf{M}) \cdot \begin{bmatrix} \delta\tilde{n} \\ \delta\tilde{p} \\ \delta\tilde{n}_{\mathrm{t}} \end{bmatrix} = \begin{bmatrix} \tilde{\gamma}_n \\ \tilde{\gamma}_p \\ \tilde{\gamma}_{\mathrm{D}} \end{bmatrix}, \tag{1.112}$$

where the phenomenological relaxation matrix [16, 34, 39] is

$$\mathsf{M} = \begin{bmatrix} \frac{1}{\tau_n}\left(1 - \frac{n_{\mathrm{t}0}}{N_{\mathrm{t}}}\right) & 0 & -\frac{1}{\tau_n}\frac{n_0 + n_1}{N_{\mathrm{t}}} \\ 0 & \frac{1}{\tau_p}\frac{n_{\mathrm{t}0}}{N_{\mathrm{t}}} & \frac{1}{\tau_p}\frac{p_0 + p_1}{N_{\mathrm{t}}} \\ -\frac{1}{\tau_n}\left(1 - \frac{n_{\mathrm{t}0}}{N_{\mathrm{t}}}\right) & \frac{1}{\tau_p}\frac{n_{\mathrm{t}0}}{N_{\mathrm{t}}} & \frac{1}{\tau_n}\frac{n_0 + n_1}{N_{\mathrm{t}}} + \frac{1}{\tau_p}\frac{p_0 + p_1}{N_{\mathrm{t}}} \end{bmatrix}. \tag{1.113}$$

Equation (1.112) can be inverted to yield the correlation matrix of the carrier fluctuations to be exploited in (1.104):

$$\begin{bmatrix} K_{\delta n,\delta n} & K_{\delta n,\delta p} & K_{\delta n,\delta n_\mathrm{t}} \\ K_{\delta p,\delta n} & K_{\delta p,\delta p} & K_{\delta p,\delta n_\mathrm{t}} \\ K_{\delta n_\mathrm{t},\delta n} & K_{\delta n_\mathrm{t},\delta p} & K_{\delta n_\mathrm{t},\delta n_\mathrm{t}} \end{bmatrix} = (\mathrm{i}\omega\mathsf{I}+\mathsf{M})^{-1}\cdot\begin{bmatrix} K_{\delta\gamma_n,\delta\gamma_n} & K_{\delta\gamma_n,\delta\gamma_p} & K_{\delta\gamma_n,\delta\gamma_\mathrm{D}} \\ K_{\delta\gamma_p,\delta\gamma_n} & K_{\delta\gamma_p,\delta\gamma_p} & K_{\delta\gamma_p,\delta\gamma_\mathrm{D}} \\ K_{\delta\gamma_\mathrm{D},\delta\gamma_n} & K_{\delta\gamma_\mathrm{D},\delta\gamma_p} & K_{\delta\gamma_\mathrm{D},\delta\gamma_\mathrm{D}} \end{bmatrix}\cdot(\mathrm{i}\omega\mathsf{I}+\mathsf{M})^{-\dagger}, \tag{1.114}$$

which can be shown [16] to have embedded a frequency dependence resulting from the superposition of two Lorentzians, whose corner frequencies are related to the opposite of the two non-zero eigenvalues of the relaxation matrix M, solutions of the second-order algebraic equation:

$$\lambda^2+\lambda\left[\frac{1}{\tau_n}\left(1-\frac{n_{\mathrm{t}0}}{N_\mathrm{t}}+\frac{n_0+n_1}{N_\mathrm{t}}\right)+\frac{1}{\tau_p}\left(\frac{n_{\mathrm{t}0}}{N_\mathrm{t}}+\frac{p_0+p_1}{N_\mathrm{t}}\right)\right] + \frac{1}{\tau_n\tau_p}\left[\frac{n_{\mathrm{t}0}}{N_\mathrm{t}}\frac{n_0+n_1}{N_\mathrm{t}}+\left(1-\frac{n_{\mathrm{t}0}}{N_\mathrm{t}}\right)\frac{n_{\mathrm{t}0}}{N_\mathrm{t}}+\left(1-\frac{n_{\mathrm{t}0}}{N_\mathrm{t}}\right)\frac{p_0+p_1}{N_\mathrm{t}}\right]=0. \tag{1.115}$$

Notice that, in many practical cases, trap-assisted GR noise can be approximated by carrier density fluctuations as in (1.111), i.e. fully correlated and equal in magnitude for electrons and holes, at least if deep traps (recombination centers) are assumed and the free carrier concentration is high enough with respect to the trap-level density [39]. Let us consider a uniformly n-doped silicon sample ($N_\mathrm{D}=10^{15}$ cm^{-3}), wherein traps are characterized by $E_\mathrm{t}=E_\mathrm{Fi}$ (the intrinsic Fermi level), the trap level density is $N_\mathrm{t}=10^{11}$ cm^{-3}, and the recombination lifetimes are $\tau_n=100$ ps and $\tau_p=200$ ps. Figure 1.6 shows the frequency dependence of the carrier fluctuations power spectra $K_{\delta n,\delta n}$ and $K_{\delta p,\delta p}$, and the correlation coefficient

$$C_{\delta n,\delta p}=\frac{K_{\delta n,\delta p}}{K_{\delta n,\delta n}K_{\delta p,\delta p}},$$

normalized to $K_{\delta n,\delta n}(0)$. The electron and hole density fluctuations are fully correlated and Lorentzian in shape, and therefore equivalent to the behavior exhibited in the band-to-band case, up to frequencies of the order of 100 GHz. This corresponds to the minority carrier lifetime τ_p, which in turn is the lower eigenvalue of M. The second Lorentzian observed in the majority carrier fluctuations is related to the higher eigenvalue of M (corresponding to a time constant equal to 0.01 ps), while for minority carriers a zero almost exactly compensates such a pole. From a physical standpoint, the frequency behavior can be explained as follows. For low-frequency components of hole

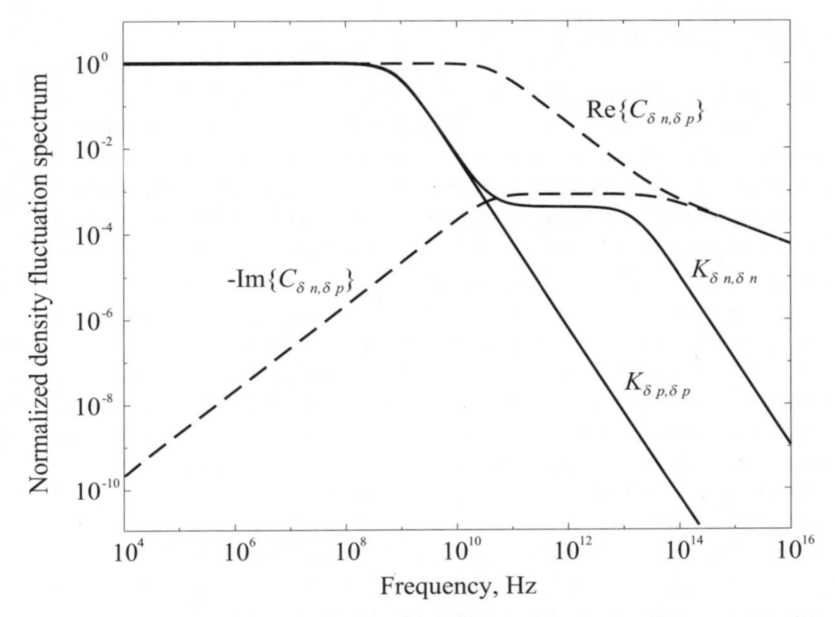

Fig. 1.6. Frequency dependence of the carrier density fluctuations spectra and correlation coefficient, normalized to $K_{\delta n,\delta n}(0)$, for a uniformly doped n-type sample. The material data are $N_{\mathrm{D}} = 10^{15}$ cm^{-3}, $E_{\mathrm{t}} = E_{\mathrm{Fi}}$, $N_{\mathrm{t}} = 10^{11}$ cm^{-3}, $\tau_n = 100$ ps and $\tau_p = 200$ ps (from [39])

and electron fluctuations, the trap level introduces negligible delay, and therefore strong, fully correlated fluctuations arise. On increasing the fluctuation frequency the first corner frequency is met, which corresponds to the minority carrier lifetime. Beyond that frequency, band-to-band transitions through the trap gradually disappear because the interaction between the trap level and minority carriers becomes increasingly inefficient, implying that the density fluctuation spectra of both minority and majority carriers decrease. However, since the interaction between the trap level and the majority carrier is still efficient, correlation remains high. After the zero corner frequency, the exchange between the trap level and the valence band becomes negligible, and fluctuations are almost entirely due to the interaction between the trap level and the conduction band: this corresponds to a decreasing correlation. The second pole is related to the time constant of a system including the trap and the conduction band only, which is very low, owing to the large majority carrier concentration; after the second pole, the exchange between the trap and the conduction band also becomes increasingly inefficient, thus causing a final decrease of the majority carrier fluctuation spectrum.

1.2.3 $1/f$-Like Fluctuations

Unlike diffusion and GR noise, the physical origin of $1/f$ and $1/f$-like fluctuations, often called *flicker* noise, is not well established yet: several models have been proposed to explain their occurrence, but up to now no theory exists to interpret exhaustively all the available experimental data. Nevertheless, $1/f$ noise has paramount importance in practical applications both with respect to the noise performance of MOS and bipolar transistors operated in linear conditions (e.g. see [40–43] and [44–46], respectively, and references therein), and for the noise analysis of RF semiconductor devices operated in the nonlinear quasi-periodic regime. In the latter case, although for RF operation diffusion noise is the only component directly affecting device performance, the nonlinear regime determines up-conversion of low-frequency noise components (i.e. the frequency band where flicker noise has a major impact) around the central frequencies of the harmonic noiseless steady-state. This phenomenon has been studied for a long time at a circuit level description, e.g. see [47–49], but the analysis has been only recently extended to physics-based noise models. More details on this can be found in Chap. 5.

Experimental data collected by several researchers have led to the following conclusions. Let us apply a constant voltage v to a homogeneous semiconductor sample with resistance R. The current i flowing through the sample undergoes fluctuations δi whose power spectrum is proportional to $1/f$, where $\omega = 2\pi f$. More precisely, experiments yield[5]

$$\frac{S_{\delta i,\delta i}}{|i|^2} = \frac{\text{const.}}{f}. \tag{1.116}$$

Since the applied voltage is held fixed, current fluctuations can be ascribed only to fluctuations in the material conductivity:

$$v = Ri = \text{const.} \Longrightarrow \frac{\delta i}{i} = -\frac{\delta R}{R} \Longrightarrow \frac{S_{\delta i,\delta i}}{|i|^2} = \frac{S_{\delta R,\delta R}}{|R|^2}.$$

In order to give a unified interpretation of several experimental data, Hooge proposed [23, 50] the following phenomenological expression:

$$\frac{S_{\delta i,\delta i}}{|i|^2} = \frac{S_{\delta R,\delta R}}{|R|^2} = \frac{\alpha_{\mathrm{H}}}{Nf}, \tag{1.117}$$

where α_{H} is called the *Hooge constant*, a quality parameter for the $1/f$ noise material properties, and N is the number of carriers included in the sample. The only assumption underlying (1.117) is that the single carrier fluctuations yielding the $1/f$ spectrum are uncorrelated.

[5] Notice that experiments often result in power spectra proportional to $1/f^\gamma$ with $\gamma = 1 \div 2$. We shall explicitly account for the case $\gamma = 1$ only, since very often the other cases can be interpreted in terms of diffusion processes [23].

Notice that the very assumption of $1/f$ power spectrum results in difficulties (see, for example, the discussion in [51]): this frequency behavior can not hold rigorously for any frequency, since this would give rise to a process with an unphysically divergent energy (i.e. variance). Although this does not provide an insuperable difficulty for the high-frequency behavior (at some higher frequency $f_{\rm h}$ the spectrum slope must be steeper than -1, though this $f_{\rm h}$ has never been experimentally observed because at this frequency $1/f$ fluctuations are hidden by the white velocity fluctuations), at low frequency the situation is much more involved. The simplest solution would be to assume the $1/f$ behavior breaks down at very low frequency, so as to yield a convergent integral. Unfortunately, experiments have shown no sign of slope change down to 10^{-6} Hz [23, 52]. Another possible solution to this contradiction has been proposed by Keshner [53]: fluctuations causing $1/f$ noise are assumed to be represented by a *non-stationary* random process, where the autocorrelation function does not depend only on the difference τ between the two observation times, but also on the time t elapsed since the process began. Therefore, the power spectrum (defined by Fourier transforming the autocorrelation function with respect to τ) depends on t as well. This dependency has not been experimentally observed up to now, but this is ascribed to the fact that the non-stationary process is a very slowly varying function of t, and that the time intervals exploited in the measurements are short with respect to t itself. An example of a non-stationary random process amenable to be considered as a model for general $1/f$ fluctuations is introduced in [53].

Even neglecting these mathematical difficulties, the very nature of $1/f$ microscopic fluctuations is still an open problem. Experiments show that, in homogeneous semiconductor samples, $\alpha_{\rm H}$ values are spread between 2×10^{-3} and 5×10^{-3} [23, 54]; therefore, it is quite reasonable to assume that $1/f$ noise results from some fundamental physical phenomenon. Since the resistance of a homogeneous sample depends on the product between carrier mobility μ and carrier number N, resistance fluctuations δR can be due to two different, uncorrelated phenomena:

$$\frac{\delta R}{R} = -\frac{\delta\mu}{\mu} - \frac{\delta N}{N} \Longrightarrow \frac{S_{\delta R,\delta R}}{|R|^2} = \frac{S_{\delta\mu,\delta\mu}}{|\mu|^2} + \frac{S_{\delta N,\delta N}}{|N|^2},$$

i.e. $1/f$ noise can be due to *mobility* fuctuations or *number* fluctuations, or both, depending on material composition and purity. In bulk pure materials, experiments have shown the occurrence of mobility fluctuations with $\alpha_{\rm H} \approx 10^{-4}$. On the other hand, in MOS devices experimental results can be effectively explained assuming carrier number fluctuations related to surface GR processes (see [23, 40] and references therein).

From a theoretical standpoint, at least three different frameworks have been proposed yielding a $1/f$ microscopic noise source. Concerning carrier number fluctuations, the $1/f$ spectrum is produced by superimposing a continuous distribution of Lorentzian spectra obtained from uncorrelated GR processes characterized by a proper distribution of lifetimes [23, 51]. For example, McWhorter [55] proposed a model for surface GR transitions. Mobility fluctuations, on the other hand, require a more detailed analysis. Two theories have been developed: the first, named *local interference noise* [56–58], attains the correct frequency dependence as a superposition of Lorentzians not related to GR mechanisms (i.e. no carrier number fluctuations are involved). The second is Handel's *quantum* $1/f$ *noise* [24, 59, 60], based on an interference phenomenon between the bremsstrahlung and non-bremsstrahlung parts of the single-electron wavefunction.

In order to account for $1/f$ fluctuations in physics-based noise models, a microscopic noise source in terms of local current-density fluctuations must be provided to be included in the proper continuity equation as a Langevin source. Since, from first principles, white sources only can be rigorously derived (see Sect. A.6), the $1/f$ noise source must be considered as purely phenomenological. Let us consider a volume V, small enough to assume constant electron[6] density n. Then, $N = nV$ and $i = JA$, where A is the volume cross section with respect to the electron flow. In such conditions, (1.117) yields

$$S_{\delta i,\delta i} = \frac{\alpha_{\mathrm{H}}}{f} \frac{|JA|^2}{nV}, \tag{1.118}$$

where $S_{\delta i,\delta i} = (A)^2 S_{\delta J,\delta J}$. Therefore one has

$$S_{\delta J,\delta J} = \frac{\alpha_{\mathrm{H}}}{f} \frac{|J|^2}{nV}. \tag{1.119}$$

Taking the limit $V \to 0$, for spatially uncorrelated sources one obtains the microscopic noise source for $1/f$ noise proposed by Nougier in [28]:

$$\mathsf{S}_{\delta \boldsymbol{J}_n,\delta \boldsymbol{J}_n}(\boldsymbol{r},\boldsymbol{r}';\omega) = \mathsf{K}_{\delta \boldsymbol{J}_n,\delta \boldsymbol{J}_n}(\boldsymbol{r};\omega)\delta(\boldsymbol{r}-\boldsymbol{r}'), \tag{1.120a}$$

$$\mathsf{S}_{\delta \boldsymbol{J}_p,\delta \boldsymbol{J}_p}(\boldsymbol{r},\boldsymbol{r}';\omega) = \mathsf{K}_{\delta \boldsymbol{J}_p,\delta \boldsymbol{J}_p}(\boldsymbol{r};\omega)\delta(\boldsymbol{r}-\boldsymbol{r}'), \tag{1.120b}$$

where the local noise source for $1/f$ noise is

$$\mathsf{K}_{\delta \boldsymbol{J}_n,\delta \boldsymbol{J}_n}(\boldsymbol{r};\omega) = \frac{\alpha_{\mathrm{H}n}}{f} \frac{|\boldsymbol{J}_n(\boldsymbol{r})|^2}{n(\boldsymbol{r})}\mathsf{I}, \tag{1.121a}$$

$$\mathsf{K}_{\delta \boldsymbol{J}_p,\delta \boldsymbol{J}_p}(\boldsymbol{r};\omega) = \frac{\alpha_{\mathrm{H}p}}{f} \frac{|\boldsymbol{J}_p(\boldsymbol{r})|^2}{p(\boldsymbol{r})}\mathsf{I}. \tag{1.121b}$$

1.3 Back to Basics: the Fundamental Approach

In the previous section, the mesoscopic treatment of the noise sources has been developed exploiting ad hoc, quasi-classical models. In particular, no

[6] The same considerations apply if hole conduction is involved.

unified, first-principle model including fluctuations has been presented to justify the Langevin approach. The purpose of this section is to present, without any attempt at completeness, a short review of the *fundamental approach* to the development of a semiconductor noise model. This approach is shown to lead to a rigorous quasi-classical model for carrier transport including fluctuations, which is given by a Boltzmann transport equation (BTE) with a stochastic forcing term. Such an equation can be considered as the starting point for the derivation of all other PDE-based transport models including fluctuations. For the sake of simplicity, only electron diffusion noise will be considered, i.e. we shall not include interband transitions.

The fundamental approach is based on a collective, many-body quantum description of the free electrons participating in electrical conduction. The system Hamiltonian is decomposed according to [19, 20]

$$H = H_0 + \lambda V - \boldsymbol{A} \cdot \boldsymbol{F}(t), \tag{1.122}$$

where:

- H_0 is the Hamiltonian of the noninteracting particle systems, electrons and phonons in our case, which can be diagonalized according to the eigenstates $|\gamma\rangle$ with energy E_γ:

  $$H_0|\gamma\rangle = E_\gamma|\gamma\rangle.$$

  The eigenstates can be expressed in terms of the occupation numbers $\boldsymbol{n}_\zeta$ and $\boldsymbol{N}_\eta$ for the electron and phonon states, respectively; therefore

  $$|\gamma\rangle = |\boldsymbol{n}_\zeta, \boldsymbol{N}_\eta\rangle.$$

- λV represents the interactions between the two particle subsystems, and is therefore responsible for the random transitions between the H_0 eigenstates; in other words, this term is responsible for the dissipative phenomena included in the description [20].
- $-\boldsymbol{A} \cdot \boldsymbol{F}(t)$ includes the interaction of the quantum particles with an external applied field; for instance, for a set of charged particles of charge q_i submitted to an electric field $\boldsymbol{\mathcal{E}}(t)$ one has:

  $$\boldsymbol{A} \cdot \boldsymbol{F}(t) = \sum_i q_i \left(\boldsymbol{r}_i - \boldsymbol{r}_{i,\mathrm{eq}}\right) \cdot \boldsymbol{\mathcal{E}}(t),$$

  where $\boldsymbol{r}_i$ is the i-th particle position, equal to $\boldsymbol{r}_{i,\mathrm{eq}}$ before the external field is switched on.

The transition rate from state $|\gamma\rangle$ to $|\overline{\gamma}\rangle$ due to the dissipative process λV is given by the Fermi golden rule:

$$W_{\gamma\overline{\gamma}} = \frac{2\pi\lambda^2}{\hbar} \left|\langle\gamma|V|\overline{\gamma}\rangle\right|^2 \delta(E_\gamma - E_{\overline{\gamma}}) \tag{1.123}$$

and satisfies microscopic reversibility since $W_{\gamma\overline{\gamma}} = W_{\overline{\gamma}\gamma}$.

The statistical properties of the particle system are defined by the density operator $\rho_D(t)$, whose dynamics satisfies the *quantum Liouville* equation, or *von Neumann* equation:

$$\frac{\partial \rho_D}{\partial t} + \frac{1}{i\hbar}[\rho_D, H] = 0, \tag{1.124}$$

where $[\cdot,\cdot]$ is the commutator. This equation has to be approximated in order to obtain relationships between the macroscopic system properties (e.g. electrical conductivity) and the microscopic scattering events which cause them. Two paths can be followed, both leading to the same results. The first is based on deriving the Kubo–Green relations [61] relating the particle transport coefficients and the correlation function of fluctuations around an equilibrium state, and then applying the van Hove limit [19], i.e. the following set of conditions:

- $\lambda \longrightarrow 0$, representing a weak coupling between electrons and phonons;
- $t/\tau_t \longrightarrow +\infty$, where τ_t is the transition time (coarse graining in time);
- $\lambda^2 t$ finite.

According to the second methodology, Zwanzig's projection operator technique is applied to (1.124) to derive the dynamics of the diagonal and off-diagonal components of the density operator, and then the van Hove limit is evaluated [20]. The resulting equation for the probability density$f(\gamma,t) = \langle\gamma|\rho_D^R|\gamma\rangle$ (here, superscript R denotes the reduced operator after the van Hove limit) is typical of a Markov process, and has the form of a nonhomogeneous *many-body master equation*, where the forcing term is proportional to the external field [20, 22]:

$$\frac{\partial f}{\partial t} = \sum_{\overline{\gamma}}[W_{\overline{\gamma}\gamma}f(\overline{\gamma},t) - W_{\gamma\overline{\gamma}}f(\gamma,t)] + \frac{1}{k_B T}f_{eq}(\gamma)\boldsymbol{F}(t)\cdot\langle\gamma|\dot{\boldsymbol{A}}^R|\gamma\rangle, \tag{1.125}$$

where $f_{eq}(\gamma)$ is the stationary, zero-field solution, and $\dot{\boldsymbol{A}}^R$ is the time derivative of $\boldsymbol{A}^R$. The discussion in [20] shows that the diagonal part of $\rho_D^R(t)$ is adequate to properly describe the system macroscopic transport properties provided the observation time t is large enough with respect to the characteristic time of the microscopic transitions, i.e. for low-frequency operation.

As described in Sect. A.6, (1.125) can be transformed by exploiting the Kramers–Moyal expansion (A.92), therefore yielding an infinite hierarchy of moment equations (the *master hierarchy* [16,21]) involving the Fokker–Planck moments:

$$F_k(\boldsymbol{n}_\zeta) = \sum_{\overline{\gamma}}(\overline{n}_{\zeta 1} - n_{\zeta 1})\ldots(\overline{n}_{\zeta k} - n_{\zeta k})\,W_{\gamma\overline{\gamma}}, \tag{1.126}$$

where $\boldsymbol{n}_\zeta$ is a vector of k components $n_{\zeta 1},\ldots,n_{\zeta k}$. To the first order, the *quantum Boltzmann equation* [21] is obtained:

$$\frac{\partial \langle n_\zeta \rangle_t}{\partial t} = \langle F_1(n_\zeta) \rangle_t + \frac{1}{k_\mathrm{B}T} \langle n_\zeta \rangle_\mathrm{eq} \boldsymbol{F}(t) \cdot \langle \gamma | \dot{\boldsymbol{A}}^\mathrm{R} | \gamma \rangle. \tag{1.127}$$

The first term in the right-hand side (r.h.s.) is the collision term [21]:

$$\langle F_1(n_\zeta) \rangle_t = -\sum_{\zeta_1} \left[w_{\zeta\zeta_1} \langle n_\zeta \rangle_t \left(1 - \langle n_{\zeta_1} \rangle_t\right) - w_{\zeta_1\zeta} \langle n_{\zeta_1} \rangle_t \left(1 - \langle n_\zeta \rangle_t\right) \right], \tag{1.128}$$

where $w_{\zeta\zeta_1}$ is the transition rate from the electron state ζ to ζ_1. The second Fokker–Planck moment can be evaluated as well, see [16, 62]:

$$\begin{aligned} B\left(n_{\zeta_1}, n_{\zeta_2}\right) = & \langle F_2\left(n_{\zeta_1}, n_{\zeta_2}\right) \rangle_t = - \left[w_{\zeta_1\zeta_2} \langle n_{\zeta_1} \rangle_t \left(1 - \langle n_{\zeta_2} \rangle_t\right) \right. \\ & \left. + w_{\zeta_1\zeta_2} \langle n_{\zeta_2} \rangle_t \left(1 - \langle n_{\zeta_1} \rangle_t\right) \right] \\ & + \delta_{\zeta_1,\zeta_2} \sum_{\zeta} \left[w_{\zeta_1\zeta} \langle n_{\zeta_1} \rangle_t \left(1 - \langle n_\zeta \rangle_t\right) \right. \\ & \left. + w_{\zeta\zeta_1} \langle n_\zeta \rangle_t \left(1 - \langle n_{\zeta_1} \rangle_t\right) \right], \end{aligned} \tag{1.129}$$

where δ_{ζ_1,ζ_2} is the Kronecker symbol.

As shown in [21], the quantum Boltzmann equation (1.127) reduces, exploiting the Wigner function, to the classical BTE for the free electrons in a crystalline structure whose states are characterized by the wave vector $\boldsymbol{k}$ (Bloch states). Further, the master equation (1.125) can be converted into an equivalent Langevin equation, as described in Sect. A.6, for the free electron distribution function, wherein the deterministic part is fully equivalent to the BTE:

$$\left[\frac{\partial}{\partial t} + \boldsymbol{v}_{\boldsymbol{k}} \cdot \nabla_{\boldsymbol{r}} + \frac{1}{\hbar} \boldsymbol{F}(t) \cdot \nabla_{\boldsymbol{k}} \right] f_{\boldsymbol{r},\boldsymbol{k}}(\boldsymbol{r}, \boldsymbol{k}, t) = F_1(n) + \xi(\boldsymbol{r}, \boldsymbol{k}, t), \tag{1.130}$$

where F_1 is the first Fokker–Planck moment (1.128), $\nabla_{\boldsymbol{\alpha}}$ acts on variable $\boldsymbol{\alpha}$, and the stochastic forcing term is white with correlation spectrum proportional to the second Fokker–Planck moment (see (A.95) and (1.129)):

$$\begin{aligned} S_{\xi,\xi}(\boldsymbol{k}_1, \boldsymbol{k}_2) = & -2 \left[w_{\boldsymbol{k}_1\boldsymbol{k}_2} \langle n_{\boldsymbol{k}_1} \rangle_t \left(1 - \langle n_{\boldsymbol{k}_2} \rangle_t\right) \right. \\ & \left. + w_{\boldsymbol{k}_1\boldsymbol{k}_2} \langle n_{\boldsymbol{k}_2} \rangle_t \left(1 - \langle n_{\boldsymbol{k}_1} \rangle_t\right) \right] \\ & + \delta(\boldsymbol{k}_1 - \boldsymbol{k}_2) \int \left[w_{\boldsymbol{k}_1\boldsymbol{k}} \langle n_{\boldsymbol{k}_1} \rangle_t \left(1 - \langle n_{\boldsymbol{k}} \rangle_t\right) \right. \\ & \left. + w_{\boldsymbol{k}\boldsymbol{k}_1} \langle n_{\boldsymbol{k}} \rangle_t \left(1 - \langle n_{\boldsymbol{k}_1} \rangle_t\right) \right] \, \mathrm{d}\boldsymbol{k}. \end{aligned} \tag{1.131}$$

Notice that in these equations $f_{\boldsymbol{r},\boldsymbol{k}}(\boldsymbol{r}, \boldsymbol{k}, t) \, \mathrm{d}\boldsymbol{r} \, \mathrm{d}\boldsymbol{k}$ has the usual meaning of number of electrons which, at time t, can be found between $\boldsymbol{r}$ and $\boldsymbol{r} + \mathrm{d}\boldsymbol{r}$, and between $\boldsymbol{k}$ and $\boldsymbol{k} + \mathrm{d}\boldsymbol{k}$, i.e. the electron distribution function.

Taking moments of the Langevin BTE (1.130), one obtains the hydrodynamic (drift-diffusion, in the simplest case; see the discussion in Sect. 2.1) transport models including the microscopic noise sources.

1.4 Equivalent Representation of Noisy Devices

Velocity or population fluctuations taking place within a semiconductor device cause induced current or voltage fluctuations to appear at the device terminals. Such an *external* noise can be detected by means of measurements and its characterization is of paramount importance in the circuit-oriented modeling of the noisy device, aimed at assessing its interaction with the circuit this is connected to by means of standard circuit analysis technique.

From a circuit standpoint, noise can be assimilated to a small-amplitude random signal. According to a Langevin picture of noisy circuit operation, small-amplitude voltage or current fluctuations associated to noisy device ports are impressed sources whose response corresponds to voltage or current fluctuations in the network connected to the noisy elements. The assumption of small amplitude generally implies that such a response can be derived by linearizing the circuit under consideration either around a DC working point (if the noiseless operation is small-signal) or around an AC dynamic working point (if the noiseless operation is large-signal, (quasi-) periodic). This latter case will be examined in detail in Chap. 5, while the present treatment will be confined to small-signal operation.

Since the physically meaningful quantity at a circuit level is the noise power exchanged by the elements, circuit noise characterizations are entirely based on second-order statistical properties, in particular power or correlation spectra of the external noise sources. Such spectra can be experimentally characterized, and only in particular cases are they directly related to the small-signal device response namely, if operation takes place at or near equilibrium, in which case the Nyquist theorem holds. In the following sections a brief review is presented on the circuit-oriented noise characterization of a multi-port in small-signal conditions; then, the Nyquist theorem will be extended to noisy multi-ports at or near equilibrium. Finally, system noise parameters such as the noise figure and the optimum noise impedance of a two-port noisy device will be reviewed.

1.4.1 Circuit-Oriented Device Noise Parameters

Let us consider a linear, passive (i.e. without internal generators), noisy N-port and denote with $\boldsymbol{v}(t)$ the array of instantaneous port voltages and with $\boldsymbol{i}(t)$ the array of instantaneous currents entering the device ports; let $\tilde{\boldsymbol{v}}$ and $\tilde{\boldsymbol{\imath}}$ be the Fourier transforms of $\boldsymbol{v}$ and $\boldsymbol{i}$. The Norton and Thévenin theorems yield, respectively:

$$\tilde{\boldsymbol{\imath}} = \mathsf{Y}(\omega) \cdot \tilde{\boldsymbol{v}} + \tilde{\boldsymbol{\imath}}_0, \tag{1.132}$$

$$\tilde{\boldsymbol{v}} = \mathsf{Z}(\omega) \cdot \tilde{\boldsymbol{\imath}} + \tilde{\boldsymbol{v}}_0, \tag{1.133}$$

where $\boldsymbol{i}_0$ is the array of the short-circuit port currents, $\boldsymbol{v}_0$ the array of the open-circuit port voltages, Y the admittance matrix, and Z the impedance

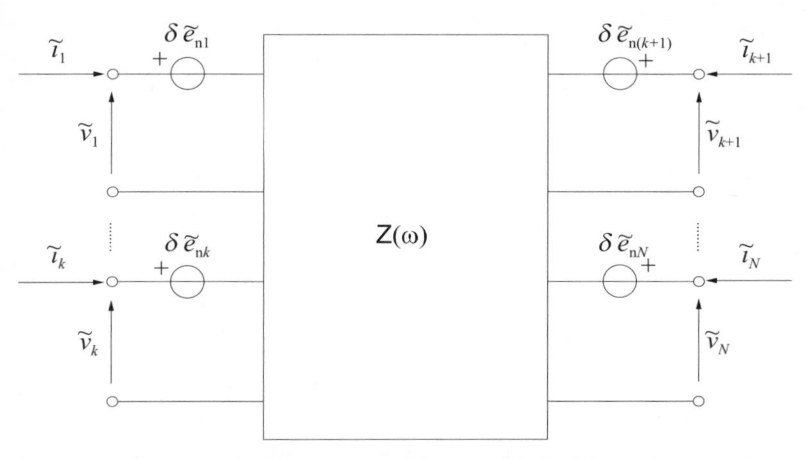

Fig. 1.7. Series equivalent circuit of a noisy linear N-port

matrix of the N-port. The parallel and series representations (1.132) and (1.133) will be used for further treatment, although others exist, such as the hybrid or transmission representations [63]. We have supposed that both representations exist, i.e. both the admittance and impedance matrices can be inverted. In a noisy N-port, the open-circuit voltages $\boldsymbol{v}_0$ and the short circuit currents $\boldsymbol{i}_0$ are zero-average random processes, expressing the voltage or current fluctuations induced at the device ports by the internal noise sources. Supposing that fluctuations are independent of the average values of voltages and currents, we can identify the processes $\boldsymbol{v}_0$ and $\boldsymbol{i}_0$ as the *noise generators* appearing in the series and parallel equivalent circuits shown in Fig. 1.7 and Fig. 1.8, respectively, as:

$$\delta\boldsymbol{e} \equiv \boldsymbol{v}_0, \tag{1.134}$$

$$\delta\boldsymbol{i} \equiv \boldsymbol{i}_0. \tag{1.135}$$

The formulation also holds if $\delta\boldsymbol{e}$ ($\delta\boldsymbol{i}$) depend on $\boldsymbol{v}$ or $\boldsymbol{i}$, although in this case voltage (current) fluctuations cannot be identified with open-circuit voltage (short-circuit current) generators.

From a circuit standpoint we are interested in the second-order statistical properties of $\delta\boldsymbol{e}$ and $\delta\boldsymbol{i}$, since these enable us to estimate the noise power delivered to the network by the noisy N-port. The second-order characterization reduces to the correlation matrices $\mathsf{S}_{\delta\boldsymbol{e},\delta\boldsymbol{e}}(\omega)$ and $\mathsf{S}_{\delta\boldsymbol{i},\delta\boldsymbol{i}}(\omega)$, whose elements are defined, respectively, as:

$$\left(\mathsf{S}_{\delta\boldsymbol{e},\delta\boldsymbol{e}}\right)_{i,j} = S_{\delta e_i,\delta e_j}, \tag{1.136}$$

$$\left(\mathsf{S}_{\delta\boldsymbol{i},\delta\boldsymbol{i}}\right)_{i,j} = S_{\delta i_i,\delta i_j}, \tag{1.137}$$

i.e. as the power and correlation spectra of the voltage (current) port noise generators. Since $S_{\delta e_i,\delta e_j} = S^*_{\delta e_j,\delta e_i}$ and $S_{\delta i_i,\delta i_j} = S^*_{\delta i_j,\delta i_i}$, see (A.69), a linear,

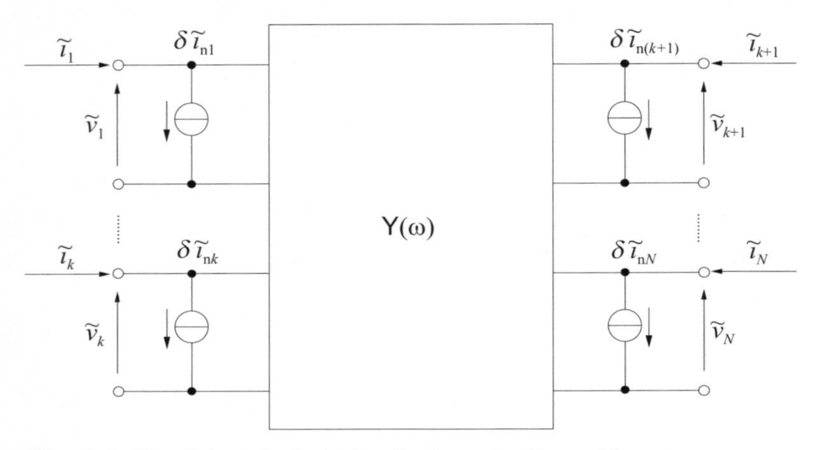

Fig. 1.8. Parallel equivalent circuit of a noisy linear N-port

noisy N-port is characterized by a set of N (real) power spectra and of $N(N-1)/2$ (complex) correlation spectra that generally depend on the operating frequency. Note that the previous treatment formally applies also to a linear, passive, noisy $(N+1)$-pole wherein one of the poles is taken as the common voltage reference; for the particular (but important) case $N = 2$, we can thus conclude that the noise of a two-port or three-pole is characterized by the power spectra of the input and output noise generators and by their correlation spectrum. Since (1.132) and (1.133) yield:

$$\delta\widetilde{\boldsymbol{\imath}} = -\mathsf{Y} \cdot \delta\widetilde{\boldsymbol{e}}, \tag{1.138}$$

$$\delta\widetilde{\boldsymbol{e}} = -\mathsf{Z} \cdot \delta\widetilde{\boldsymbol{\imath}}, \tag{1.139}$$

by exploiting (A.75) the correlation spectra of the voltage and current noise generators are related as follows:

$$\mathsf{S}_{\delta \boldsymbol{i},\delta \boldsymbol{i}} = \mathsf{Y} \cdot \mathsf{S}_{\delta \boldsymbol{e},\delta \boldsymbol{e}} \cdot \mathsf{Y}^{\dagger}, \tag{1.140a}$$

$$\mathsf{S}_{\delta \boldsymbol{e},\delta \boldsymbol{e}} = \mathsf{Z} \cdot \mathsf{S}_{\delta \boldsymbol{i},\delta \boldsymbol{i}} \cdot \mathsf{Z}^{\dagger}. \tag{1.140b}$$

The circuit models developed for linear, passive, noisy N-ports can be readily extended to nonlinear N-ports operating under small-signal conditions. In this case voltages and currents are small-signal values (i.e. small perturbations with respect to a DC operating point), while the (differential) impedance and admittance matrices originate from linearization, around the DC operating point, of the current or voltage-controlled constitutive relationships of the nonlinear N-port. In turn, the noise generators will generally depend on the DC operating point. Since fluctuations are expected to have small amplitude, noise generators are compatible with the small-signal regime.

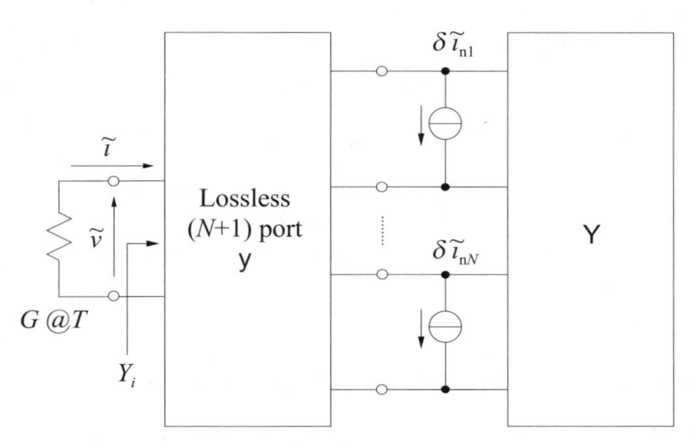

Fig. 1.9. Connecting a noisy one-port to a noisy multi-port through an embedding lossless network.

1.4.2 Nyquist Theorem for Linear Passive Multi-Ports

Following a definition that is customary in circuit theory, albeit not entirely rigorous, we shall define as *linear passive multi-port* any (reciprocal or non-reciprocal) multi-port operating at or near thermodynamic equilibrium. Assume that the multi-port is characterized through its admittance matrix Y. The correlation spectrum of the short-circuit noise currents can be shown to take the form:

$$\mathsf{S}_{\delta \boldsymbol{i},\delta \boldsymbol{i}} = 2k_{\mathrm{B}}T\left(\mathsf{Y} + \mathsf{Y}^{\dagger}\right). \tag{1.141}$$

The generalized Nyquist theorem (1.141) can be readily extended to evaluating the correlation spectrum of the open-circuit noise voltage taking into account (1.140); one immediately has:

$$\mathsf{S}_{\delta \boldsymbol{e},\delta \boldsymbol{e}} = 2k_{\mathrm{B}}T\left(\mathsf{Z} + \mathsf{Z}^{\dagger}\right). \tag{1.142}$$

where Z is the multi-port impedance matrix.

The proof of (1.141) can be carried out, for example, as shown in [64]. Consider a noisy N-port and a noisy resistor in thermodynamic equilibrium, and suppose we connect them through a reactive, noiseless $(N+1)$-port whose port 1 is connected to the resistor, while all remaining ports are connected to the noisy N-port, as shown in Fig. 1.9. Let Y be the $N \times N$ admittance matrix of the N-port, y the $(N+1) \times (N+1)$ admittance matrix of the embedding lossless network, which will be decomposed as follows:

$$\mathsf{y} = \begin{bmatrix} y_{11} & \boldsymbol{y}_{12}^{\mathrm{T}} \\ \boldsymbol{y}_{21} & \mathsf{y}_{22} \end{bmatrix}. \tag{1.143}$$

The input admittance seen from port 1 can be easily shown to be:

$$Y_\mathrm{i} = y_{11} - \boldsymbol{y}_{12}^\mathrm{T} \cdot (\mathsf{y}_{22} + \mathsf{Y}) \cdot \boldsymbol{y}_{21}, \tag{1.144}$$

while the short-circuit noise current spectrum at port 1 is:

$$S_{\delta i,\delta i} = \boldsymbol{y}_{12}^\mathrm{T} \cdot (\mathsf{y}_{22} + \mathsf{Y}) \cdot \mathsf{S}_{\boldsymbol{\delta i},\boldsymbol{\delta i}} \cdot (\mathsf{y}_{22} + \mathsf{Y})^{-\dagger} \cdot \boldsymbol{y}_{21}^\dagger. \tag{1.145}$$

Since the system is in thermodynamic equilibrium, the net noise power exchanged at port 1 must be zero, i.e. the Nyquist theorem must hold for the one-port made by the connected noisy N-port and embedding reactive network. Therefore one also has:

$$S_{\delta i,\delta i} = 2k_\mathrm{B}T\,(Y_\mathrm{i} + Y_\mathrm{i}^*)\,. \tag{1.146}$$

Inserting (1.144) into (1.146) and comparing with (1.145) one obtains the following relationship:

$$\begin{aligned} &\boldsymbol{y}_{12}^\mathrm{T} \cdot (\mathsf{y}_{22} + \mathsf{Y}) \cdot \mathsf{S}_{\boldsymbol{\delta i},\boldsymbol{\delta i}} \cdot (\mathsf{y}_{22} + \mathsf{Y})^{-\dagger} \cdot \boldsymbol{y}_{21}^* = \\ &2k_\mathrm{B}T\boldsymbol{y}_{12}^\mathrm{T} \cdot \left[(\mathsf{y}_{22} + \mathsf{Y})^{-1} + (\mathsf{y}_{22} + \mathsf{Y})^{-\dagger}\right] \cdot \boldsymbol{y}_{12}^*, \end{aligned} \tag{1.147}$$

which must hold for arbitrary y; taking into account that $\mathsf{y} = \mathsf{y}^\dagger$ one finally obtains (1.141).

1.4.3 System-Oriented Device Noise Parameters: the Noise Figure

Short-circuit current or open-circuit voltage fluctuation power and correlation spectra are a convenient device-level model, which enables us to evaluate the noise power exchanged in a network through circuit analysis. Nevertheless, these parameters are difficult to measure directly and do not provide direct information on system performance. Thus, other, mathematically equivalent, noise parameters are customarily introduced to express the noise properties of a (usually two-port or three-pole) device.

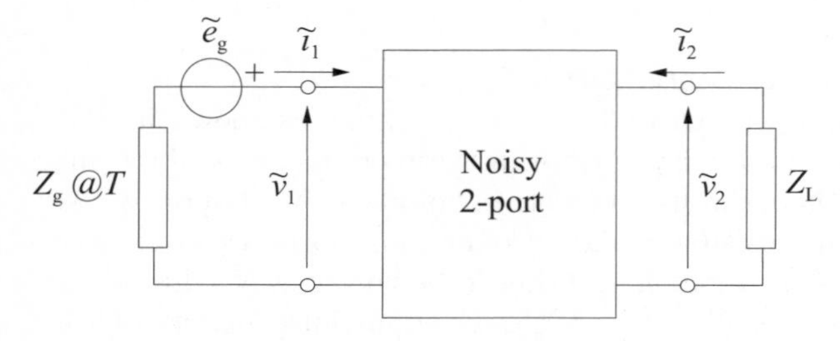

Fig. 1.10. Loaded two-port

Let us consider a noisy two-port connected at port 1 (the input) to a noisy generator and at port 2 (the output) to a load, see Fig. 1.10. From a system design standpoint, we want to assess:

1. The relative effect of the noisy two-port and of the noisy generator on the noise power delivered to the load, through the ratio:

$$F = \frac{P_{n1}}{P_{n2}}, \tag{1.148}$$

where P_{n1} is the total noise power on the load originating from both the noisy generator and the noisy two-port, while P_{n2} is the total noise power that would be delivered to the load if the generator were noisy, but the two-port were noiseless. The parameter F is called the *noise figure* of the two-port;[7] if the noise powers are interpreted per unit frequency, F (also called spot noise figure) depends on frequency. Notice that F is a relative, rather than absolute, measure of the two-port noise.
2. The generator impedance Z_o (or admittance Y_o) for which the noise figure is minimum; the minimum noise figure will be referred to as F_{min}.
3. The sensitivity of the noise figure to changes in the input impedance or admittance with respect to Z_o (Y_o), which will be shown to be proportional to a parameter called the noise conductance g_n (resistance R_n).

The analysis follows the classical method in [66]. As a first step, we define for a two-port a series or parallel equivalent circuit whose noise generators all appear at the input port, see Fig. 1.11. Circuit inspection yields:

$$\delta\tilde{e} = \delta\tilde{e}_1 - \delta\tilde{e}_2 Z_{11}/Z_{21} = -\delta\tilde{\imath}_2/Y_{21}, \tag{1.149}$$

$$\delta\tilde{\imath} = -\delta\tilde{e}_2/Z_{21} = \delta\tilde{\imath}_1 - \delta\tilde{\imath}_2 Y_{11}/Y_{21} \,. \tag{1.150}$$

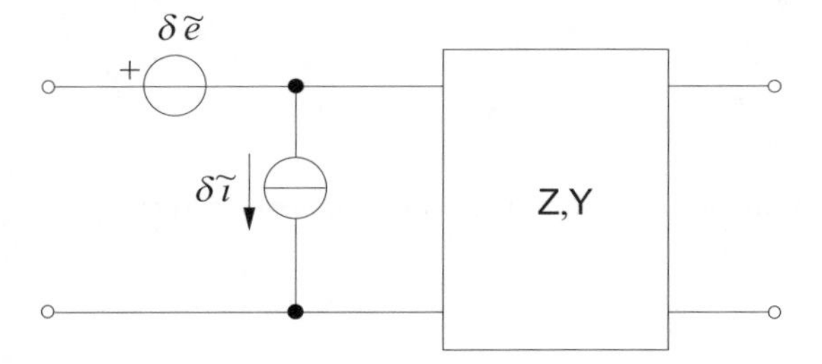

Fig. 1.11. Series (parallel) representation with input generators

We now decompose the noise generators so as to obtain a representation where noise generators are uncorrelated. In the series case, we define:

$$\delta\tilde{e} = \delta\tilde{e}_c + \delta\tilde{e}_u, \tag{1.151}$$

[7] In fact, in the standard definition of F the generator noise must be thermal, at a reference temperature, see [65].

where $\delta\tilde{e}_c$ ($\delta\tilde{e}_u$) is fully correlated (uncorrelated) with $\delta\tilde{\imath}$. Full correlation can be expressed by defining:

$$\delta\tilde{e}_c = Z_c \delta\tilde{\imath} \, . \tag{1.152}$$

The *correlation impedance* $Z_c = R_c + iX_c$ can be evaluated by imposing that $S_{\delta e_u,\delta i}$ be zero. The correlation impedance Z_c and the power spectra of δe_u and δi result as:

$$Z_c = Z_{11} - Z_{21} \frac{S_{\delta e_1,\delta e_2}}{S_{\delta e_2,\delta e_2}}, \tag{1.153}$$

$$S_{\delta e_u,\delta e_u} = S_{\delta e_1,\delta e_1} \left(1 - \frac{|S_{\delta e_1,\delta e_2}|^2}{|S_{\delta e_1,\delta e_1}||S_{\delta e_2,\delta e_2}|} \right) \equiv 4k_B T r_n(\omega), \tag{1.154}$$

$$S_{\delta i,\delta i} = \frac{S_{\delta e_2,\delta e_2}}{|Z_{21}|^2} \equiv 4k_B T g_n(\omega), \tag{1.155}$$

where the series noise resistance r_n and conductance g_n are defined by formally applying the Nyquist relationship to the power spectra of δe_u and δi. The series representation with input uncorrelated generators can be shown to admit the circuit interpretation in Fig. 1.12.

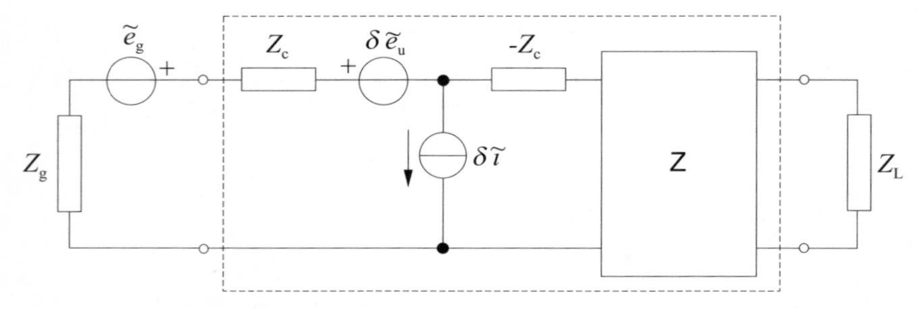

Fig. 1.12. Loaded two-port with uncorrelated input generators, series representation

In the parallel case, we have:

$$\delta\tilde{\imath} = \delta\tilde{\imath}_c + \delta\tilde{\imath}_u, \tag{1.156}$$

where $\delta\tilde{\imath}_c$ ($\delta\tilde{\imath}_u$) is fully correlated (uncorrelated) with $\delta\tilde{e}$. Full correlation can again be expressed as:

$$\delta\tilde{\imath}_c = Y_c \delta\tilde{e} \, . \tag{1.157}$$

The *correlation admittance* $Y_c = G_c + iB_c$ can be evaluated from the condition $S_{\delta i_u,\delta e} = 0$; one obtains for Y_c and the power spectra of $\delta\tilde{\imath}_u$ and $\delta\tilde{e}$:

$$Y_{\rm c} = Y_{11} - Y_{21}\frac{S_{\delta i_1,\delta i_2}}{S_{\delta i_2,\delta i_2}}, \tag{1.158}$$

$$S_{\delta i_{\rm u},\delta i_{\rm u}} = S_{\delta i_1,\delta i_1}\left(1 - \frac{\left|S_{\delta i_1,\delta i_2}\right|^2}{\left|S_{\delta i_1,\delta i_1}\right|\left|S_{\delta i_2,\delta i_2}\right|}\right) \equiv 4k_{\rm B}TG_{\rm n}(\omega), \tag{1.159}$$

$$S_{\delta e,\delta e} = \frac{S_{\delta i_2,\delta i_2}}{|Y_{21}|^2} \equiv 4k_{\rm B}TR_{\rm n}(\omega), \tag{1.160}$$

where the parallel noise conductance $G_{\rm n}$ and resistance $R_{\rm n}$ are again defined by formally applying the Nyquist relationship. The parallel representation with input uncorrelated generators admits the circuit representation of Fig. 1.13. Finally, the series and parallel parameters are related as follows [66]:

$$Y_{\rm c} = \frac{Z_{\rm c}^*}{|Z_{\rm c}|^2 + r_{\rm n}/g_{\rm n}}, \tag{1.161}$$

$$G_{\rm n} = \frac{r_{\rm n}}{|Z_{\rm c}|^2 + r_{\rm n}/g_{\rm n}}, \tag{1.162}$$

$$R_{\rm n} = r_{\rm n} + g_{\rm n}|Z_{\rm c}|^2. \tag{1.163}$$

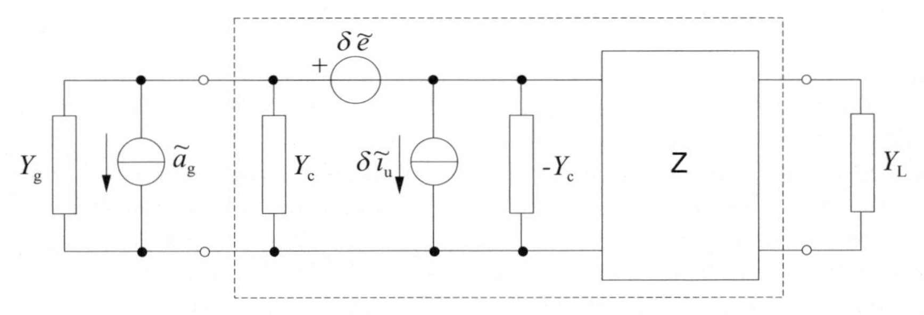

Fig. 1.13. Loaded two-port with uncorrelated input generators, parallel representation

Having placed all noise sources (originating from the two-pole and from the generator internal impedance or admittance) at the device input, the noise figure can be simply evaluated as the ratio between the total input noise available power, and the input available power originating from the input generator only. Defining $Z_{\rm g} = R_{\rm g} + {\rm i}X_{\rm g}$, $Y_{\rm g} = G_{\rm g} + {\rm i}B_{\rm g}$, we assume that the input generator is affected by thermal noise only, and that its temperature coincides with the two-port temperature. In this case one has for the noise figure:

$$F = 1 + \frac{r_{\rm n}}{R_{\rm g}} + \frac{g_{\rm n}}{R_{\rm g}}\left[(R_{\rm g} + R_{\rm c})^2 + (X_{\rm g} + X_{\rm c})^2\right], \tag{1.164}$$

or, equivalently:

$$F = 1 + \frac{G_\mathrm{n}}{G_\mathrm{g}} + \frac{R_\mathrm{n}}{G_\mathrm{g}} \left[(G_\mathrm{g} + G_\mathrm{c})^2 + (B_\mathrm{g} + B_\mathrm{c})^2 \right] . \tag{1.165}$$

The minimum noise figure F_min can be obtained, for a given device, by properly chosing the generator impedance or admittance. The optimum generator impedance $Z_\mathrm{o} = R_\mathrm{o} + \mathrm{i}X_\mathrm{o}$ and admittance $Y_\mathrm{o} = G_\mathrm{o} + \mathrm{i}B_\mathrm{o}$ can be obtained as:

$$Z_\mathrm{o} = \sqrt{\frac{r_\mathrm{n}}{g_\mathrm{n}} + R_\mathrm{c}^2} - \mathrm{i}X_\mathrm{c}, \tag{1.166}$$

$$Y_\mathrm{o} = \sqrt{\frac{G_\mathrm{n}}{R_\mathrm{n}} + G_\mathrm{c}^2} - \mathrm{i}B_\mathrm{c}, \tag{1.167}$$

and the resulting minimum noise figure is:

$$F_\mathrm{min} = 1 + 2g_\mathrm{n}(R_\mathrm{c} + R_\mathrm{o}) = 1 + 2R_\mathrm{n}(G_\mathrm{c} + G_\mathrm{o}) . \tag{1.168}$$

As expected, the optimum source is uniquely defined, i.e. $Z_\mathrm{o} = 1/Y_\mathrm{o}$. The noise figure with arbitrary source can be finally expressed as:

$$F = F_\mathrm{min} + \frac{g_\mathrm{n}}{R_\mathrm{g}} |Z_\mathrm{g} - Z_\mathrm{o}|^2 = F_\mathrm{min} + \frac{R_\mathrm{n}}{G_\mathrm{g}} |Y_\mathrm{g} - Y_\mathrm{o}|^2 ; \tag{1.169}$$

thus, the series noise conductance g_n and the parallel noise resistance R_n express the *sensitivity* of the noise figure to a change of the source impedance (admittance) with respect to its optimum value.

We conclude that, while at a circuit level a full noise characterization for a two-port is given by the (frequency-dependent) power and correlation spectra of its input noise voltage or current generators, at a system level the same information (four real numbers for each frequency) can be conveyed by the equivalent sets $(F_\mathrm{min}, Z_\mathrm{o}, g_\mathrm{n})$ or $(F_\mathrm{min}, Y_\mathrm{o}, R_\mathrm{n})$, again four real numbers for each frequency. Besides having a more direct impact on the circuit or system designer, the system-oriented parameters can be measured directly.

2. Noise Analysis Techniques

2.1 Semiconductor Device Physical Models for Noise Modeling

Physics-based noise modeling can be, as a matter of principle, obtained as a by-product of a direct evaluation of the carrier distribution function carried out through the Monte Carlo technique. In fact, this approach provides a direct mimicry of the microscopic scattering mechanisms, and therefore yields both microscopic variables (such as carrier velocities and free charge densities) and external variables (such as terminal voltages and currents) as realizations of random processes, thus allowing averages to be separated from fluctuations. Although fluctuations are amplified by the limited number of particles included in the simulation, their physical value can be recovered through a scaling process. Thus, full information on any kind of statistical properties related to device noise can be recovered by inspection.

The Monte Carlo approach to noise simulation has been successfully pursued by many research groups [67–69]. In particular, Reggiani's group has carried out extensive investigations on the fluctuation properties of semiconductors. However, the Monte Carlo technique is still deemed too computationally intensive to allow the simulation of multidimensional (e.g. 3D) devices. Thus, in order to provide a more computationally efficient tool for device analysis and optimization, noise simulation should also be founded on the more conventional framework of PDE-based models, such as the full hydrodynamic, the energy balance or the drift-diffusion models. As a matter of principle, such models only provide ensemble averages of quantities such as the carrier density, velocity, and energy; the discussion will therefore be confined, as a first step, to a fluctuation-free system.

Starting from the Boltzmann equation, whose unknown is the distribution function of carriers, and from the quasi-static relationship between charges and electric fields provided by the Poisson equation, the so-called method of moments provides a general strategy for deriving partial differential equations involving the *central moments* (in a statistical sense) of the carrier distribution functions. The formal procedure can be found in several papers and books, e.g. see [70]. From a physical standpoint, the first three central moments can be related, in increasing moment order, to the carrier concentration, and the carrier average momentum, the carrier average energy. Unfortunately, the PDE system thus derived is not, strictly speaking, closed if a truncation is performed. This means that the first three moment equations derived (corresponding to the charge conservation equation or continuity equation, the momentum conservation equation, and the energy conservation equation) also involve the fourth moment, connected to the so-called heat flow of the carrier distribution. Thus, not only has the system to be suitably truncated, but approximate closure relationships have to be provided in order to eliminate the higher-order moments by means of approximate relationships with lower-order ones.

Although the technique can be exploited to derive PDE systems of arbitrarily high order, the computational complexity makes the approach hardly advantageous with respect to the full Monte Carlo method when the number of moments involved is larger than three. Three moment equations expressing the carrier, momentum and energy conservation for each carrier species involved yield, together with the Poisson equation, the so-called *full hydrodynamic* model (e.g. see [70, 71]). This model, albeit actually exploited for device simulation, shows a number of computational problems which suggest simplifications. The main one consists in assuming that the time scale of the average momentum and energy dynamics can be separated, meaning that the average momentum dynamics is fast, while the average energy dynamics is slow. This assumption, added to neglecting the so-called convective terms in the momentum equation and assuming that the collision term in the same equation follows the relaxation-time approximation, leads to turning the momentum equation into a generalized, energy-dependent constitutive relationship for the carrier velocity (and therefore the carrier current density). In this way, the carrier current density can be separated into a generalized drift contribution (with energy-dependent mobility) and a generalized diffusion contribution (with energy-dependent diffusivity and carrier temperature different from the lattice temperature T) [71]. The resulting model is referred to as the *energy balance* model, or, if the kinetic (collective) term in the average carrier energy is neglected with respect to the thermal (disordered) term, proportional to the carrier temperature, as the *temperature* model. The final step of this simplifying process consists in assuming that time-dependent and space-dependent variations in the energy transport equation can be neglected, thus relating the average carrier energy to the local electric field. Therefore, the carrier mobility and diffusivity become functions of the local and instantaneous electric field, and the customary *drift-diffusion* model is obtained, including as usual the Poisson equation. According to this model, the electron and hole transport current densities are expressed as the sum of a drift (conduction) and a diffusion term:

$$\boldsymbol{J}_n = -qn\mu_n \nabla\varphi + qD_n \nabla n \tag{2.1a}$$

$$\boldsymbol{J}_p = -qp\mu_p \nabla\varphi - qD_p \nabla p. \tag{2.1b}$$

The problem now arises as to how fluctuations can be treated within the framework of PDE-based models. In general, the strategy followed is to apply a Langevin approach, i.e. by treating fluctuations as the effect of a proper stochastic forcing term added to the model equations. For the drift-diffusion approach, such a stochastic forcing term is directly related to carrier-velocity and population fluctuations, which can be expressed in turn according to a mesoscopic formulation as outlined in Sect. 1.2. However, the extension of the Langevin approach to higher-order transport models requires exploition of a more rigorous method.

As a matter of principle, PDE-based models of any order including fluctuations can be formally derived from the BTE with the Langevin stochastic forcing term (1.130) by application of the method of moments. In this way, a self-consistent hyerarchy of Langevin forcing terms can be obtained for the transport equations arising in the full hydrodynamic or energy-balance models. An example of such a derivation is outlined in [18]. Although higher-order models are increasingly important whenever the devices under examination exhibit submicrometer features inducing significant non-stationary transport effects, concerning noise analysis the derivation of a coherent set of Langevin sources and their actual implementation within the framework of numerical device simulators is still under development, see Sect. 2.2.5 for a short review of the most recent literature.

We can therefore conclude that the drift-diffusion approach is still useful in noise analysis, at least for a first-approximation solution, also in the field of submicrometer devices.

2.1.1 The Drift-Diffusion Model

In this section a short summary will be provided of the drift-diffusion approach, mainly to establish a notation to be exploited in the rest of the book. The bipolar drift-diffusion model includes Poisson's equation, the electron and hole continuity equations, and a set of rate equations describing the evolution of the population of energy levels (acceptor or donor, shallow or deep) falling into the forbidden band. Notice that such rate equations are ordinary differential equations rather than partial, unless hopping conduction models are included in the formulation. In DC or small-signal periodic operation (in the frequency domain) such equations become algebraic relationships and can be eliminated by substitution. The bipolar drift-diffusion model therefore reads:

$$\nabla^2 \varphi = -\frac{q}{\varepsilon}\left(\sum_k C^+_{\mathrm{D}k} + p - n\right) \tag{2.2a}$$

$$\frac{\partial n}{\partial t} + \nabla \cdot (n\mu_n \nabla\varphi - D_n \nabla n) + U_n = 0 \tag{2.2b}$$

$$\frac{\partial p}{\partial t} - \nabla \cdot (p\mu_p \nabla \varphi + D_p \nabla p) + U_p = 0 \tag{2.2c}$$

$$\frac{\partial (C^+_{\mathrm{D}k})}{\partial t} + \sum_{j \neq k} (p_{jk} - p_{kj}) = 0 \quad \text{for all } k, \tag{2.2d}$$

where n and p are the electron and hole concentrations, μ_α and D_α ($\alpha = n, p$) are the field-dependent mobilities and diffusivities, respectively, U_n and U_p are the net recombination rates for electrons and holes. For any trap level k, p_{kj} describes the electron transition rate from k to level j (see the discussion in Sect. 1.2.2), while $C^+_{\mathrm{D}k}$ is the net equivalent ionized donor concentration, i.e. $C^+_{\mathrm{D}k} = N^+_{\mathrm{D}k}$ for donor levels, $C^+_{\mathrm{D}k} = -N^-_{\mathrm{A}k}$ for acceptor levels, with N_α, $\alpha = \mathrm{D}, \mathrm{A}$ the donor and acceptor concentration, respectively.

Boundary Conditions. The physical model described above is associated with a suitable set of boundary conditions (BC), e.g. see [35]. A first kind of BC concerns ideal insulating boundaries, for which the normal component of the current density is zero. This condition corresponds to zero normal derivative (also called homogeneous Neumann BC) for the potential and both carrier concentrations:

$$\frac{\partial \varphi}{\partial \hat{n}} = 0, \tag{2.3}$$

$$\frac{\partial n}{\partial \hat{n}} = 0, \tag{2.4}$$

$$\frac{\partial p}{\partial \hat{n}} = 0. \tag{2.5}$$

Interfaces with dielectric media are treated likewise, but the normal derivative of the potential must satisfy the continuity of the normal component of electric displacement across the boundary.

Metallic contacts fall into two main classes: ohmic and rectifying (Schottky) contacts. In both cases metallic contacts are assumed to be made of ideal conductors, i.e. they are equipotential. Ohmic contacts ideally behave as short-circuits, i.e. do not support any potential drop, while rectifying Schottky contacts exhibit a diode-like behavior.

For ohmic contacts, the local electron and hole concentrations assume the equilibrium values corresponding to the local doping, yielding a Dirichlet BC (i.e. the unknown is assigned):

$$n = \frac{1}{2}\left(\sqrt{C^{+2}_{\mathrm{D}} + 4n_{\mathrm{i}}^2} + C^+_{\mathrm{D}}\right) \tag{2.6}$$

$$p = \frac{1}{2}\left(\sqrt{C^{+2}_{\mathrm{D}} + 4n_{\mathrm{i}}^2} - C^+_{\mathrm{D}}\right) \tag{2.7}$$

where n_{i} is the intrinsic concentration and C^+_{D} is the local effective ionized donor concentration. If the contact is controlled by an ideal voltage source, a Dirichlet BC also arises for the potential:

$$\varphi = e_{\mathrm{s}} + \varphi_{\mathrm{rif}} \tag{2.8}$$

where e_{s} is the applied (source) voltage and φ_{rif} is a reference shift depending on the definition of the internal potential (e.g. as the vacuum level rather than the intrinsic Fermi level). Ohmic contacts driven by a real source (or, in the limit case, by an ideal current generator) correspond to a nonlinear BC; supposing, for the sake of simplicity, a resistive internal generator admittance, one has:

$$i_{\mathrm{s}} + Gv + i = 0 \tag{2.9}$$

where i_{s} is the current source connected to the contact, G the internal generator conductance, i the total current entering the contact. The contact (external) potential v is related to the internal potential φ as $\varphi = v + \varphi_{\mathrm{rif}}$ and is unknown. Since i depends nonlinearly on φ, the BC is nonlinear.

The treatment of Schottky contacts is more involved, since several approximations can be exploited. A fairly accurate set of BC can be derived from the Sze–Bethe thermionic-drift diffusion theory [72]. According to this model, the normal component of the electron and hole total current densities $\boldsymbol{J}_{n\mathrm{t}}$, $\boldsymbol{J}_{p\mathrm{t}}$ follow an equivalent surface recombination BC:

$$\boldsymbol{J}_{n\mathrm{t}} \cdot \hat{n} = q v_{\mathrm{s}n}(n - n_{\mathrm{s}}), \tag{2.10}$$

$$\boldsymbol{J}_{p\mathrm{t}} \cdot \hat{n} = q v_{\mathrm{s}p}(p - p_{\mathrm{s}}), \tag{2.11}$$

where $v_{\mathrm{s}n}$ and $v_{\mathrm{s}p}$ are equivalent surface recombination velocities, and n_{s}, p_{s} can be derived from the Schottky contact barrier height according to the equilibrium statistics, see [72]. If the contact is driven by an ideal voltage source, the potential follows a simple Dirichlet BC:

$$\varphi = e_{\mathrm{s}} + v_{\mathrm{bi}} + \varphi_{\mathrm{rif}}, \tag{2.12}$$

where v_{bi} is the Schottky barrier built-in voltage. If, on the other hand, the contact is driven by a real generator, (2.12) becomes:

$$\varphi = v + v_{\mathrm{bi}} + \varphi_{\mathrm{rif}} \tag{2.13}$$

and (2.9) must be added to evaluate the unknown contact voltage v.

Heterojunctions are characterized by continuity conditions across the material discontinuity whose treatment is beyond the scope of this book; for a thorough discussion of the available models, see [73].

2.1.2 The Model Solution

Following a formal approach, the drift-diffusion model equations (Poisson, electron and hole continuity equations in this order) can be expressed as the vector system

$$\boldsymbol{F}(\varphi, n, p, \dot{n}, \dot{p}) = \mathbf{0}, \tag{2.14a}$$
$$\boldsymbol{b}(\varphi, n, p, \dot{n}, \dot{p}, \boldsymbol{s}_{\mathrm{e}}) = \mathbf{0}, \tag{2.14b}$$

where $\dot{x} = \partial x / \partial t$ and $\boldsymbol{b}$ represents the boundary conditions, including the external electrical sources $\boldsymbol{s}_{\mathrm{e}}$ applied to the device terminals. The system (2.14) can be solved in several different operating conditions, arising in the analysis of a circuit including the device under consideration. In the most general case, the set of applied sources is time-varying and has arbitrary amplitude. This case is often referred to as the transient analysis of the device; on the other hand, if the source set is (quasi-)periodic,[1] the corresponding regime is called large-signal (LS). In both cases, the analysis requires the solution of a non-stationary, nonlinear PDE system. Notice that in the LS case the solution can be referred to as a (quasi-)periodic steady state.

In the simplest case, the set of external sources is timeindependent, i.e. $\boldsymbol{s}_{\mathrm{e}}(t) = \boldsymbol{s}_{\mathrm{e}0}$. This analysis corresponds to the so-called DC or working-point simulation of the device. The stationary, generally nonlinear PDE system to be solved is then:

$$\boldsymbol{F}(\varphi_0, n_0, p_0, 0, 0) = \mathbf{0}, \tag{2.15a}$$
$$\boldsymbol{b}(\varphi_0, n_0, p_0, 0, 0, \boldsymbol{s}_{\mathrm{e}0}) = \mathbf{0}. \tag{2.15b}$$

Suppose now that a set of time-varying, small-amplitude sources $\boldsymbol{s}_{\mathrm{e}}^{\mathrm{ss}}(t)$ is added to the DC excitation $\boldsymbol{s}_{\mathrm{e}0}$, so that:

$$\boldsymbol{s}_{\mathrm{e}}(t) = \boldsymbol{s}_{\mathrm{e}0} + \boldsymbol{s}_{\mathrm{e}}^{\mathrm{ss}}(t). \tag{2.16}$$

As a consequence, the DC solution is perturbed so as to become:

$$\varphi(t) = \varphi_0 + \varphi^{\mathrm{ss}}(t), \tag{2.17}$$
$$n(t) = n_0 + n^{\mathrm{ss}}(t), \tag{2.18}$$
$$p(t) = p_0 + p^{\mathrm{ss}}(t). \tag{2.19}$$

The device model now becomes:

$$\boldsymbol{F}(\varphi_0 + \varphi^{\mathrm{ss}}, n_0 + n^{\mathrm{ss}}, p_0 + p^{\mathrm{ss}}, \dot{n}^{\mathrm{ss}}, \dot{p}^{\mathrm{ss}}) = \mathbf{0},$$
$$\boldsymbol{b}(\varphi_0 + \varphi^{\mathrm{ss}}, n_0 + n^{\mathrm{ss}}, p_0 + p^{\mathrm{ss}}, \dot{n}^{\mathrm{ss}}, \dot{p}^{\mathrm{ss}}, \boldsymbol{s}_{\mathrm{e}0} + \boldsymbol{s}_{\mathrm{e}}^{\mathrm{ss}}) = \mathbf{0}.$$

Under the condition of small-amplitude time-varying perturbation of the source term, the device equations can be linearized around the DC working point thus yielding the small-signal (SS) system:

[1] A quasi-periodic function of time is the superposition of periodic waveforms, whose periods are uncommensurate [74].

$$\left.\frac{\partial \boldsymbol{F}}{\partial \varphi}\right|_0 \varphi^{\mathrm{ss}} + \left.\frac{\partial \boldsymbol{F}}{\partial n}\right|_0 n^{\mathrm{ss}} + \left.\frac{\partial \boldsymbol{F}}{\partial p}\right|_0 p^{\mathrm{ss}} + \left.\frac{\partial \boldsymbol{F}}{\partial \dot{n}}\right|_0 \dot{n}^{\mathrm{ss}} + \left.\frac{\partial \boldsymbol{F}}{\partial \dot{p}}\right|_0 \dot{p}^{\mathrm{ss}} = \mathbf{0} \tag{2.20a}$$

$$\left.\frac{\partial \boldsymbol{b}}{\partial \varphi}\right|_0 \varphi^{\mathrm{ss}} + \left.\frac{\partial \boldsymbol{b}}{\partial n}\right|_0 n^{\mathrm{ss}} + \left.\frac{\partial \boldsymbol{b}}{\partial p}\right|_0 p^{\mathrm{ss}} + \left.\frac{\partial \boldsymbol{b}}{\partial \dot{n}}\right|_0 \dot{n}^{\mathrm{ss}} + \left.\frac{\partial \boldsymbol{b}}{\partial \dot{p}}\right|_0 \dot{p}^{\mathrm{ss}} + \left.\frac{\partial \boldsymbol{b}}{\partial \boldsymbol{s}_{\mathrm{e}}}\right|_0 \cdot \boldsymbol{s}_{\mathrm{e}}^{\mathrm{ss}} = \mathbf{0}, \tag{2.20b}$$

which can be effectively expressed in the frequency domain as the linear PDE system:

$$\begin{aligned}\left.\frac{\partial \boldsymbol{F}}{\partial \varphi}\right|_0 \tilde{\varphi}^{\mathrm{ss}} &+ \left(\left.\frac{\partial \boldsymbol{F}}{\partial n}\right|_0 + \mathrm{i}\omega \left.\frac{\partial \boldsymbol{F}}{\partial \dot{n}}\right|_0 \right) \tilde{n}^{\mathrm{ss}} \\ &+ \left(\left.\frac{\partial \boldsymbol{F}}{\partial p}\right|_0 + \mathrm{i}\omega \left.\frac{\partial \boldsymbol{F}}{\partial \dot{p}}\right|_0 \right) \tilde{p}^{\mathrm{ss}} = \mathbf{0}\end{aligned} \tag{2.21a}$$

$$\begin{aligned}\left.\frac{\partial \boldsymbol{b}}{\partial \varphi}\right|_0 \tilde{\varphi}^{\mathrm{ss}} &+ \left(\left.\frac{\partial \boldsymbol{b}}{\partial n}\right|_0 + \mathrm{i}\omega \left.\frac{\partial \boldsymbol{b}}{\partial \dot{n}}\right|_0 \right) \tilde{n}^{\mathrm{ss}} \\ &+ \left(\left.\frac{\partial \boldsymbol{b}}{\partial p}\right|_0 + \mathrm{i}\omega \left.\frac{\partial \boldsymbol{b}}{\partial \dot{p}}\right|_0 \right) \tilde{p}^{\mathrm{ss}} + \left.\frac{\partial \boldsymbol{b}}{\partial \boldsymbol{s}_{\mathrm{e}}}\right|_0 \cdot \tilde{\boldsymbol{s}}_{\mathrm{e}}^{\mathrm{ss}} = \mathbf{0}.\end{aligned} \tag{2.21b}$$

We stress that in all of the above cases the device is assumed to be noiseless, and therefore the solutions are deterministic, i.e. fluctuation-free.

2.2 Langevin Approach to Noise Analysis

Classical physics-based noise analysis through PDE-based models is founded on a two-step approach; in the first step, microscopic noise sources are identified in terms of fundamental fluctuations of carrier velocity (diffusion noise) and carrier number (GR noise) and added to the relevant conservation equation [6, 16]. This means that the PDE-based physical model describing the noiseless device behavior is converted into a Langevin equation [75], wherein the microscopic noise sources are the Langevin forces. Such forces are included in the carrier continuity equation and, for GR noise induced by trap-assisted phenomena, in the trap-occupancy rate equation (see Sect. 2.2.4). For the sake of simplicity, we shall consider in this section a simple bipolar drift-diffusion physical model (2.2), neglecting trap-assisted transitions.

2.2.1 Green's Function Solution Techniques

We start again from (2.14), which, from a purely mathematical standpoint, expresses the three drift-diffusion equations together with their boundary conditions. The noiseless steadystate solution (DC or LS) depends on physical and geometrical parameters and on the boundary conditions, particularly on the external electrical sources. If these are timeindependent (DC), i.e. the device is driven by constant voltages and/or currents, the steadystate solution (φ_0, n_0, p_0) is constant as well. On the other hand, if the device is operated under (quasi-periodic) LS conditions, i.e. is driven by time-varying (quasi-periodic) large-amplitude generators, the steadystate solution will be (quasi-periodic) time-dependent; this deeply affects the statistical nature of fluctuations. Throughout this chapter, we shall always assume the noisy device is operated in small-signal conditions around a stationary steadystate, as happens for a linear amplifier, so that all the stochastic processes representing fluctuations will be wide-sense stationary [27]. For a treatment of noise in devices operated in quasi-periodic large-signal condition see Chap. 5.

The microscopic noise sources $\boldsymbol{s}$ are then added to the right-hand side of the physical model, thereby perturbing the DC steadystate solution by the corresponding fluctuations:

$$\boldsymbol{F}\left(\varphi_0 + \delta\varphi, n_0 + \delta n, p_0 + \delta p, \delta\dot{n}, \delta\dot{p}\right) = \boldsymbol{s},$$
$$\boldsymbol{b}\left(\varphi_0 + \delta\varphi, n_0 + \delta n, p_0 + \delta p, \delta\dot{n}, \delta\dot{p}, \boldsymbol{s}_{\mathrm{e}0}\right) = \boldsymbol{0}.$$

Notice that this implicitly assumes no noise sources are present on the device boundary; otherwise, a further source term should be added to the boundary conditions $\boldsymbol{b}$.

Since the sources $\boldsymbol{s}$ are assumed to be small enough to linearly excite the device, the previous equation can be linearized around the noiseless steadystate yielding the linear Langevin equation for the fluctuations:

$$\left.\frac{\partial \boldsymbol{F}}{\partial \varphi}\right|_0 \delta\varphi + \left.\frac{\partial \boldsymbol{F}}{\partial n}\right|_0 \delta n + \left.\frac{\partial \boldsymbol{F}}{\partial p}\right|_0 \delta p + \left.\frac{\partial \boldsymbol{F}}{\partial \dot{n}}\right|_0 \delta\dot{n} + \left.\frac{\partial \boldsymbol{F}}{\partial \dot{p}}\right|_0 \delta\dot{p} = \boldsymbol{s}, \tag{2.22a}$$

$$\left.\frac{\partial \boldsymbol{b}}{\partial \varphi}\right|_0 \delta\varphi + \left.\frac{\partial \boldsymbol{b}}{\partial n}\right|_0 \delta n + \left.\frac{\partial \boldsymbol{b}}{\partial p}\right|_0 \delta p + \left.\frac{\partial \boldsymbol{b}}{\partial \dot{n}}\right|_0 \delta\dot{n} + \left.\frac{\partial \boldsymbol{b}}{\partial \dot{p}}\right|_0 \delta\dot{p} = \boldsymbol{0}, \tag{2.22b}$$

where the derivatives are evaluated in the noiseless steadystate (φ_0, n_0, p_0). Equation (2.22) is then completed by additional boundary conditions imposed on the linearized system only (see the later discussion).

The linear PDE (2.22) can be solved through a Green's function approach as follows. To fix the ideas, suppose the model under consideration is a drift-diffusion bipolar model, so that $\boldsymbol{F} = (F_\varphi, F_n, F_p)$ where F_φ is Poisson's equation, F_n and F_p the electron and hole continuity equations, respectively. Let us define the Green's function $G_{\alpha,\beta}(\boldsymbol{r}, \boldsymbol{r}_1; t, t_1)$ $(\alpha, \beta = \varphi, n, p)$ for equation α and input variable β as the response in variable α to a unit source $\delta(\boldsymbol{r} - \boldsymbol{r}_1)\delta(t - t_1)$ (δ here is Dirac's delta function) injected in equation β. Green's theorem allows the fluctuation $\delta\alpha$ induced by the vector source $\boldsymbol{s}$ to be evaluated as the spatial and temporal convolution integrals:

$$\delta\alpha(\boldsymbol{r}, t) = \sum_{\beta=\varphi,n,p} \int_\Omega \int_{-\infty}^{t} G_{\alpha,\beta}(\boldsymbol{r}, \boldsymbol{r}_1; t, t_1) s_\beta(\boldsymbol{r}_1, t_1)\, \mathrm{d}t_1 \mathrm{d}\boldsymbol{r}_1, \tag{2.23}$$

where Ω is the device volume and $\alpha = \varphi, n, p$.

Equation (2.23) can be simplified provided the noiseless steadystate is stationary, since this assumption leads to a time-invariant linear system

$$G_{\alpha,\beta}(\boldsymbol{r},\boldsymbol{r}_1;t,t_1) = G_{\alpha,\beta}(\boldsymbol{r},\boldsymbol{r}_1;t-t_1),$$

which, in turn, allows for a frequency-domain analysis of the linear system (2.22). By introducing the Fourier transform $\tilde{f}(\boldsymbol{r},\omega)$ of function $f(\boldsymbol{r},t)$, (2.23) simplifies to:

$$\delta\tilde{\alpha}(\boldsymbol{r},\omega) = \sum_{\beta=\varphi,n,p} \int_{\Omega} \tilde{G}_{\alpha,\beta}(\boldsymbol{r},\boldsymbol{r}_1;\omega)\tilde{s}_{\beta}(\boldsymbol{r}_1,\omega)\,\mathrm{d}\boldsymbol{r}_1. \tag{2.24}$$

Once the Green's functions have been evaluated, the correlation spectrum of the fluctuations $\delta\alpha(\boldsymbol{r})$, $\delta\beta(\boldsymbol{r}')$ $(\alpha,\beta=\varphi,n,p)$ can therefore be computed as:

$$\begin{aligned} S_{\delta\alpha,\delta\beta}(\boldsymbol{r},\boldsymbol{r}';\omega) = \sum_{\gamma,\delta=\varphi,n,p} \int_{\Omega}\int_{\Omega} &\tilde{G}_{\alpha,\gamma}(\boldsymbol{r},\boldsymbol{r}_1;\omega) \\ &S_{s_\gamma,s_\delta}(\boldsymbol{r}_1,\boldsymbol{r}_2;\omega)\tilde{G}^*_{\beta,\delta}(\boldsymbol{r}',\boldsymbol{r}_2;\omega)\,\mathrm{d}\boldsymbol{r}_1\mathrm{d}\boldsymbol{r}_2, \end{aligned} \tag{2.25}$$

where $S_{s_\gamma,s_\delta}(\boldsymbol{r}_1,\boldsymbol{r}_2;\omega)$ is the correlation spectrum of the microscopic noise sources $s_\gamma(\boldsymbol{r}_1)$ and $s_\delta(\boldsymbol{r}_2)$, and x^* is the conjugate part of x. According to the customary assumption of spatially uncorrelated microscopic noise sources (cf. Sect. 1.2), the *local* microscopic noise source K_{s_γ,s_δ} is often introduced

$$S_{s_\gamma,s_\delta}(\boldsymbol{r}_1,\boldsymbol{r}_2;\omega) = K_{s_\gamma,s_\delta}(\boldsymbol{r}_1;\omega)\delta(\boldsymbol{r}_1-\boldsymbol{r}_2),$$

so that (2.25) reads

$$\begin{aligned} S_{\delta\alpha,\delta\beta}(\boldsymbol{r},\boldsymbol{r}';\omega) = \sum_{\gamma,\delta=\varphi,n,p} \int_{\Omega} &\tilde{G}_{\alpha,\gamma}(\boldsymbol{r},\boldsymbol{r}_1;\omega) \\ &K_{s_\gamma,s_\delta}(\boldsymbol{r}_1;\omega)\tilde{G}^*_{\beta,\delta}(\boldsymbol{r}',\boldsymbol{r}_1;\omega)\,\mathrm{d}\boldsymbol{r}_1. \end{aligned} \tag{2.26}$$

2.2.2 Application of the Green's Function Technique to the Drift-Diffusion Model

As discussed in Sect. 1.4, from a circuit standpoint a noisy $N_\mathrm{c}+1$ terminal device operated in stationary (small-signal) conditions can be represented as a linear circuit wherein one of the terminals is taken as reference for potential, namely terminal $N_\mathrm{c}+1$, plus N_c independent noise generators connected to the ungrounded terminals. The aim of noise analysis is to evaluate the second-order statistical properties, i.e. the correlation spectra, of these noise generators, which can be either open-circuit noise voltage sources, short-circuit noise current sources or any combination of these, at least for a device with more than two terminals, i.e. a device with more than one port. The choice of external noise generators to be evaluated is reflected in the additional boundary conditions to be imposed on (2.22):

- to evaluate the noise current generator connected to terminal i, no potential fluctuation is allowed on such a terminal (i.e. small-signal short-circuit conditions are imposed):

$$\delta\tilde{\varphi}(\boldsymbol{r}) = 0, \tag{2.27}$$

for any point $\boldsymbol{r}$ pertaining to terminal i;

- to compute the noise voltage generator connected to terminal j, no current fluctuation occurs on such a terminal (i.e. small-signal open-circuit conditions are enforced):

$$\delta\tilde{\imath}_{\mathrm{c},j} = 0, \tag{2.28}$$

where the total fluctuation current $\delta\tilde{\imath}_{\mathrm{c},j}$ flowing through terminal j must be considered, i.e. electron, hole and displacement components have to be added, to properly apply (2.28).

In both cases, $N_{\mathrm{c}} + 1$ auxiliary equations (N_{c} short-circuit or open-circuit conditions for the ungrounded terminals and one condition to enforce terminal $N_{\mathrm{c}} + 1$ is grounded) need to be added to the small-change equation (2.22) to properly evaluate the relevant Green's function.

We shall hereafter consider the simplest case of transport analysis: the bipolar drift-diffusion model (2.2) neglecting trap-assisted transitions. In this case, the microscopic noise sources appear in the electron and hole continuity equations only (cf. Sect. 1.2), i.e.

$$s_{\varphi} = 0, \tag{2.29a}$$

$$s_{n} = \gamma_{n} + \frac{1}{q}\nabla \cdot \delta \boldsymbol{J}_{n}, \tag{2.29b}$$

$$s_{p} = \gamma_{p} - \frac{1}{q}\nabla \cdot \delta \boldsymbol{J}_{p}, \tag{2.29c}$$

where $\gamma_{\alpha}(\boldsymbol{r};t)$ and $\delta\boldsymbol{J}_{\alpha}(\boldsymbol{r};t)$ represent, respectively, GR and diffusion noise sources for carrier α ($\alpha = n,p$). The stochastic PDEs defining the fuctuations around the stationary steadystate can therefore be written, in the frequency domain, as:

$$\nabla^{2}\delta\tilde{\varphi} = -\Lambda_{\varphi}\left(\delta\tilde{\varphi}, \delta\tilde{n}, \delta\tilde{p}\right), \tag{2.30a}$$

$$\mathrm{i}\omega\delta\tilde{n} = -\Lambda_{n}\left(\delta\tilde{\varphi}, \delta\tilde{n}, \delta\tilde{p}\right) + \tilde{\gamma}_{n} + \frac{1}{q}\nabla \cdot \delta\tilde{\boldsymbol{J}}_{n}, \tag{2.30b}$$

$$\mathrm{i}\omega\delta\tilde{p} = -\Lambda_{p}\left(\delta\tilde{\varphi}, \delta\tilde{n}, \delta\tilde{p}\right) + \tilde{\gamma}_{p} - \frac{1}{q}\nabla \cdot \delta\tilde{\boldsymbol{J}}_{p}, \tag{2.30c}$$

where i is the imaginary unit and Λ_{α} ($\alpha = \varphi, n, p$) are linear functions of their arguments.

Without loss in generality, we shall evaluate the open-circuit noise voltage fluctuations induced at the N_{c} ungrounded terminals, which is the classical configuration exploited in Shockley's impedance field method [1]. The N_c auxiliary equations to be added to enforce the N_c ungrounded terminals to be open-circuited are therefore akin to (2.28). Moreover, only two (namely, $\tilde{G}_{\varphi,n}$ and $\tilde{G}_{\varphi,p}$) of the nine Green's functions defined on the bipolar drift-diffusion physical model are exploited to evaluate terminal fluctuations: sources are added to the electron and hole continuity equations only, and the open-circuit boundary conditions imposed at the ungrounded device terminals assure potential fluctuations only are induced at the terminals themselves. According to the definition of such Green's functions, we have

$$\nabla^2_{\boldsymbol{r}}\tilde{G}_{\varphi,\alpha}(\boldsymbol{r},\boldsymbol{r}_1;\omega) = -\Lambda_\varphi\left(\tilde{G}_{\varphi,\alpha},\delta\tilde{n},\delta\tilde{p}\right), \tag{2.31a}$$

$$\mathrm{i}\omega\delta\tilde{n}(\boldsymbol{r},\boldsymbol{r}_1;\omega) = -\Lambda_n\left(\tilde{G}_{\varphi,\alpha},\delta\tilde{n},\delta\tilde{p}\right) + \delta_{\alpha,n}\delta(\boldsymbol{r}-\boldsymbol{r}_1), \tag{2.31b}$$

$$\mathrm{i}\omega\delta\tilde{p}(\boldsymbol{r},\boldsymbol{r}_1;\omega) = -\Lambda_p\left(\tilde{G}_{\varphi,\alpha},\delta\tilde{n},\delta\tilde{p}\right) + \delta_{\alpha,p}\delta(\boldsymbol{r}-\boldsymbol{r}_1), \tag{2.31c}$$

where $\alpha = n,p$, $\delta_{\alpha,\beta}$ is the Kronecker symbol and $\nabla_{\boldsymbol{r}}$ indicates derivatives with respect to $\boldsymbol{r}$.

From (2.24) and (2.29), the potential fluctuations induced within the device volume can therefore be computed as

$$\begin{aligned}\delta\tilde{\varphi}(\boldsymbol{r},\omega) = &\sum_{\alpha=n,p}\int_\Omega \tilde{G}_{\varphi,\alpha}(\boldsymbol{r},\boldsymbol{r}_1;\omega)\tilde{\gamma}_\alpha(\boldsymbol{r}_1;\omega)\,\mathrm{d}\boldsymbol{r}_1\\ &-\frac{1}{q}\sum_{\alpha=n,p}\kappa_\alpha\int_\Omega \tilde{G}_{\varphi,\alpha}(\boldsymbol{r},\boldsymbol{r}_1;\omega)\nabla_{\boldsymbol{r}_1}\cdot\delta\tilde{\boldsymbol{J}}_\alpha(\boldsymbol{r}_1;\omega)\,\mathrm{d}\boldsymbol{r}_1,\end{aligned} \tag{2.32}$$

where $\kappa_n = -1$ and $\kappa_p = 1$. The convolution integral involving the diffusion (vector) noise source $\delta\tilde{\boldsymbol{J}}_\alpha$ can be simplified as follows. Since the following vector identity holds

$$\begin{aligned}\nabla_{\boldsymbol{r}_1}\cdot\left[\tilde{G}_{\varphi,\alpha}(\boldsymbol{r},\boldsymbol{r}_1;\omega)\delta\tilde{\boldsymbol{J}}_\alpha(\boldsymbol{r}_1;\omega)\right] = &\nabla_{\boldsymbol{r}_1}\tilde{G}_{\varphi,\alpha}(\boldsymbol{r},\boldsymbol{r}_1;\omega)\cdot\delta\tilde{\boldsymbol{J}}_\alpha(\boldsymbol{r}_1;\omega)\\ &+\tilde{G}_{\varphi,\alpha}(\boldsymbol{r},\boldsymbol{r}_1;\omega)\nabla_{\boldsymbol{r}_1}\cdot\delta\tilde{\boldsymbol{J}}_\alpha(\boldsymbol{r}_1;\omega),\end{aligned}$$

by applying Gauss's theorem the second integral in (2.32) becomes

$$\begin{aligned}&\int_\Omega \tilde{G}_{\varphi,\alpha}(\boldsymbol{r},\boldsymbol{r}_1;\omega)\nabla_{\boldsymbol{r}_1}\cdot\delta\tilde{\boldsymbol{J}}_\alpha(\boldsymbol{r}_1;\omega)\,\mathrm{d}\boldsymbol{r}_1\\ &\qquad=\int_\Omega \nabla_{\boldsymbol{r}_1}\cdot\left[\tilde{G}_{\varphi,\alpha}(\boldsymbol{r},\boldsymbol{r}_1;\omega)\nabla_{\boldsymbol{r}_1}\delta\tilde{\boldsymbol{J}}_\alpha(\boldsymbol{r}_1;\omega)\right]\,\mathrm{d}\boldsymbol{r}_1\\ &\qquad\quad-\int_\Omega \nabla_{\boldsymbol{r}_1}\tilde{G}_{\varphi,\alpha}(\boldsymbol{r},\boldsymbol{r}_1;\omega)\cdot\delta\tilde{\boldsymbol{J}}_\alpha(\boldsymbol{r}_1;\omega)\,\mathrm{d}\boldsymbol{r}_1,\\ &\qquad=\oint_\Sigma \tilde{G}_{\varphi,\alpha}(\boldsymbol{r},\boldsymbol{r}_1;\omega)\delta\tilde{\boldsymbol{J}}_\alpha(\boldsymbol{r}_1;\omega)\cdot\hat{n}\,\mathrm{d}\sigma\\ &\qquad\quad-\int_\Omega \nabla_{\boldsymbol{r}_1}\tilde{G}_{\varphi,\alpha}(\boldsymbol{r},\boldsymbol{r}_1;\omega)\cdot\delta\tilde{\boldsymbol{J}}_\alpha(\boldsymbol{r}_1;\omega)\,\mathrm{d}\boldsymbol{r}_1,\end{aligned}$$

where Σ is the boundary surface of device volume Ω, and $\hat{n}$ its external normal versor. Since we shall neglect surface microscopic noise sources [76], the surface integral is zero. Therefore, defining the *vector* Green's function $\tilde{\boldsymbol{G}}_{\varphi,\alpha}$ for carrier $\alpha = n, p$ as the gradient with respect to the injection point $\boldsymbol{r}_1$ of the *scalar* Green's function $\tilde{G}_{\varphi,\alpha}$:

$$\tilde{\boldsymbol{G}}_{\varphi,\alpha}(\boldsymbol{r}, \boldsymbol{r}_1; \omega) = \frac{1}{q} \nabla_{\boldsymbol{r}_1} \tilde{G}_{\varphi,\alpha}(\boldsymbol{r}, \boldsymbol{r}_1; \omega) \tag{2.33}$$

the potential fluctuation distribution due to the microscopic noise sources within the device volume is finally given by:

$$\begin{aligned} \delta\tilde{\varphi}(\boldsymbol{r}, \omega) = &\sum_{\alpha=n,p} \int_\Omega \tilde{G}_{\varphi,\alpha}(\boldsymbol{r}, \boldsymbol{r}_1; \omega) \tilde{\gamma}_\alpha(\boldsymbol{r}_1; \omega) \, \mathrm{d}\boldsymbol{r}_1 \\ &+ \sum_{\alpha=n,p} \kappa_\alpha \int_\Omega \tilde{\boldsymbol{G}}_{\varphi,\alpha}(\boldsymbol{r}, \boldsymbol{r}_1; \omega) \cdot \delta\tilde{\boldsymbol{J}}_\alpha(\boldsymbol{r}_1; \omega) \, \mathrm{d}\boldsymbol{r}_1 . \end{aligned} \tag{2.34}$$

Evaluation of the Correlation Matrix. The potential fluctuations induced within the device volume by the microscopic noise sources (2.29) can be evaluated from (2.34) once the relevant Green's functions have been computed. From a circuit standpoint, the noisy device is completely characterized by its small-signal impedance matrix Z and by the correlation matrix $\mathsf{S}_{\delta\boldsymbol{e},\delta\boldsymbol{e}}$ of the open-circuit noise voltage generators corresponding to the potential fluctuations induced on the N_c open-circuited terminals themselves. Since all the points of the portion $\Sigma_{\mathrm{c},k}$ of the device boundary Σ pertaining to terminal k ($k = 1, \ldots, N_\mathrm{c}$) are equipotential, the open-circuit voltage δe_k induced on terminal k can be unambiguously evaluated as

$$\delta e_k(\omega) = \delta\varphi(\boldsymbol{r}_k; \omega) \qquad \boldsymbol{r}_k \in \Sigma_{\mathrm{c},k},$$

where $\boldsymbol{r}_k$ is one of the points belonging to $\Sigma_{\mathrm{c},k}$.

Within the customary assumption of spatially uncorrelated noise sources, the local noise sources for GR (K) and diffusion (K) noise are defined as in Sect. 1.2:

$$S_{\gamma_\alpha,\gamma_\beta}(\boldsymbol{r}_1, \boldsymbol{r}_2; \omega) = K_{\gamma_\alpha,\gamma_\beta}(\boldsymbol{r}_1; \omega) \delta(\boldsymbol{r}_1 - \boldsymbol{r}_2), \tag{2.35a}$$

$$\mathsf{S}_{\delta\boldsymbol{J}_\alpha,\delta\boldsymbol{J}_\beta}(\boldsymbol{r}_1, \boldsymbol{r}_2; \omega) = \mathsf{K}_{\delta\boldsymbol{J}_\alpha,\delta\boldsymbol{J}_\beta}(\boldsymbol{r}_1; \omega) \delta(\boldsymbol{r}_1 - \boldsymbol{r}_2), \tag{2.35b}$$

where $\alpha, \beta = n, p$. The element i, j of the correlation matrix $\mathsf{S}_{\delta\boldsymbol{e},\delta\boldsymbol{e}}$ is then evaluated exploiting (2.34) to obtain a relation akin to (2.26):

$$\begin{aligned} S_{\delta e_i,\delta e_j}(\omega) = &\sum_{\alpha,\beta=n,p} \int_\Omega \tilde{G}_{\varphi,\alpha}(\boldsymbol{r}_i, \boldsymbol{r}; \omega) K_{\gamma_\alpha,\gamma_\beta}(\boldsymbol{r}; \omega) \tilde{G}^*_{\varphi,\beta}(\boldsymbol{r}_j, \boldsymbol{r}; \omega) \, \mathrm{d}\boldsymbol{r} \\ &+ \sum_{\alpha,\beta=n,p} \kappa_\alpha \kappa_\beta \int_\Omega \tilde{\boldsymbol{G}}_{\varphi,\alpha}(\boldsymbol{r}_i, \boldsymbol{r}; \omega) \cdot \mathsf{K}_{\delta\boldsymbol{J}_\alpha,\delta\boldsymbol{J}_\beta}(\boldsymbol{r}; \omega) \\ &\cdot \tilde{\boldsymbol{G}}^\dagger_{\varphi,\beta}(\boldsymbol{r}_j, \boldsymbol{r}; \omega) \, \mathrm{d}\boldsymbol{r}, \end{aligned} \tag{2.36}$$

which is the classical result of noise analysis through physics-based models [6,16]. In (2.36) $\boldsymbol{x}^\dagger$ is the Hermitian conjugate of $\boldsymbol{x}$. Notice that, for spatially uncorrelated noise sources, the contributions to the correlation matrix in (2.36) arising from GR or diffusion noise are expressed as the volume integral of the functions:

$$P_{i,j}^{(\mathrm{GR})}(\boldsymbol{r}) = \sum_{\alpha,\beta=n,p} \tilde{G}_{\varphi,\alpha}(\boldsymbol{r}_i,\boldsymbol{r};\omega) K_{\gamma_\alpha,\gamma_\beta}(\boldsymbol{r};\omega) \tilde{G}^*_{\varphi,\beta}(\boldsymbol{r}_j,\boldsymbol{r};\omega), \tag{2.37}$$

$$P_{i,j}^{(\mathrm{D})}(\boldsymbol{r}) = \sum_{\alpha,\beta=n,p} \kappa_\alpha \kappa_\beta \tilde{\boldsymbol{G}}_{\varphi,\alpha}(\boldsymbol{r}_i,\boldsymbol{r};\omega) \cdot \mathsf{K}_{\delta \boldsymbol{J}_\alpha,\delta \boldsymbol{J}_\beta}(\boldsymbol{r};\omega) \cdot \tilde{\boldsymbol{G}}^\dagger_{\varphi,\beta}(\boldsymbol{r}_j,\boldsymbol{r};\omega), \tag{2.38}$$

for GR and diffusion noise, respectively. The functions $P_{i,j}^{(\mathrm{D})}(\boldsymbol{r})$ and $P_{i,j}^{(\mathrm{GR})}(\boldsymbol{r})$ yielding the spatial distribution of the noise power or correlation spectra within the device volume can be generally referred to as the *spatial noise densities* for diffusion or GR noise. Notice that for $i \neq j$ the spatial noise density is complex, while it is real and positive for $i = j$. Assuming that the local noise source for diffusion noise is diagonal (i.e. electron and hole velocity fluctuations are uncorrelated), the corresponding spatial noise density can be written as:

$$P_{i,j}^{(\mathrm{D})}(\boldsymbol{r}) = P_{ni,j}^{(\mathrm{D})}(\boldsymbol{r}) + P_{pi,j}^{(\mathrm{D})}(\boldsymbol{r}), \tag{2.39}$$

where the two terms separately express the electron and hole contributions to the noise spectrum:

$$P_{\alpha i,j}^{(\mathrm{D})}(\boldsymbol{r}) = \tilde{\boldsymbol{G}}_{\varphi,\alpha}(\boldsymbol{r}_i,\boldsymbol{r};\omega) \cdot \mathsf{K}_{\delta \boldsymbol{J}_\alpha,\delta \boldsymbol{J}_\alpha}(\boldsymbol{r};\omega) \cdot \tilde{\boldsymbol{G}}^\dagger_{\varphi,\alpha}(\boldsymbol{r}_j,\boldsymbol{r};\omega) \qquad \alpha = n,p. \tag{2.40}$$

2.2.3 The Green's Function Approach and Other Noise Analysis Techniques

The approach to noise analysis outlined in the previous section can be easily connected to other classical methods introduced in the literature: Shockley's *impedance field* method [1] and the *transfer impedance field* approach proposed by van Vliet et al. [75]. The connection will be here described directly for a bipolar model, although originally both techniques were introduced for monopolar devices: such a simplification can be readily obtained neglecting one of the continuity equations, and therefore the corresponding noise sources and Green's functions in the previous (and following) formulae.

The source term qs_α ($\alpha = n,p$) is homogeneous to a *scalar* current source (per unit volume) injected into the device, since the continuity equation for carrier α yields

$$\nabla \cdot \boldsymbol{J}_\alpha = \frac{q}{\kappa_\alpha}\left(s_\alpha - U_\alpha - \frac{\partial \alpha}{\partial t}\right) \qquad \alpha = n,p. \tag{2.41}$$

Therefore, one can easily understand that a direct connection exists between the scalar Green's function $\tilde{G}_{\varphi,\alpha}$ and the scalar impedance field $\tilde{Z}_\alpha$. The scalar impedance field was defined in [1] as the open-circuit small-change potential induced on a device external terminal by a unit (i.e. $\delta\tilde{\imath}_\alpha = 1$) scalar current source $\delta\tilde{\imath}_\alpha \delta(\boldsymbol{r} - \boldsymbol{r}_1)$ for carrier α injected into the device volume. In a straightforward extension of the impedance field concept, the observation point for the induced potential was placed anywhere in the device volume, not only on the external contacts; according to this extended definition, one immediately has from (2.41):

$$\tilde{Z}_\alpha(\boldsymbol{r}, \boldsymbol{r}_1; \omega) = \frac{\kappa_\alpha}{q} \tilde{G}_{\varphi,\alpha}(\boldsymbol{r}, \boldsymbol{r}_1; \omega) \qquad \alpha = n, p. \tag{2.42}$$

For a more detailed discussion on this topic see [75, 77].

Small-change open-circuit conditions for the ungrounded device terminals are exploited in [1] to define the *vector* impedance field $\tilde{\boldsymbol{Z}}_\alpha$ as well, which is a vector function connecting the small-change potential induced within the device to a vector unit (i.e. $\delta\tilde{\boldsymbol{J}}_\alpha = \boldsymbol{1}$) current density source $\delta\tilde{\boldsymbol{J}}_\alpha \delta(\boldsymbol{r} - \boldsymbol{r}_1)$ for carrier α injected into the device. $\tilde{\boldsymbol{Z}}_\alpha$ can be related to the vector Green's function exploiting (2.33):

$$\tilde{\boldsymbol{Z}}_\alpha(\boldsymbol{r}, \boldsymbol{r}_1; \omega) = \kappa_\alpha \tilde{\boldsymbol{G}}_{\varphi,\alpha}(\boldsymbol{r}, \boldsymbol{r}_1; \omega) \qquad \alpha = n, p \tag{2.43}$$

since the following relation holds [1]:

$$\tilde{\boldsymbol{Z}}_\alpha(\boldsymbol{r}, \boldsymbol{r}_1; \omega) = \nabla_{\boldsymbol{r}_1} \tilde{Z}_\alpha(\boldsymbol{r}, \boldsymbol{r}_1; \omega).$$

A physical interpretation of the scalar and vector impedance fields for carrier α is depicted in Fig. 2.1.

Finally, the *transfer impedance field* tensor $\tilde{\mathsf{T}}_\alpha$ ($\alpha = n, p$), introduced by van Vliet et al. [75] as the Green's function connecting a current density source for carrier α to the small-change electric field $\delta\boldsymbol{\mathcal{E}}$ induced within the device volume, can be calculated from the scalar Green's function $\tilde{G}_{\varphi,\alpha}$ as well. Since $\delta\boldsymbol{\mathcal{E}}(\boldsymbol{r}) = -\nabla_{\boldsymbol{r}} \delta\varphi(\boldsymbol{r})$, one clearly has:

$$\begin{aligned} \tilde{\mathsf{T}}_\alpha(\boldsymbol{r}, \boldsymbol{r}_1; \omega) &= -\kappa_\alpha \nabla_{\boldsymbol{r}} \tilde{\boldsymbol{G}}_{\varphi,\alpha}(\boldsymbol{r}, \boldsymbol{r}_1; \omega) \\ &= -\frac{\kappa_\alpha}{q} \nabla_{\boldsymbol{r}} \nabla_{\boldsymbol{r}_1} \tilde{G}_{\varphi,\alpha}(\boldsymbol{r}, \boldsymbol{r}_1; \omega) \qquad \alpha = n, p. \end{aligned} \tag{2.44}$$

2.2.4 Trap-Assisted GR Noise

The treatment of trap-assisted GR noise can be carried out within a Green's function approach by extending, in a natural way, the discussion in Sect. 2.2.1 and Sect. 2.2.2. The starting point is again the drift-diffusion system[2] (2.2),

[2] As previously noted, this assumption is by no means restrictive. Provided the proper microscopic noise sources have been identified, the present discussion can be immediately extended to a PDE-based model of any degree.

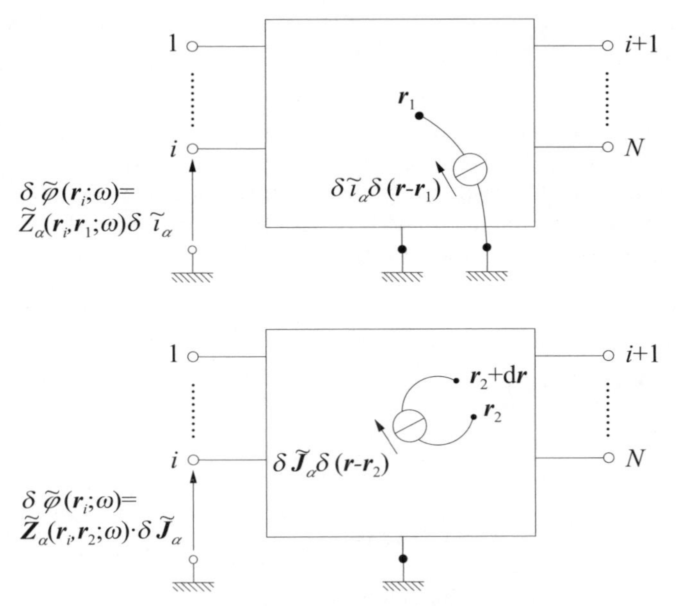

Fig. 2.1. Physical interpretation of the scalar (*above*) and vector (*below*) impedance field for carrier α in a $N_c + 1$ terminal device

where now the three basic equations (Poisson and free carrier continuity equations) are complemented by a number of rate equations (2.2d) equal to the number of trap levels. Such equations are homogeneous, i.e. do not contain any differential operator in space, and therefore can, in principle, be eliminated from the system by substitution, at least in AC small-signal operation.

For what concerns noise analysis, a Langevin source is added to the rate equations as well as to the free carrier continuity equations (see Sect. 1.2.2). The linearized and frequency-transformed Langevin rate equations can be cast in matrix form as:

$$\mathrm{i}\omega\delta\tilde{\boldsymbol{C}}_\mathrm{D}^+ = -\mathsf{M}\cdot\delta\tilde{\boldsymbol{C}}_\mathrm{D}^+ - \boldsymbol{L}_\varphi\delta\tilde{\varphi} - \boldsymbol{L}_n\delta\tilde{n} - \boldsymbol{L}_p\delta\tilde{p} + \tilde{\boldsymbol{s}}_\mathrm{D}, \tag{2.45}$$

where $\boldsymbol{C}_\mathrm{D}^+$ is the collection of trap-level occupancies $C_{\mathrm{D}k}^+$, and $\tilde{\boldsymbol{s}}_\mathrm{D}$ is the collection of microscopic noise sources $\tilde{s}_k$ to be included in the rate equations, whose correlation spectra are evaluated as in (1.84) and (1.85). Further, the elements of M are related to the first derivative, evaluated in the noiseless steady state, of the transition rates with respect to the trap occupancy:

$$(\mathsf{M})_{m,n} = \sum_{j\neq m}\left[\left.\frac{\partial p_{jm}}{\partial C_{\mathrm{D}n}^+}\right|_0 - \left.\frac{\partial p_{mj}}{\partial C_{\mathrm{D}n}^+}\right|_0\right],$$

and the m-th element of vectors $\boldsymbol{L}_\alpha$ ($\alpha = \varphi, n, p$) is given by:

$$(\boldsymbol{L}_\alpha)_m = \sum_{j \neq m} \left[\left. \frac{\partial p_{jm}}{\partial \alpha} \right|_0 - \left. \frac{\partial p_{mj}}{\partial \alpha} \right|_0 \right].$$

Since no microscopic noise source appears in the Poisson equation (compare the discussion in Sects. 1.2 and 2.2.2), for the evaluation of the open-circuit noise voltage generators the two scalar Green's functions corresponding to a source term in the continuity equations must be complemented by a scalar Green's function for each trap level rate equation. In the frequency domain, the linearized stochastic PDEs defining fluctuations around the stationary steadystate result as:

$$\nabla^2 \delta\tilde{\varphi} = -\Lambda_\varphi \left(\delta\tilde{\varphi}, \delta\tilde{n}, \delta\tilde{p}, \delta\tilde{\boldsymbol{C}}_{\mathrm{D}}^+ \right), \tag{2.46a}$$

$$\mathrm{i}\omega\delta\tilde{n} = -\Lambda_n \left(\delta\tilde{\varphi}, \delta\tilde{n}, \delta\tilde{p}, \delta\tilde{\boldsymbol{C}}_{\mathrm{D}}^+ \right) + \tilde{s}_n, \tag{2.46b}$$

$$\mathrm{i}\omega\delta\tilde{p} = -\Lambda_p \left(\delta\tilde{\varphi}, \delta\tilde{n}, \delta\tilde{p}, \delta\tilde{\boldsymbol{C}}_{\mathrm{D}}^+ \right) + \tilde{s}_p, \tag{2.46c}$$

$$\mathrm{i}\omega\delta\tilde{\boldsymbol{C}}_{\mathrm{D}}^+ = -\mathsf{M} \cdot \delta\tilde{\boldsymbol{C}}_{\mathrm{D}}^+ - \boldsymbol{L}_\varphi\delta\tilde{\varphi} - \boldsymbol{L}_n\delta\tilde{n} - \boldsymbol{L}_p\delta\tilde{p} + \tilde{\boldsymbol{s}}_{\mathrm{D}}. \tag{2.46d}$$

Therefore (2.24), (2.25) and (2.26) can be generalized to:

$$\delta\tilde{\varphi}(\boldsymbol{r}, \omega) = \sum_{\beta=n,p,k} \int_\Omega \tilde{G}_{\varphi,\beta}(\boldsymbol{r}, \boldsymbol{r}_1; \omega) \tilde{s}_\beta(\boldsymbol{r}_1, \omega) \, \mathrm{d}\boldsymbol{r}_1, \tag{2.47}$$

$$\begin{aligned} S_{\delta\varphi,\delta\varphi}(\boldsymbol{r}, \boldsymbol{r}'; \omega) = & \sum_{\gamma,\delta=n,p,k} \int_\Omega \int_\Omega \tilde{G}_{\varphi,\gamma}(\boldsymbol{r}, \boldsymbol{r}_1; \omega) \\ & S_{s_\gamma,s_\delta}(\boldsymbol{r}_1, \boldsymbol{r}_2; \omega) \tilde{G}^*_{\varphi,\delta}(\boldsymbol{r}', \boldsymbol{r}_2; \omega) \, \mathrm{d}\boldsymbol{r}_1 \mathrm{d}\boldsymbol{r}_2, \end{aligned} \tag{2.48}$$

$$\begin{aligned} S_{\delta\varphi,\delta\varphi}(\boldsymbol{r}, \boldsymbol{r}'; \omega) = & \sum_{\gamma,\delta=n,p,k} \int_\Omega \tilde{G}_{\varphi,\gamma}(\boldsymbol{r}, \boldsymbol{r}_1; \omega) \\ & K_{s_\gamma,s_\delta}(\boldsymbol{r}_1; \omega) \tilde{G}^*_{\varphi,\delta}(\boldsymbol{r}', \boldsymbol{r}_1; \omega) \, \mathrm{d}\boldsymbol{r}_1. \end{aligned} \tag{2.49}$$

An alternative, albeit equivalent, formulation makes use of three Green's functions only. This simplification is based on eliminating the linearized algebraic equation (2.45) from the linearized drift-diffusion system (2.46) by substitution. From (2.45) one has:

$$\begin{aligned} \delta\tilde{\boldsymbol{C}}_{\mathrm{D}}^+ = & - (\mathrm{i}\omega\mathsf{I} + \mathsf{M})^{-1} \cdot \boldsymbol{L}_\varphi\delta\tilde{\varphi} - (\mathrm{i}\omega\mathsf{I} + \mathsf{M})^{-1} \cdot \boldsymbol{L}_n\delta\tilde{n} \\ & - (\mathrm{i}\omega\mathsf{I} + \mathsf{M})^{-1} \cdot \boldsymbol{L}_p\delta\tilde{p} + (\mathrm{i}\omega\mathsf{I} + \mathsf{M})^{-1} \cdot \tilde{\boldsymbol{s}}_{\mathrm{D}}, \end{aligned} \tag{2.50}$$

which can be substituted in the Poisson equation (2.2a) making use of the identity:

$$\sum_k \delta\tilde{C}_{\mathrm{D}k}^+ = \boldsymbol{e}^{\mathrm{T}} \cdot \delta\tilde{\boldsymbol{C}}_{\mathrm{D}}^+$$

where $\boldsymbol{e}$ is a column vector whose elements are 1. This procedure results in two modifications on the linearized drift-diffusion model described in Sect. 2.2.2. First, the Langevin sources in the continuity equations are modified by an additive term corresponding to the last term in (2.50), and a nonwhite noise source appears in the Poisson equation (cf. (2.2a) and (2.29)):

$$\tilde{s}'_\varphi = -\frac{q}{\varepsilon}\boldsymbol{e}^{\mathrm{T}} \cdot (\mathrm{i}\omega\mathsf{I} + \mathsf{M})^{-1} \cdot \tilde{\boldsymbol{s}}_{\mathrm{D}}. \tag{2.51}$$

Second, the linearized stochastic equations (2.46) become:

$$\nabla^2\delta\tilde{\varphi} = -\Lambda'_\varphi\,(\delta\tilde{\varphi}, \delta\tilde{n}, \delta\tilde{p}) + \tilde{s}'_\varphi \tag{2.52a}$$

$$\mathrm{i}\omega\delta\tilde{n} = -\Lambda'_n\,(\delta\tilde{\varphi}, \delta\tilde{n}, \delta\tilde{p}) + \tilde{s}'_n \tag{2.52b}$$

$$\mathrm{i}\omega\delta\tilde{p} = -\Lambda'_p\,(\delta\tilde{\varphi}, \delta\tilde{n}, \delta\tilde{p}) + \tilde{s}'_p, \tag{2.52c}$$

where $(\alpha = \varphi, n, p)$

$$\begin{aligned}\Lambda'_\alpha\,(\delta\tilde{\varphi}, \delta\tilde{n}, \delta\tilde{p}) = \Lambda_\alpha \Big[&\delta\tilde{\varphi}, \delta\tilde{n}, \delta\tilde{p}, -(\mathrm{i}\omega\mathsf{I} + \mathsf{M})^{-1} \cdot \boldsymbol{L}_\varphi\delta\tilde{\varphi} \\ &- (\mathrm{i}\omega\mathsf{I} + \mathsf{M})^{-1} \cdot \boldsymbol{L}_n\delta\tilde{n} - (\mathrm{i}\omega\mathsf{I} + \mathsf{M})^{-1} \cdot \boldsymbol{L}_p\delta\tilde{p}\Big].\end{aligned} \tag{2.53}$$

As an example of application, let us apply the previous formalism to the case of single-level trap-assisted transitions, modeled according to the theory developed by Shockley, Read and Hall as discussed in Sect. 1.2.2. The four jump processes exploited to derive the expressions for the transition rates are depicted in Fig. 1.5. According to (2.2d), the rate equation makes use of the positively ionized trap density C_{D}^{+}. Therefore, (1.95)–(1.98) are still valid provided the following relationship holds:

$$n_{\mathrm{t}} = -C_{\mathrm{D}}^{+}.$$

The rate equation of the trap level reads:

$$\frac{\partial C_{\mathrm{D}}^{+}}{\partial t} = R_p^{\mathrm{srh}} - G_p^{\mathrm{srh}} + G_n^{\mathrm{srh}} - R_n^{\mathrm{srh}}. \tag{2.54}$$

Taking into account that in DC stationary conditions the time derivative in (2.54) must be zero, one easily obtains:

$$\frac{C_{\mathrm{D0}}^{+}}{N_{\mathrm{t}}} = -\frac{p_1\tau_n + n_0\tau_p}{\tau_p(n_0 + n_1) + \tau_n(p_0 + p_1)}, \tag{2.55}$$

so that the steadystate expressions for the four transition rates are given by [35, 39]:

$$R_{n0}^{\mathrm{srh}} = \frac{n_0}{\tau_n} \frac{n_1\tau_p + p_0\tau_n}{\tau_p(n_0+n_1)+\tau_n(p_0+p_1)}, \tag{2.56a}$$

$$G_{n0}^{\mathrm{srh}} = \frac{n_1}{\tau_n} \frac{n_0\tau_p + p_1\tau_n}{\tau_p(n_0+n_1)+\tau_n(p_0+p_1)}, \tag{2.56b}$$

$$R_{p0}^{\mathrm{srh}} = \frac{p_0}{\tau_p} \frac{n_0\tau_p + p_1\tau_n}{\tau_p(n_0+n_1)+\tau_n(p_0+p_1)}, \tag{2.56c}$$

$$G_{p0}^{\mathrm{srh}} = \frac{p_1}{\tau_p} \frac{n_1\tau_p + p_0\tau_n}{\tau_p(n_0+n_1)+\tau_n(p_0+p_1)}. \tag{2.56d}$$

The linearized rate equation (2.45) is therefore:

$$\begin{aligned}\mathrm{i}\omega\delta\tilde{C}_{\mathrm{D}}^{+} = &-\left(\frac{p_0+p_1}{N_{\mathrm{t}}\tau_p}+\frac{n_0+n_1}{N_{\mathrm{t}}\tau_n}\right)\delta\tilde{C}_{\mathrm{D}}^{+} \\ &-\frac{1}{\tau_n}\frac{n_1\tau_p+p_0\tau_n}{\tau_p(n_0+n_1)+\tau_n(p_0+p_1)}\delta\tilde{n} \\ &+\frac{1}{\tau_p}\frac{n_0\tau_p+p_1\tau_n}{\tau_p(n_0+n_1)+\tau_n(p_0+p_1)}\delta\tilde{p}+\tilde{s}_{\mathrm{D}}.\end{aligned} \tag{2.57}$$

The substitution procedure (2.51) yields for the noise sources [39]:

$$\tilde{s}'_{\varphi} = -\Gamma_{\varphi\mathrm{D}}\tilde{s}_{\mathrm{D}}, \tag{2.58a}$$

$$\tilde{s}'_{n} = \tilde{s}_n - \Gamma_{n\mathrm{D}}\tilde{s}_{\mathrm{D}}, \tag{2.58b}$$

$$\tilde{s}'_{p} = \tilde{s}_p - \Gamma_{p\mathrm{D}}\tilde{s}_{\mathrm{D}}, \tag{2.58c}$$

thus obtaining:

$$S_{s'_\varphi,s'_\varphi} = |\Gamma_{\varphi\mathrm{D}}|^2 S_{s_\mathrm{D},s_\mathrm{D}}, \tag{2.59a}$$

$$S_{s'_n,s'_n} = S_{s_n,s_n} + |\Gamma_{n\mathrm{D}}|^2 S_{s_\mathrm{D},s_\mathrm{D}} - 2\mathrm{Re}\left\{\Gamma_{n\mathrm{D}}^{*} S_{s_n,s_\mathrm{D}}\right\}, \tag{2.59b}$$

$$S_{s'_p,s'_p} = S_{s_p,s_p} + |\Gamma_{p\mathrm{D}}|^2 S_{s_\mathrm{D},s_\mathrm{D}} - 2\mathrm{Re}\left\{\Gamma_{p\mathrm{D}}^{*} S_{s_p,s_\mathrm{D}}\right\}, \tag{2.59c}$$

$$S_{s'_n,s'_\varphi} = \Gamma_{n\mathrm{D}}\Gamma_{\varphi\mathrm{D}}^{*} S_{s_\mathrm{D},s_\mathrm{D}} - \Gamma_{\varphi\mathrm{D}}^{*} S_{s_n,s_\mathrm{D}}, \tag{2.59d}$$

$$S_{s'_n,s'_p} = \Gamma_{n\mathrm{D}}\Gamma_{p\mathrm{D}}^{*} S_{s_\mathrm{D},s_\mathrm{D}} - \Gamma_{p\mathrm{D}}^{*} S_{s_n,s_\mathrm{D}} - \Gamma_{n\mathrm{D}}^{*} S_{s_p,s_\mathrm{D}}, \tag{2.59e}$$

$$S_{s'_p,s'_\varphi} = \Gamma_{p\mathrm{D}}\Gamma_{\varphi\mathrm{D}}^{*} S_{s_\mathrm{D},s_\mathrm{D}} - \Gamma_{\varphi\mathrm{D}}^{*} S_{s_p,s_\mathrm{D}}, \tag{2.59f}$$

where the coefficients Γ are given by:

$$\Gamma_{\varphi\mathrm{D}} = \frac{qN_{\mathrm{t}}\tau_n\tau_p/\varepsilon}{[\tau_p(n_0+n_1)+\tau_n(p_0+p_1)]+\mathrm{i}\omega N_{\mathrm{t}}\tau_n\tau_p}, \tag{2.60a}$$

$$\Gamma_{n\mathrm{D}} = \frac{(n_0+n_1)\tau_p}{[\tau_p(n_0+n_1)+\tau_n(p_0+p_1)]+\mathrm{i}\omega N_{\mathrm{t}}\tau_n\tau_p}, \tag{2.60b}$$

$$\Gamma_{p\mathrm{D}} = \frac{-(p_0+p_1)\tau_n}{[\tau_p(n_0+n_1)+\tau_n(p_0+p_1)]+\mathrm{i}\omega N_{\mathrm{t}}\tau_n\tau_p}. \tag{2.60c}$$

On the other hand, (2.53) yields ($\alpha = \varphi, n, p$)

$$\Lambda'_{\alpha}(\delta\tilde{\varphi}, \delta\tilde{n}, \delta\tilde{p}) = \Lambda_{\alpha}(\delta\tilde{\varphi}, \delta\tilde{n}, \delta\tilde{p}) + \Gamma_{\alpha n}\delta\tilde{n} + \Gamma_{\alpha p}\delta\tilde{p}, \tag{2.61}$$

where $\Lambda_{\alpha}(\delta\tilde{\varphi}, \delta\tilde{n}, \delta\tilde{p})$ corresponds to the DD model linearized with respect to potential and free carrier fluctuations only, and

$$\Gamma_{\varphi n} = \frac{-qN_{\mathrm{t}}\tau_p/\varepsilon}{[\tau_p(n_0+n_1)+\tau_n(p_0+p_1)]+\mathrm{i}\omega N_{\mathrm{t}}\tau_n\tau_p} \times \frac{n_1\tau_p+p_0\tau_n}{\tau_p(n_0+n_1)+\tau_n(p_0+p_1)}, \tag{2.62a}$$

$$\Gamma_{\varphi p} = \frac{qN_{\mathrm{t}}\tau_n/\varepsilon}{[\tau_p(n_0+n_1)+\tau_n(p_0+p_1)]+\mathrm{i}\omega N_{\mathrm{t}}\tau_n\tau_p} \times \frac{n_0\tau_p+p_1\tau_n}{\tau_p(n_0+n_1)+\tau_n(p_0+p_1)}, \tag{2.62b}$$

$$\Gamma_{nn} = \frac{-(n_0+n_1)\tau_p/\tau_n}{[\tau_p(n_0+n_1)+\tau_n(p_0+p_1)]+\mathrm{i}\omega N_{\mathrm{t}}\tau_n\tau_p} \times \frac{n_1\tau_p+p_0\tau_n}{\tau_p(n_0+n_1)+\tau_n(p_0+p_1)}, \tag{2.62c}$$

$$\Gamma_{np} = \frac{(n_0+n_1)}{[\tau_p(n_0+n_1)+\tau_n(p_0+p_1)]+\mathrm{i}\omega N_{\mathrm{t}}\tau_n\tau_p} \times \frac{n_0\tau_p+p_1\tau_n}{\tau_p(n_0+n_1)+\tau_n(p_0+p_1)}, \tag{2.62d}$$

$$\Gamma_{pn} = \frac{(p_0+p_1)}{[\tau_p(n_0+n_1)+\tau_n(p_0+p_1)]+\mathrm{i}\omega N_{\mathrm{t}}\tau_n\tau_p} \times \frac{n_1\tau_p+p_0\tau_n}{\tau_p(n_0+n_1)+\tau_n(p_0+p_1)}, \tag{2.62e}$$

$$\Gamma_{pp} = \frac{-(p_0+p_1)\tau_n/\tau_p}{[\tau_p(n_0+n_1)+\tau_n(p_0+p_1)]+\mathrm{i}\omega N_{\mathrm{t}}\tau_n\tau_p} \times \frac{n_0\tau_p+p_1\tau_n}{\tau_p(n_0+n_1)+\tau_n(p_0+p_1)}. \tag{2.62f}$$

2.2.5 Noise Analysis Through Non-stationary Transport Models

As already recalled, the application of the method of moments to the BTE with Langevin forcing term, as outlined in [18], to numerical noise modeling, is still under development. Recent contributions to this topic are found in [78–80]. While the nature of the Langevin source term arising in the momentum transport equations is fairly well established, since this must be related to carrier velocity fluctuations, a greater uncertainity seems to arise with respect to the stochastic source term for the energy transport equation. In an early approch, such a term was assumed to be proportional to the covariance of carrier velocity and carrier energy fluctuations [78]; later, a more conventional term, only dependent on energy fluctuations, was postulated, see [79, 80]. Through a straightforward generalization of the Green's function approach, velocity and energy fluctuations were propagated to the device terminals and the resulting current or voltage fluctuations were evaluated by means of an extension of (2.25) or (2.26), according to whether spatial correlation was accounted for. This approach obviously requires us to evaluate not only the power spectrum of velocity fluctuations, but also the power spectrum of energy fluctuations and their correlation spectrum. This was performed in [79, 80] directly by means of a Monte Carlo simulation.

While all the previously mentioned non-stationary transport models are based on the full hydrodynamic approach, no attempt has yet been made, to our knowledge, to implement a noise energy-balance model including energy fluctuations. On the other hand, examples of Green's function based numerical noise models exploiting full hydrodynamic or energy-balance approaches considering diffusion noise only (i.e. velocity fluctuations), can be found in [81] and [82], respectively.

In any case, the noise modeling approach based on Green's function techniques and the numerical methods described in Chap. 3 can be exploited, for the solution of non-stationary models, in much the same way as is done for the drift-diffusion approach, only with increased computational intensity.

2.3 Applications of the Green's Function Approach: Compact Device Noise Modeling

In this section, a few simple applications of the Green's function technique will be discussed. Suitable approximations in the geometry and/or device physics allow the method to be applied in analytical form, thus laying the basis for the so-called *compact noise models* of electron devices. To overcome the limitations involved in the analytical approach, a full numerical implementation of the Green's function technique must be performed, as discussed in detail in Chap. 3

2.3.1 Thermal Noise in a Linear Semiconductor Resistor

Consider a majority-carrier (e.g. n-type), uniform sample of area A and length L, grounded in $x = 0$ and left open in $x = L$ (Fig. 2.2). A simple, one-dimensional analysis yields the open-circuit voltage induced in $x = L$ by a current source $\delta\tilde{\imath}$ injected in point x as $\delta\tilde{\varphi} = Z(x)\delta\tilde{\imath}$. The impedance of the sample between x and 0 reads:

$$Z(x) = \frac{x}{A\sigma}\frac{1}{1 + \mathrm{i}\omega\tau_{\mathrm{d}}}, \tag{2.63}$$

where $\tau_{\mathrm{d}} = \varepsilon/\sigma$ is the semiconductor dielectric relaxation time, and σ is the semiconductor conductivity. Clearly, $Z(x) \equiv \tilde{Z}_n(L, x; \omega)$, where $\tilde{Z}_n$ is the

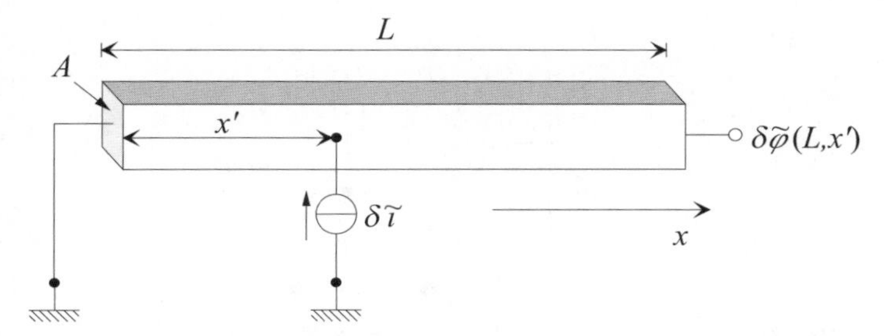

Fig. 2.2. Evaluating the impedance field in a linear resistor

electron scalar impedance field. The vector impedance field is therefore:

$$\tilde{\boldsymbol{Z}}_n(L,x;\omega) = \frac{\mathrm{d}}{\mathrm{d}x}\tilde{Z}_n(L,x;\omega)\hat{x} = \frac{\hat{x}}{A\sigma}\frac{1}{1+\mathrm{i}\omega\tau_{\mathrm{d}}} \,. \tag{2.64}$$

Taking into account that for thermal noise $\mathsf{K}_{\delta J_n,\delta J_n} = 4q^2 D_n N_{\mathrm{D}}\mathsf{I}$ where N_{D} is the ionized net donor concentration, D_n the electron diffusivity, and I the identity matrix, one has from (2.36) that the power spectrum of the open-circuit voltage fluctuations reads:

$$S_{\delta e,\delta e}(\omega) = A\int_0^L \frac{4q^2 D_n N_{\mathrm{D}}}{A^2\sigma^2}\frac{1}{1+\omega^2\tau_{\mathrm{d}}^2}\,\mathrm{d}x = 4k_{\mathrm{B}}T\frac{R}{1+\omega^2\tau_{\mathrm{d}}^2}, \tag{2.65}$$

where $R = L/(A\sigma)$ is the sample DC resistance and T is the ambient temperature. In (2.65) the Einstein relationship $D_n = \mu_n k_{\mathrm{B}}T/q$ has been used together with $\sigma = q\mu_n N_{\mathrm{D}}$, and μ_n is the electron mobility. The Lorentzian dependence in (2.65) originates from the geometrical capacitance of the resistor, and can be neglected at low frequency yielding the customary Nyquist law. Similarly, one has:

$$S_{\delta i,\delta i}(\omega) = 4k_{\mathrm{B}}TG \tag{2.66}$$

where δi are short-circuit current fluctuations and $G = 1/R$ is the sample conductance. The relevant series and parallel equivalent circuits are shown in Fig. 2.3; the resistor geometrical capacitance is $C = \varepsilon A/L$.

2.3.2 Compact Noise Models for Field-Effect Transistors

A huge effort has been spent during the past years in the compact noise modeling of several classes of field-effect transistors (FETs). Most approaches originate from Shockley's gradual channel approximation, with modifications aimed at accounting for field-dependent mobilities, velocity-saturated channels, short-channel and hot-electron effects. Owing to the similarities between JFETs, MESFETs, MOSFETs and HEMTs, a unified treatment can be attempted, at least at a simple, first-approximation level, e.g. see Sect. 5.2

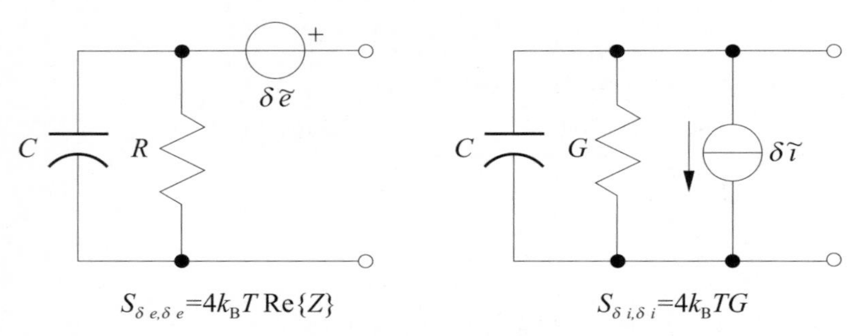

Fig. 2.3. Series and parallel equivalent circuits of a noisy resistive sample, including the effect of the geometrical capacitance

of [83] and references therein. More specialized (and hopefully accurate) compact models have been developed for specific devices, see, for example, for the MOSFET Sect. 7.3.4 of [12] and references therein; for the MESFET [84,85], and [86] for the HEMT. As usually happens in FET compact models, long-channel devices are comparatively easy to model, while short-channel devices are more challenging and generally require numerical approaches, see Chap. 3. In what follows, a simple unified theory will be presented for field-effect transistor noise, again exploiting the Green's function approach in the impedance-field form (see [87] for a similar approach based on the transfer-impedance method).

Drain Noise. As a first step, we briefly recall some elements of the gradual-channel FET theory; for simplicity, we restrict ourselves to the n-channel device. In this case, the drain current at a section x of the channel is:

$$i_{\mathrm{D}}(x) = W q \mu_n N(x) \frac{\mathrm{d}\varphi}{\mathrm{d}x} \equiv G(x) \frac{\mathrm{d}\varphi}{\mathrm{d}x}, \tag{2.67}$$

where W is the channel width, μ_n the electron mobility, $N(x)$ the total channel electron concentration per unit surface, defined as:

$$N(x) = \int_{\mathrm{ch}} n(\boldsymbol{r}) \, \mathrm{d}y, \tag{2.68}$$

where n is the electron concentration in the channel, and the integral extends to the channel depth (see Fig. 2.4). In (2.67) φ is the channel potential, G the channel conductance per unit length along the channel (i.e. along x). Since the channel electron concentration depends on the potential difference between the channel and the gate (we neglect the effect of bulk or backgate contacts), we have $G(x) = G(\varphi(x) - v_{\mathrm{GS}}) = G(\varphi)$. The source is taken at the reference potential.

By integrating (2.67) from $x = 0$ (the source) and $x = L$ (the drain) and taking into account that the drain current is constant along the channel, one obtains:

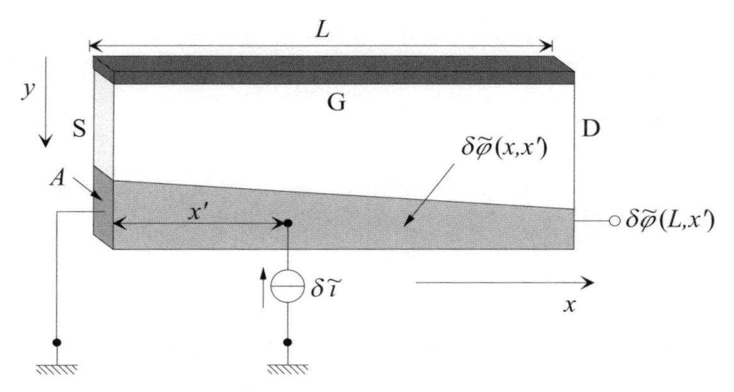

Fig. 2.4. Evaluating the FET scalar impedance field

$$i_{\mathrm{D}} = \frac{1}{L}\int_0^L G(x)\frac{\mathrm{d}\varphi}{\mathrm{d}x}\,\mathrm{d}x = \frac{1}{L}\int_0^{v_{\mathrm{DS}}} G(\varphi)\,\mathrm{d}\varphi. \tag{2.69}$$

Suppose now that a constant DC drain current is forced throughout the channel, while a small-signal, harmonic scalar current generator δi is connected between point x' in the channel and the ground (see Fig. 2.4). The induced potential fluctuation along the channel will be $\delta\varphi$. By linearizing (2.67) one readily obtains, in the frequency domain:

$$\delta\tilde{\imath} = \frac{\mathrm{d}\left[G(x)\delta\tilde{\varphi}(x)\right]}{\mathrm{d}x}\ . \tag{2.70}$$

The induced current fluctuation is constant from 0 to x' and zero from x' to L; by integrating from 0 and L, one obtains

$$\frac{\delta\tilde{\varphi}(x,x')}{\delta\tilde{\imath}} = \begin{cases} \dfrac{x}{G(x)} & x \le x' \\ \dfrac{x'}{G(x)} & x > x'\ . \end{cases} \tag{2.71}$$

Therefore, the open-circuit voltage fluctuation in L is:

$$\delta\tilde{\varphi}(L,x') = \frac{x'}{G(L)}\delta\tilde{\imath}\ ; \tag{2.72}$$

thus, the electron scalar impedance field reads:

$$\tilde{Z}_n(L,x;\omega) = \frac{x}{G(L)}, \tag{2.73}$$

where $G(L) = G(v_{\mathrm{DS}})$. By differentiating with respect to the injection point, one readily obtains:

$$\tilde{\boldsymbol{Z}}_n(L,x;\omega) = \frac{\hat{x}}{G(L)}\ . \tag{2.74}$$

From (2.73) the small-signal drain (channel) impedance can also be derived as:

$$Z_{\mathrm{D}} = \tilde{Z}_n(L, L; \omega) = \frac{L}{G(L)} \,. \tag{2.75}$$

Finally, the spectrum of the open-circuit voltage fluctuations on the drain results as:

$$S_{\delta e_{\mathrm{DS}}, \delta e_{\mathrm{DS}}}(\omega) = \frac{1}{G^2(L)} \int_{\Omega} \hat{x} \cdot \mathsf{K}_{\delta J_n, \delta J_n}(\boldsymbol{r}, \omega) \cdot \hat{x} \; \mathrm{d}\boldsymbol{r}, \tag{2.76}$$

where Ω is the channel volume. Assuming that diffusion noise occurs in the channel, we have:

$$\hat{x} \cdot \mathsf{K}_{\delta J_n, \delta J_n}(\boldsymbol{r}, \omega) \cdot \hat{x} = 4q^2 D_n n(\boldsymbol{r}) = 4k_{\mathrm{B}} T_n(\boldsymbol{r}) q \mu_n n(\boldsymbol{r}), \tag{2.77}$$

where a generalized Einstein relationship with electron temperature T_n has been postulated. Therefore:

$$\begin{aligned} &\int_{\Omega} \hat{x} \cdot \mathsf{K}_{\delta J_n, \delta J_n}(\boldsymbol{r}, \omega) \cdot \hat{x} \; \mathrm{d}\boldsymbol{r} \\ &\qquad = W \int_{\mathrm{ch}} \int_0^L 4k_{\mathrm{B}} T_n q \mu_n n \; \mathrm{d}x \mathrm{d}y = 4k_{\mathrm{B}} T \int_0^L \frac{T_n(x)}{T} G(x) \; \mathrm{d}x, \end{aligned} \tag{2.78}$$

where T is the ambient temperature; one finally obtains:

$$S_{\delta e_{\mathrm{DS}}, \delta e_{\mathrm{DS}}}(\omega) = \frac{4k_{\mathrm{B}} T}{G^2(L)} \int_0^L \frac{T_n(x)}{T} G(x) \; \mathrm{d}x. \tag{2.79}$$

Neglecting, as in long-channel devices, electron heating, the variable change $\mathrm{d}x = (G/i_{\mathrm{D}}) \; \mathrm{d}\varphi$ yields the standard expression (e.g. see [87]):

$$S_{\delta e_{\mathrm{DS}}, \delta e_{\mathrm{DS}}}(\omega) = \frac{4k_{\mathrm{B}} T}{i_{\mathrm{D}} G^2(L)} \int_0^{v_{\mathrm{DS}}} G^2(\varphi) \; \mathrm{d}\varphi. \tag{2.80}$$

From (2.80) we can also express short-circuit drain current fluctuations as:

$$S_{\delta i_{\mathrm{D}}, \delta i_{\mathrm{D}}}(\omega) = \frac{S_{\delta e_{\mathrm{DS}}, \delta e_{\mathrm{DS}}}}{|Z_{\mathrm{D}}|^2} = \frac{4k_{\mathrm{B}} T}{i_{\mathrm{D}} L^2} \int_0^{v_{\mathrm{DS}}} G^2(\varphi) \; \mathrm{d}\varphi. \tag{2.81}$$

To account for electron heating, we have to introduce suitable models for the dependence of T_n on the electric field, e.g. see p. 812 of [88].

Up to now, we made no assumptions on G. Suppose now that the linear approximation holds:

$$G(\varphi) \approx W \mu_n C_{\mathrm{ch}} (v_{\mathrm{GS}} - \varphi - v_{\mathrm{th}}) \tag{2.82}$$

for $v_{\mathrm{GS}} - \varphi - v_{\mathrm{th}} \geq 0$, where v_{th} is the threshold voltage, C_{ch} is the equivalent specific (per unit surface) channel capacitance. Notice that for $\varphi = 0$, i.e. on the source end of the channel, (2.82) simply states that the channel conductance linearly depends on the difference between v_{GS} and the threshold voltage, and is zero below threshold. The approximation (2.82) is often exploited in describing the channel control in MOSFETs and HEMTs, but is also a first-order model for low-threshold, implanted JFETs and MESFETs. In fact, in MOSFETs we have:

$$C_{\mathrm{ch}} = C_{\mathrm{ox}} = \varepsilon_{\mathrm{ox}}/t_{\mathrm{ox}} \tag{2.83}$$

the oxide capacitance per unit surface, while in MESFETs and JFETs:

$$C_{\mathrm{ch}} \approx \varepsilon/a, \tag{2.84}$$

where ε is the semiconductor permittivity, a the open channel thickness (e.g. the thickness of the epitaxial layer in epitaxial devices). Finally, the channel capacitance of HEMTs can be approximated as:

$$C_{\mathrm{ch}} \approx \frac{\varepsilon}{a+\Delta}, \tag{2.85}$$

where a is the thickness of the large-bandgap supply layer, Δ the equivalent thickness of the two-dimensional electron gas (e.g. see [89]).

Using (2.82) in (2.67) we obtain:

$$\begin{aligned} i_{\mathrm{D}} &= \frac{W\mu_n C_{\mathrm{ch}}}{L}\int_0^{v_{\mathrm{DS}}} (v_{\mathrm{GS}} - \varphi - v_{\mathrm{th}})\,\mathrm{d}\varphi \\ &= \frac{W\mu_n C_{\mathrm{ch}}}{L}\left[(v_{\mathrm{GS}} - v_{\mathrm{th}})v_{\mathrm{DS}} - \frac{v_{\mathrm{DS}}^2}{2}\right] . \end{aligned} \tag{2.86}$$

As is well known, (2.86) is valid up to $v_{\mathrm{DS}} = v_{\mathrm{DSS}} = v_{\mathrm{GS}} - v_{\mathrm{th}}$; for larger drain voltages the drain current saturates to an almost constant value:

$$i_{\mathrm{D,s}} = \frac{W\mu_n C_{\mathrm{ch}}}{2L}(v_{\mathrm{GS}} - v_{\mathrm{th}})^2 \tag{2.87}$$

and the channel is pinched off at the drain end. The saturation transconductance g_m is therefore:

$$g_m = \frac{\partial i_{\mathrm{D,s}}}{\partial v_{\mathrm{GS}}} = \frac{W\mu_n C_{\mathrm{ch}}}{L}(v_{\mathrm{GS}} - v_{\mathrm{th}}) . \tag{2.88}$$

Notice that, somewhat accidentally, the linear charge control model with saturation due to channel pinch-off yields $g_m = G_{\mathrm{DS0}}$, where G_{DS0} is the drain conductance evaluated in $v_{\mathrm{DS}} = 0$.

Equations (2.82) and (2.81) yield, after carrying out the integration:

$$S_{\delta i_{\mathrm{D}},\delta i_{\mathrm{D}}} = 4k_{\mathrm{B}}T\frac{(W\mu_n C_{\mathrm{ch}})^2}{3L^2 i_{\mathrm{D}}}\left[(v_{\mathrm{GS}} - v_{\mathrm{th}})^3 - (v_{\mathrm{GS}} - v_{\mathrm{DS}} - v_{\mathrm{th}})^3\right] . \tag{2.89}$$

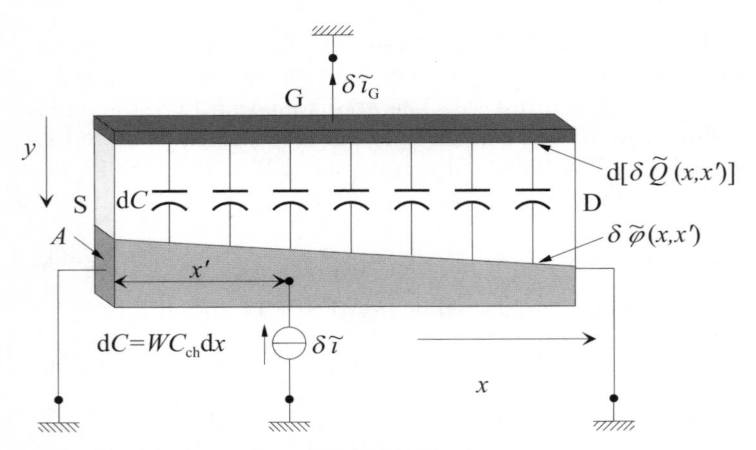

Fig. 2.5. Evaluating the FET induced gate noise

Beyond saturation, i.e. for $v_{DS} \geq v_{DSS} = v_{GS} - v_{th}$, (2.89) becomes, taking into account (2.87):

$$S_{\delta i_D,\delta i_D}(\omega) = 4k_B T \frac{2W\mu_n C_{ch}}{3L}(v_{GS} - v_{th}), \tag{2.90}$$

which can be expressed, through (2.88), in the common form:

$$S_{\delta i_D,\delta i_D}(\omega) = 4k_B T\gamma G_{DS0} = 4k_B T\gamma g_m, \tag{2.91}$$

where $\gamma = 2/3$. According to this simple model, the short-circuit drain current fluctuation spectrum therefore increases linearly in $v_{GS} - v_{th}$.

Induced Gate Noise. Potential fluctuations arising along the channel cause a net charge to be induced on the gate electrode, thus leading to induced potential fluctuations on the gate, or, if the gate is short-circuited, to induced short-circuit gate current fluctuations. Owing to the capacitive coupling mechanism, the short-circuit gate current noise spectrum is characterized by a ω^2-dependence. Moreover, gate and drain noise are partially correlated, with a ω frequency dependence. Several simplified models have been proposed to analytically estimate the induced gate noise, see [83]. A first-order solution can be derived as follows.

First, we evaluate the small-signal potential fluctuation $\delta\tilde{\varphi}(x, x')$ induced along the channel by a scalar harmonic current source $\delta\tilde{\imath}$ injected in point x' (see Fig. 2.5). The problem is similar to that leading to (2.72), but now we assume, as boundary condition, that the drain is connected to the source, i.e. $\delta\tilde{\varphi}(L, x') = 0$. From $\delta\tilde{\varphi}(x, x')$ we can estimate the charge induced on the gate as $\mathrm{d}[\delta\tilde{Q}(x, x')] = WC_{ch}\,\mathrm{d}x\,\delta\tilde{\varphi}(x, x')$. Integrating (2.70) with proper boundary conditions we obtain:

$$\frac{1}{\delta\tilde{\imath}}\frac{\mathrm{d}\delta\tilde{Q}(x,x')}{\mathrm{d}x} = \begin{cases} WC_{\mathrm{ch}}\dfrac{x}{G(x)}\left[1-\dfrac{x'}{L}\right] & x \le x' \\ WC_{\mathrm{ch}}\dfrac{x'}{G(x)}\left[1-\dfrac{x}{L}\right] & x > x' \, . \end{cases} \tag{2.92}$$

The total gate charge fluctuation $\delta\tilde{Q}_{\mathrm{G}}$ can be recovered by integration from $x = 0$ to $x = L$; defining the scalar Green's function $G_Q(x', \omega)$ such as:

$$\delta\tilde{Q}_{\mathrm{G}} = G_Q(x', \omega)\delta\tilde{\imath}, \tag{2.93}$$

where the current fluctuation is injected in x', we have:

$$G_Q(x', \omega) = \frac{1}{\delta\tilde{\imath}}\int_0^L \frac{\mathrm{d}\delta\tilde{Q}(x,x')}{\mathrm{d}x}\,\mathrm{d}x \, . \tag{2.94}$$

From (2.94) the corresponding vector Green's function can be obtained:

$$\boldsymbol{G}_Q(x', \omega) = \frac{\hat{x}}{\delta\tilde{\imath}}\int_0^L \frac{\mathrm{d}^2\delta\tilde{Q}(x,x')}{\mathrm{d}x\mathrm{d}x'}\,\mathrm{d}x \, ; \tag{2.95}$$

a straightforward computation yields:

$$\boldsymbol{G}_Q(x', \omega) = WC_{\mathrm{ch}}\hat{x}\left\{\int_{x'}^L \frac{1}{G(x)}\,\mathrm{d}x - \int_0^L \frac{x}{LG(x)}\,\mathrm{d}x\right\} . \tag{2.96}$$

As usual, the integrals in (2.96) can be solved through the variable change:

$$x(\varphi) = \frac{1}{i_{\mathrm{D}}}\int_0^{\varphi} G(\varphi)\,\mathrm{d}\varphi \, .$$

For the sake of simplicity, suppose the device is in saturation. With the assumption of linear charge control, one has:

$$\boldsymbol{G}_Q(x', \omega) = \frac{WC_{\mathrm{ch}}\hat{x}}{i_{\mathrm{D}}}\left[(v_{\mathrm{GS}} - \varphi(x') - v_{\mathrm{th}}) - \frac{2}{3}(v_{\mathrm{GS}} - v_{\mathrm{th}})\right] . \tag{2.97}$$

The charge fluctuation spectrum can be finally expressed as:

$$S_{\delta Q_{\mathrm{G}},\delta Q_{\mathrm{G}}}(\omega) = \int_0^L |\boldsymbol{G}_Q(x, \omega)|^2 \, 4k_{\mathrm{B}}TG(x)\,\mathrm{d}x \tag{2.98}$$

and the short-circuit gate current fluctuation spectrum will be $S_{\delta i_{\mathrm{G}},\delta i_{\mathrm{G}}}(\omega) = \omega^2 S_{\delta Q_{\mathrm{G}},\delta Q_{\mathrm{G}}}(\omega)$. The final result can be expressed in the form [83]:

$$S_{\delta i_{\mathrm{G}},\delta i_{\mathrm{G}}} = 4k_{\mathrm{B}}T\beta G_{\mathrm{G}}, \tag{2.99}$$

where $\beta = 4/3$ and the saturation gate conductance G_{G} reads:

$$G_{\mathrm{G}} = \frac{4}{45}\frac{\omega^2 (C_{\mathrm{ch}} W L)^2}{g_m} \; . \tag{2.100}$$

In a similar way we can evaluate the correlation spectrum between short-circuit gate and drain noise currents, as $S_{\delta i_{\mathrm{G}}, \delta i_{\mathrm{D}}} = \mathrm{i}\omega S_{\delta Q_{\mathrm{G}}, \delta i_{\mathrm{D}}}$. Since the charge gate fluctuation is (at least at low frequency) in phase with the potential channel fluctuation, which in turn is in phase with the short-circuit gate fluctuation, gate and drain short-circuit fluctuations will be in quadrature. In saturation, one has:

$$S_{\delta i_{\mathrm{G}}, \delta i_{\mathrm{D}}} = \mathrm{i}\omega \frac{C_{\mathrm{ch}} W L}{9} 4 k_{\mathrm{B}} T \; ; \tag{2.101}$$

it follows that the correlation coefficient between gate and drain current fluctuations in saturation is approximately i0.4. Such a comparatively low correlation coefficient originates from the induced charge profile on the gate from channel fluctuations, and is not expected to occur in short-gate FETs.

Short-Channel Effects. The simple, long-channel theory outlined so far is expected to be inadequate in describing short-channel devices. First, either gradual (as in Si) or abrupt (as in III-V compounds) velocity saturation takes place along the drain side of the channel, thus becoming the main cause of current saturation. Secondly, the electron temperature becomes larger than the lattice temperature at the drain end of the channel. Notice that in the velocity-saturated part of the channel microscopic diffusion noise occurs, whose effect on the external terminals can be no longer evaluated following the simple, analytical Green's function approach applied to the ohmic channel which has been discussed previously.

A simplified compact model for silicon-like velocity saturation is discussed in Sect. 5.2c of [83], with application to MOSFET devices. Although the result strongly depends on the approximations made in describing electron heating as a function of the electric field, the approach suggests that in short-channel devices the noise factor γ should increase. On the other hand, even for long-channel devices, accurate noise modeling requires an improved description of charge control, not limited to the strong inversion region, such as the one discussed in [90]. Compact noise models based on simplified approaches have been shown to successfully reproduce the experimental device behavior [91] also in the submicrometer case, provided the input physical and geometrical parameters are properly fitted. Some fundamental questions, however, such as the behavior and physical origin of excess channel noise, remain open.

Owing to their impact on low-noise, high-frequency applications, the noise behavior of III-V FETs has been perhaps the object of wider attention; a comprehensive compact model can be found in [84, 85]. Basically, the model assumes abrupt velocity saturation in a uniformly doped MESFET and analytically evaluates the noise contribution for the ohmic and saturated part of the channel. Electron heating is again accounted for approximately. The approach is able to account for the detailed noise behavior of the ohmic and saturated part of the channel; since in the latter channel noise is strongly correlated to induced gate noise, the overall correlation coefficient increases. Similar approximations are the basis of compact models for HEMTs, see [86, 89, 92]. However, compact models for short-gate devices are not expected to provide truly accurate results in the submicrometer gate range unless proper parameter tuning is made.

Generation-Recombination and $1/f$ Noise. The classical approach based on the evaluation of drain (channel) noise and on induced gate noise through simplified Green's function or IF techniques is independent of the nature of microscopic noise sources; therefore, such models can be readily extended in order to account for GR and $1/f$ noise, provided that the proper microscopic noise source is known. Approximate Lorentzian current density microscopic fluctuations describing GR or $1/f$ noise have been extensively used in the literature, e.g. see [83, 93] for a review, usually in a monopolar context.

2.3.3 Measurement-Oriented Noise Models for Field-Effect Transistors

As already recalled, a noisy two-port device (such as a FET) can be characterized in terms of a set of frequency-dependent small-signal parameters, and of the correlation matrix of noise generators. The latter information, as already noticed, amounts to four real numbers for each frequency, which can be either the input and output generator spectra and their complex correlation, or, equivalently, the noise figure, the (complex) generator impedance or admittance, and the noise resistance or conductance. From the experimental standpoint, successful attempts have been made throughout the years to reduce the number of independent noise parameters of the *intrinsic* device and to express them, as far as possible, through the help of a reduced set of frequency-independent parameters. As a first step, the following approximate expressions were proposed for the gate and drain short-circuit current spectra of the intrinsic device (i.e. neglecting the effect of external parasitic resistance, in particular of the gate resistance) and their correlation spectrum (e.g. see [85]):

$$S_{\delta i_{\mathrm{D}},\delta i_{\mathrm{D}}} \approx 4k_{\mathrm{B}}T g_m P, \tag{2.102a}$$

$$S_{\delta i_{\mathrm{G}},\delta i_{\mathrm{G}}} \approx 4k_{\mathrm{B}}T \frac{\omega^2 C_{\mathrm{GS}}^2}{g_m} R, \tag{2.102b}$$

$$S_{\delta i_{\mathrm{G}},\delta i_{\mathrm{D}}} \approx \mathrm{i}C\sqrt{S_{\delta i_{\mathrm{G}},\delta i_{\mathrm{G}}} S_{\delta i_{\mathrm{D}},\delta i_{\mathrm{D}}}}. \tag{2.102c}$$

The dimensionless, frequency-independent parameters P and R are associated with the drain and gate noise, respectively, while the correlation coefficient is assumed to be purely imaginary, with magnitude C. Notice that from (2.91) $P \equiv \gamma$ in saturation, while $R \propto \beta$ from (2.99). The assumption of a purely imaginary correlation coefficient is again in agreement with the model in (2.100). Notice that in the so-called PRC model a set of four frequency-dependent noise parameters has been replaced (or approximated) by three frequency-independent parameters. The effect of thermal noise originating from the parasitic gate resistance R_G cannot usually be neglected, but it can be easily accounted for, since R_G is in series with the device input. This yields the following expressions for the system-oriented noise parameters [94]:

$$F_{\min} = 1 + 2K_1\sqrt{g_m(R_\mathrm{S}+R_\mathrm{G})+K_2}\,\frac{f}{f_T}, \tag{2.103}$$

$$g_\mathrm{n} = g_m K_1 \left(\frac{f}{f_T}\right)^2, \tag{2.104}$$

$$Z_\mathrm{o} = \frac{\sqrt{g_m(R_\mathrm{S}+R_\mathrm{G})+K_2}}{K_1}\frac{1}{\omega C_\mathrm{GS}} + \mathrm{i}\frac{P-C\sqrt{RC}}{K_1^2}\frac{1}{\omega C_\mathrm{GS}}, \tag{2.105}$$

where $f_T = g_m/2\pi C_\mathrm{GS}$ is the FET cutoff frequency, R_S is the source resistance, and the parameters K_1 and K_2 are defined as:

$$K_1 = \sqrt{P+R-2C\sqrt{PR}}, \tag{2.106}$$

$$K_2 = PR(1-C^2)/K_1^2. \tag{2.107}$$

Neglecting the effect of intrinsic noise sources in the expression of $F_{\min}$ one finally obtains the Fukui approximation [95]:

$$F_{\min} \approx 1 + K_\mathrm{f}\sqrt{g_m(R_\mathrm{S}+R_\mathrm{G})}\,\frac{f}{f_T}, \tag{2.108}$$

where $K_\mathrm{f} = 2K_1$ is, from the experimental standpoint, a fitting factor assuming values from 1 to 3, according to the kind of device under examination.

More recently, a different approach to the measurement-oriented noise modeling was proposed by Pospieszalski [96] and Hughes [97], whereby the noise properties of the two-port are reduced to two frequency-independent noise temperatures, the gate noise temperature T_G associated with the device intrinsic resistance R_I, and the drain noise temperature T_D associated with the output conductance $G_\mathrm{DS} = 1/R_\mathrm{DS}$, see Fig. 2.6. At low frequency, the following expressions result for the short-circuit current power and correlation spectra:

$$S_{\delta i_\mathrm{D},\delta i_\mathrm{D}} \approx 4k_\mathrm{B}(T_\mathrm{G}g_m^2R_\mathrm{I} + T_\mathrm{D}G_\mathrm{DS}), \tag{2.109a}$$

$$S_{\delta i_\mathrm{G},\delta i_\mathrm{G}} \approx 4k_\mathrm{B}T_\mathrm{G}(\omega C_\mathrm{GS})^2\,R_\mathrm{I}, \tag{2.109b}$$

$$S_{\delta i_\mathrm{G},\delta i_\mathrm{D}} \approx -\mathrm{i}4k_\mathrm{B}T_\mathrm{G}\omega C_\mathrm{GS}g_m R_\mathrm{I}. \tag{2.109c}$$

Measurements carried out on several state-of-the-art MESFETs and HEMTs [96, 97] suggest that the two-temperature model is able to fit accurately enough the experimental data over a wide frequency band; moreover, T_G turns out to be close to the ambient temperature T, while typically

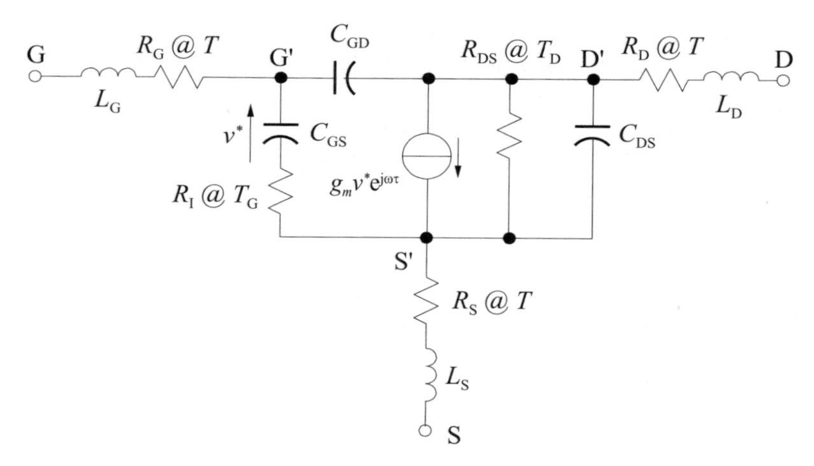

Fig. 2.6. Two-temperature FET noise model

$T_D \gg T$. Furthermore, Hughes [97] shows that a simple approximate relationship exists between the noise figure and the optimum associated gain G_{ass} of a microwave FET with low F_{min}:

$$G_{ass}(F_{min} - 1) \approx \frac{T_D}{T} . \tag{2.110}$$

A thorough discussion on the relationship between a full physics-based noise modeling approach and the outcome of measurement-oriented noise models can be found in [98], where it is shown that the gate and drain noise temperatures are, respectively, lower and higher than the average electron temperature in the channel. Thus, the interpretation initially suggested in [96] of T_G as the ambient temperature and T_D as the noise temperature is not entirely consistent with physics-based noise modeling. Nevertheless, the two-temperature approach for FET devices ultimately suggests that, at least approximately, the very complex noise behavior of such devices can be described through the help of a few frequency-independent noise parameters, associated with a lumped equivalent circuit model.

2.3.4 Compact Noise Models for Bipolar Junction Devices

The noise behavior of bipolar junction devices is deceptively simple. Although shot-like models, inherited from vacuum diode theory, suffice to explain the external noise behavior of the diode at least at low injection, the underlying interpretation of diode noise as shot noise associated with carriers crossing a space-charge region is only marginally correct [83]. As a matter of fact, a formal shot-like noise theory can be applied to any device as follows (for simplicity, we assume that current flow is one-dimensional): suppose that the device current i is split, at a given section x, into two components i_1, describing particle flow from left to right, and i_r, describing particle flow from right to left, and that fluctuations δi_l and δi_r are explicitly added to the respective mean values. Since carriers get across section x, at a microscopic level, through a ballistic free flight (i.e.they scatter left or right of x), the two currents can be physically identified. Since both currents develop full shot noise, Campbell's theorem (A.59) suggests

$$S_{\delta i,\delta i} = 2q\left(i_\mathrm{l} + i_\mathrm{r}\right). \tag{2.111}$$

Of course, the problem lies with identifying and evaluating the two current components. In a junction diode, a meaningful decomposition results from associating with i_l the inverse saturation current i_S and with i_l the injected current $i_\mathrm{S}\exp(v_\mathrm{D}/v_\mathrm{T}) = i_\mathrm{D} + i_\mathrm{S}$, where $v_\mathrm{T} = k_\mathrm{B}T/q$ and i_D is the total diode current. This leads to the standard result:

$$S_{\delta i,\delta i} = 2q\left(i_\mathrm{D} + 2i_\mathrm{S}\right) \tag{2.112}$$

reducing to $S_{\delta i,\delta i} \approx 2qi_\mathrm{D}$ in direct bias. However, the simple shot-like approach tends to obscure the fact that junction noise physically originates from thermal and GR noise taking place in the injection regions surrounding the depletion layer. GR noise in the depletion layer can be significant in reverse bias if this mechanism dominates the inverse saturation current of the junction. A Green's function based treatment of noise in bipolar devices operated under low-injection conditions is developed in [76]. For a *pn* diode, the following result is obtained

$$S_{\delta i,\delta i}(\omega) = 4k_\mathrm{B}T\,\mathrm{Re}\left\{Y_\mathrm{D}(\omega)\right\} - 2qi_\mathrm{D}, \tag{2.113}$$

which can be easily converted in the classical expression [83]

$$S_{\delta i,\delta i}(\omega) = 2q(i_\mathrm{D} + 2i_\mathrm{S}) + 4k_\mathrm{B}T\,\mathrm{Re}\left\{Y_\mathrm{D}(\omega) - Y_\mathrm{D}(0)\right\}, \tag{2.114}$$

since in low-injection conditions

$$Y_\mathrm{D}(0) = \frac{i_\mathrm{D} + i_\mathrm{S}}{v_\mathrm{T}}.$$

The high-injection behavior is analyzed by van Vliet et al. in [99, 100], where the Green's function approach already exploited for the low-injection case is extended to yield a shot-like short-circuit current noise spectrum as well. Several simplifying assumptions are made to allow for an analytical treatment, leading to the so-called *ambipolar* model [83]. Furthermore, only band-to-band and symmetric (i.e.$\tau_n = \tau_p$) Shockley-Read-Hall GR processes are considered. For a p^+n diode a noise reduction is foreseen, since

$$S_{\delta i,\delta i}(\omega) = \frac{b+1}{2(\gamma + \gamma b - 1)}\left[2q(i_\mathrm{D} + 2i_\mathrm{S}) + 4k_\mathrm{B}T\,\mathrm{Re}\left\{Y_\mathrm{D}(\omega) - Y_\mathrm{D}(0)\right\}\right], \tag{2.115}$$

where $b = \mu_n/\mu_p$ in the (lightly doped) n region and γ is the junction injection efficiency, i.e. the fraction of the diode current i_D due to the minority carriers in the lightly doped side. Since $\gamma \approx 1$ and, at least for silicon, $b > 1$, the noise spectrum is effectively reduced by the high-injection conditions. On the contrary, in a n^+p junction a noise increment is expected since the noise spectrum is now

$$S_{\delta i,\delta i}(\omega) = \frac{1/b + 1}{2(\gamma + \gamma/b - 1)} \left[2q(i_D + 2i_S) + 4k_B T \operatorname{Re}\left\{Y_D(\omega) - Y_D(0)\right\}\right], \tag{2.116}$$

where the injection efficiency γ is now the fraction of junction current carried by electrons in the p side.

As discussed in Sect. 1.2, the inclusion of $1/f$ noise in numerical physics-based models can only be performed through the help of a proper heuristic $1/f$ microscopic noise source, often expressed in terms of current density fluctuations. Owing to the non-fundamental nature of the modeling of $1/f$ noise, we shall neglect it in the following analysis; circuit models for the low-frequency noise behavior (which definitely plays a major role in many device applications) can be found in the literature, e.g. see [44] and references therein.

A straightforward Green's function approach, similar to that discussed in [83], can be exploited to obtain (2.114), at least in the low-frequency limit, as shown in the following section.

The *pn* Diode. Let us consider the simplest bipolar junction device, a monodimensional abrupt *pn* junction with constant doping levels N_A and N_D. To allow for analytical treatment, a constant carrier mobility and low-injection conditions in the quasi-neutral injection regions outside the depletion layer are assumed. Hence, neglecting the effect of GR in the depletion layer, the junction current is the sum of the minority-carrier diffusion components, evaluated at the margins of the neutral regions:

$$i_D = qAD_n \left.\frac{\partial n_p}{\partial x}\right|_{x=-x_p} - qAD_p \left.\frac{\partial p_n}{\partial x}\right|_{x=x_n}, \tag{2.117}$$

where A is the device cross section, $n_p(x)$ and $p_n(x)$ are the minority-carrier concentrations (i.e. electrons in the p side and holes in the n side, respectively), and the device geometry is reported in Fig. 2.7. Equation (2.117) suggests that minority-carrier fluctuations play a major role in inducing total current fluctuations, while majority-carrier fluctuations can only induce total current fluctuations as long as they cause minority-carrier fluctuations. This is a second-order mechanism not included in the simplified, analytical model considered; a full numerical implementation does confirm that the majority-carrier contribution is negligible (see Sect. 4.2).

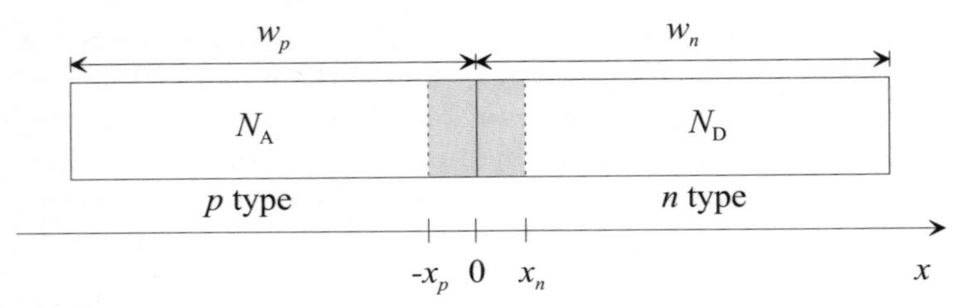

Fig. 2.7. Structure of the *pn* junction in low-injection conditions

Therefore, the microscopic noise sources included in the analysis are diffusion and GR noise due to minority carriers in the two neutral regions, where the low-injection assumption allows the (minority-carrier) continuity equations to be simplified as:

$$\frac{\partial n_p}{\partial t} = -\frac{n_p - n_{pe}}{\tau_n} + D_n \frac{\partial^2 n_p}{\partial x^2}, \tag{2.118a}$$

$$\frac{\partial p_n}{\partial t} = -\frac{p_n - p_{ne}}{\tau_p} + D_p \frac{\partial^2 p_n}{\partial x^2}, \tag{2.118b}$$

where τ_n and τ_p are the minority-carrier lifetimes, and $p_{ne} = n_i^2/N_{\mathrm{D}}$ and $n_{pe} = n_i^2/N_{\mathrm{A}}$ are the equilibrium minority-carrier concentrations in the two sides. For the sake of simplicity, we shall assume both neutral sides of the *pn* junction to be long with respect to the minority-carrier diffusion length $L_\alpha = \sqrt{D_\alpha \tau_\alpha}$ ($\alpha = n, p$), i.e.:

$$w_n - x_n \gg L_p, \qquad w_p - x_p \gg L_n;$$

the general case can be treated similarly, although calculations are quite involved. The stationary distribution of minority carriers in the two neutral regions is easily derived from (2.118):

$$n_{p0}(x) = n'_{p0}(-x_p) \exp\left(\frac{x + x_p}{L_n}\right) + n_{pe} \qquad x \le -x_p, \tag{2.119a}$$

$$p_{n0}(x) = p'_{n0}(x_n) \exp\left(-\frac{x - x_n}{L_p}\right) + p_{ne} \qquad x \ge x_n, \tag{2.119b}$$

where the excess carrier concentration at the onset of the neutral regions is given, in low-injection conditions, by the junction law:

$$n'_{p0}(-x_p) = n_{pe}\left[\exp\left(\frac{v_{\mathrm{D}}}{v_{\mathrm{T}}}\right) - 1\right], \tag{2.120}$$

$$p'_{n0}(x_n) = p_{ne}\left[\exp\left(\frac{v_{\mathrm{D}}}{v_{\mathrm{T}}}\right) - 1\right], \tag{2.121}$$

where v_{D} is the applied bias, positive on the p side.

The short-circuit noise current is then directly evaluated according to the Green's function technique discussed in Sect. 2.2.1. Imposing short-circuit conditions at the device terminals, an impulsive scalar current source is added to the minority-carrier continuity equations (2.118), which are then transformed in the frequency domain to obtain:

$$\mathrm{i}\omega\delta\tilde{n}'_p = -\frac{\delta\tilde{n}'_p}{\tau_n} + D_n\frac{\partial^2\delta\tilde{n}'_p}{\partial x^2} + \frac{\delta\tilde{\imath}_n}{qA}\delta(x-x') \qquad x' \le -x_p, \tag{2.122a}$$

$$\mathrm{i}\omega\delta\tilde{p}'_n = -\frac{\delta\tilde{p}'_n}{\tau_n} + D_p\frac{\partial^2\delta\tilde{p}'_n}{\partial x^2} + \frac{\delta\tilde{\imath}_p}{qA}\delta(x-x') \qquad x' \ge x_n. \tag{2.122b}$$

These can be analytically solved with the following boundary conditions:

- the induced charge fluctuations are zero at the onset of the neutral regions, i.e. $\delta\tilde{n}'_p(-x_p) = 0$ and $\delta\tilde{p}'_n(x_n) = 0$;
- the induced charge fluctuations tend to zero as x moves away from the injection point towards the ohmic contact;[3]
- the induced charge fluctuations are continuous at the injection point x';
- the first derivative of induced charge fluctuations is discontinuous at the injection point x'.

By integrating (2.122) between $x' - \epsilon$ and $x' + \epsilon$, taking the limit for $\epsilon \to 0$, and exploiting the continuity of the induced charge fluctuations in x', the following jump conditions are obtained:

$$\left.\frac{\partial\delta\tilde{n}'_p}{\partial x}\right|_{x=x'+} = \left.\frac{\partial\delta\tilde{n}'_p}{\partial x}\right|_{x=x'-} - \frac{\delta\tilde{\imath}_n}{qAD_n}, \tag{2.123a}$$

$$\left.\frac{\partial\delta\tilde{p}'_n}{\partial x}\right|_{x=x'+} = \left.\frac{\partial\delta\tilde{p}'_n}{\partial x}\right|_{x=x'-} - \frac{\delta\tilde{\imath}_p}{qAD_p}. \tag{2.123b}$$

According to these boundary conditions, the induced minority charge fluctuations are evaluated as follows:

$$\delta\tilde{n}'_p(x) = \begin{cases} -\delta\tilde{\imath}_n\dfrac{\tilde{L}_n}{qAD_n}\exp\left(\dfrac{x'+x_p}{\tilde{L}_n}\right)\sinh\left(\dfrac{x+x_p}{\tilde{L}_n}\right) & x' \le x \le -x_p \\ \\ -\delta\tilde{\imath}_n\dfrac{\tilde{L}_n}{qAD_n}\exp\left(\dfrac{x+x_p}{\tilde{L}_n}\right)\sinh\left(\dfrac{x'+x_p}{\tilde{L}_n}\right) & x \le x', \end{cases} \tag{2.124a}$$

[3] This approximation is justified since the long-side assumption is mathematically expressed as $w_p/L_n \to \infty$ in the p side and $w_n/L_p \to \infty$ in the n side.

$$\delta\tilde{p}'_n(x) = \begin{cases} \delta\tilde{\imath}_p \dfrac{\tilde{L}_p}{qAD_p} \exp\left(-\dfrac{x' - x_n}{\tilde{L}_p}\right) \sinh\left(\dfrac{x - x_n}{\tilde{L}_p}\right) & x_n \le x \le x' \\ \delta\tilde{\imath}_p \dfrac{\tilde{L}_p}{qAD_p} \exp\left(-\dfrac{x - x_n}{\tilde{L}_p}\right) \sinh\left(\dfrac{x' - x_n}{\tilde{L}_p}\right) & x \ge x', \end{cases} \tag{2.124b}$$

where the frequency-dependent generalized diffusion length is

$$\tilde{L}_\alpha = \frac{L_\alpha}{\sqrt{1 + \mathrm{i}\omega\tau_\alpha}} = \frac{L_\alpha}{a_\alpha + \mathrm{i}b_\alpha} \qquad \alpha = n, p$$

and the determination of the square root is chosen so as to yield $a_\alpha \ge 0$. The corresponding Green's functions relating the induced short-circuit current fluctuations to the injected current are derived from (2.124) as:

$$\begin{aligned} \tilde{G}_{np}(x', \omega) = \frac{qAD_n}{\delta\tilde{\imath}_n} \left.\frac{\partial \delta\tilde{n}'_p}{\partial x}\right|_{x=-x_p} &= -\exp\left(\frac{x' + x_p}{\tilde{L}_n}\right) \\ &= -\exp\left[\frac{x' + x_p}{L_n}(a_n + \mathrm{i}b_n)\right], \end{aligned} \tag{2.125a}$$

$$\begin{aligned} \tilde{G}_{pn}(x', \omega) = -\frac{qAD_p}{\delta\tilde{\imath}_p} \left.\frac{\partial \delta\tilde{p}'_n}{\partial x}\right|_{x=x_n} &= -\exp\left(-\frac{x' - x_n}{\tilde{L}_p}\right) \\ &= -\exp\left[-\frac{x' - x_n}{L_p}(a_p + \mathrm{i}b_p)\right]. \end{aligned} \tag{2.125b}$$

The vector Green's functions relating injected current-density sources to the short-circuit current fluctuations are then obtained from (2.125) through differentiation with respect to the injection point x':

$$\tilde{\boldsymbol{G}}_{np}(x', \omega) = -\hat{x}\frac{a_n + \mathrm{i}b_n}{L_n} \exp\left[\frac{x' + x_p}{L_n}(a_n + \mathrm{i}b_n)\right], \tag{2.126a}$$

$$\tilde{\boldsymbol{G}}_{pn}(x', \omega) = \hat{x}\frac{a_p + \mathrm{i}b_p}{L_p} \exp\left[-\frac{x' - x_n}{L_p}(a_p + \mathrm{i}b_p)\right]. \tag{2.126b}$$

Finally, the power spectrum of the short-circuit noise current generator is evaluated through superposition integrals corresponding to (minority-carrier) diffusion and GR microscopic noise sources within the neutral regions:

$$S_{\delta i, \delta i}(\omega) = S^{(\mathrm{D})}_{\delta i, \delta i}(\omega) + S^{(\mathrm{GR})}_{\delta i, \delta i}(\omega), \tag{2.127}$$

where

$$
\begin{aligned}
S^{(\mathrm{D})}_{\delta i,\delta i}(\omega) =& A \int_{x_n}^{+\infty} \tilde{\boldsymbol{G}}_{pn}(x,\omega) \cdot \mathsf{K}_{\delta \boldsymbol{J}_p,\delta \boldsymbol{J}_p}(x,\omega) \cdot \tilde{\boldsymbol{G}}_{pn}(x,\omega)\,\mathrm{d}x \\
&+ A \int_{-\infty}^{-x_p} \tilde{\boldsymbol{G}}_{np}(x,\omega) \cdot \mathsf{K}_{\delta \boldsymbol{J}_n,\delta \boldsymbol{J}_n}(x,\omega) \cdot \tilde{\boldsymbol{G}}_{np}(x,\omega)\,\mathrm{d}x
\end{aligned}
\tag{2.128}
$$

and

$$
\begin{aligned}
S^{(\mathrm{GR})}_{\delta i,\delta i}(\omega) =& q^2 A \int_{x_n}^{+\infty} \tilde{G}_{pn}(x,\omega) K_{\gamma_p,\gamma_p}(x,\omega) \tilde{G}_{pn}(x,\omega)\,\mathrm{d}x \\
&+ q^2 A \int_{-\infty}^{-x_p} \tilde{G}_{np}(x,\omega) K_{\gamma_n,\gamma_n}(x,\omega) \tilde{G}_{np}(x,\omega)\,\mathrm{d}x.
\end{aligned}
\tag{2.129}
$$

Concerning diffusion noise, one easily obtains

$$
\begin{aligned}
S^{(\mathrm{D})}_{\delta i,\delta i}(\omega) =& 4q^2 A D_p \frac{a_p^2 + b_p^2}{L_p} \left[\frac{p'_{n0}(x_n)}{1+2a_p} + \frac{p_{ne}}{2a_p} \right] \\
&+ 4q^2 A D_n \frac{a_n^2 + b_n^2}{L_n} \left[\frac{n'_{p0}(-x_p)}{1+2a_n} + \frac{n_{pe}}{2a_n} \right],
\end{aligned}
\tag{2.130}
$$

since $\hat{x} \cdot \mathsf{K}_{\delta \boldsymbol{J}_\alpha,\delta \boldsymbol{J}_\alpha} \cdot \hat{x} = 4q^2 \alpha D_\alpha$ $(\alpha = n, p)$. For the GR noise component, $K_{\gamma_p,\gamma_p}(x) = 2[p_{ne} + p_{n0}(x)]/\tau_p$ and $K_{\gamma_n,\gamma_n}(x) = 2[n_{pe} + n_{p0}(x)]/\tau_n$; therefore:

$$
\begin{aligned}
S^{(\mathrm{GR})}_{\delta i,\delta i}(\omega) =& 2q^2 A \frac{D_p}{L_p} \left[\frac{p'_{n0}(x_n)}{1+2a_p} + \frac{p_{ne}}{a_p} \right] \\
&+ 2q^2 A \frac{D_n}{L_n} \left[\frac{n'_{p0}(-x_p)}{1+2a_n} + \frac{n_{pe}}{a_n} \right].
\end{aligned}
\tag{2.131}
$$

Exploiting (2.120) and (2.121), the property $a_\alpha^2 - b_\alpha^2 = 1$ $(\alpha = n, p)$ and the expression of the small-signal diode admittance [72]:

$$
Y_\mathrm{D}(\omega) = \frac{qA}{v_\mathrm{T}} \mathrm{e}^{v_\mathrm{D}/v_\mathrm{T}} \left[p_{ne} \frac{D_p}{L_p}(a_p + \mathrm{i}b_p) + n_{pe} \frac{D_n}{L_n}(a_n + \mathrm{i}b_n) \right]
\tag{2.132}
$$

the total power spectrum of the short-circuit noise current can be easily shown to be:

$$
S_{\delta i,\delta i}(\omega) = 2q(i_\mathrm{D} + 2i_\mathrm{S}) + 4k_\mathrm{B}T \left[\mathrm{Re}\left\{ Y_\mathrm{D}(\omega) \right\} - Y_\mathrm{D}(0) \right],
\tag{2.133}
$$

where $\mathrm{Re}\{\cdot\}$ is the real part, i_D is the DC diode current and

$$
i_\mathrm{S} = qA \left[p_{ne} \frac{D_p}{L_p} + n_{pe} \frac{D_n}{L_n} \right]
$$

is the inverse saturation current of the *pn* junction. With some further effort, the analytical approach can be extended to cover the high-injection case, see [83].

The Bipolar Transistor. The noise properties of the bipolar transistor can be obtained in analogy to the calculations carried out in the previous section for the *pn* junction. Again, closed-form results can be obtained analytically in low-injection conditions [76]:

$$S_{\delta i_\alpha,\delta i_\beta} = 2k_\mathrm{B}T\left(Y_{\alpha\beta} + Y^*_{\beta\alpha}\right) - \delta_{\alpha\beta}2qi_\beta \qquad \alpha,\beta = \mathrm{B},\mathrm{C},\mathrm{E}, \tag{2.134}$$

and then extended to high-injection operation through the ambipolar approximation. From a macroscopic standpoint, the short-circuit noise current spectra exhibit a shot-like behavior, at least in the low-injection limit, since they become proportional to the DC current at the bias point. For both *pnp* and *npn* devices in the forward active region (i.e.base-emitter junction forward biased and base-collector junction reverse biased), the elements of the correlation matrix of the short-circuit noise current generators can be approximated by the following expressions [83, 101]:

- common base configuration:

$$S_{\delta i_\mathrm{E},\delta i_\mathrm{E}} \approx 2q\left|i_\mathrm{E}\right| + 4k_\mathrm{B}T\,\mathrm{Re}\left\{Y_\mathrm{EE}(\omega) - Y_\mathrm{EE}(0)\right\}, \tag{2.135a}$$

$$S_{\delta i_\mathrm{C},\delta i_\mathrm{C}} \approx 2q\left|i_\mathrm{C}\right|, \tag{2.135b}$$

$$S_{\delta i_\mathrm{E},\delta i_\mathrm{C}} \approx -2k_\mathrm{B}TY_\mathrm{CE}(\omega); \tag{2.135c}$$

- common emitter configuration:

$$S_{\delta i_\mathrm{B},\delta i_\mathrm{B}} \approx 2q\left|i_\mathrm{B}\right| + 4k_\mathrm{B}T\,\mathrm{Re}\left\{Y_\mathrm{EE}(\omega) - Y_\mathrm{EE}(0)\right\} - 4k_\mathrm{B}T\,\mathrm{Re}\left\{Y_\mathrm{CE}(\omega) - Y_\mathrm{CE}(0)\right\}, \tag{2.136a}$$

$$S_{\delta i_\mathrm{C},\delta i_\mathrm{C}} \approx 2q\left|i_\mathrm{C}\right|, \tag{2.136b}$$

$$S_{\delta i_\mathrm{B},\delta i_\mathrm{C}} \approx 2k_\mathrm{B}T\left\{Y^*_\mathrm{CE}(\omega) - Y_\mathrm{CE}(0)\right\}, \tag{2.136c}$$

where all the currents are assumed to flow inside the device terminals and the small-signal admittances Y_EE and Y_CE are evaluated at the bias point.

High-injection effects introduce correction factors in the current power and correlation spectra analogous to the ones exploited in (2.115) and (2.116); see [76, 83, 101] for a more detailed discussion.

3. Physics-Based Small-Signal Noise Simulation

3.1 Numerical Treatment of Physics-Based Device Models: a Review

Classical physical models (e.g. the drift-diffusion model) are generally formulated as systems of coupled partial differential equations (PDE). In realistic conditions, the solution of such models requires the use of numerical techniques. The numerical treatment of PDE systems, and therefore of physics-based device models, includes two logical steps: the model *discretization*, and the *solution* of the discretized model. Such issues are reviewed in a number of reference books, e.g. see [35].

In the first step, the PDE system is approximated, through spatial discretization, by a discrete system of ordinary differential equations (ODE). Spatial discretization can, in principle, be performed through classical approaches, such as the finite element method (FEM) or the finite difference (FD) technique. Nevertheless, special spatial discretization schemes have to be exploited to avoid spurious spatial oscillations of the solution, see [102]; among these, the most popular is the so-called Scharfetter–Gummel finite-box scheme.

The solution step depends on the kind of device operation which is being considered. In the steady state, DC problem time derivatives are zero and the discretized model becomes a system of nonlinear algebraic equations. Iterative methods derived from the standard Netwon approach can be conveniently exploited, even if in 3D problems their computational intensity suggests reliance on other block iterative schemes. Small-signal operation implies a small-amplitude, time-varying perturbation with respect to a DC operating point. The discretized ODE system can be conveniently linearized, and the solution found directly in the frequency domain. In transient conditions the ODE system must be solved through time-stepping algorithms, not dissimilar from those exploited in circuit simulation through SPICE-like programs. Finally, (quasi-) periodic large-signal operation can be investigated through harmonic-balance (HB) techniques [103], in analogy with large-signal nonlinear circuit simulation [74, 104].

Other, less conventional kinds of analyses are the sensitivity and the noise analysis. In sensitivity analysis, a device parameter undergoes a (small) change and the corresponding variation in the device response is evaluated, often by model linearization. Sensitivity analysis can concern the DC and small-signal parameters [105] or a set of large-signal parameters [106]. Finally, in full analogy with the case of a continuous PDE model, noise analysis corresponds to exciting the discretized model by means of a stochastic current or charge fluctuation, and evaluating the statistical properties of the induced voltage or current fluctuations on the device external electrodes. Noise analysis can be carried out either with respect to a DC steady state (small-signal noise) or with respect to a large-signal steady state (large-signal noise). This last case will be discussed in Chap. 5.

3.1.1 Finite-Box Discretization of Physics-Based Models

The finite-box (or control volume, or generalized finite-differences) approach is a flexible strategy that can be exploited to discretize conservation equations written in divergence form, either dynamic (such as the continuity equation) or static (as the Poisson equation). Owing to its popularity in the field of device simulation, it will be used as a basis for discussion. For the sake of simplicity, we shall consider a two-dimensional domain, although the same formulation can be easily extended to a three-dimensional volume.

Consider a partial differential equation in divergence form:

$$\frac{\partial f}{\partial t} + \nabla \cdot \boldsymbol{F} = s, \tag{3.1}$$

where s is a known scalar source term, f and $\boldsymbol{F}$ are an unknown scalar and vector, respectively. Suppose that the spatial domain Ω (which, for the sake of simplicity, will be assumed as two-dimensional) has been discretized through a proper, e.g. triangular, grid, whose vertices are referred to as *discretization nodes*, and whose triangular subregions are usually defined as *elements*. Consider node i and define as Σ (called the *finite box* or *control volume*) a subdomain surrounding node i. On a triangular grid made of non-obtuse triangles (also called a Delaunay grid [36]) the control volume can be conveniently defined as shown in Fig. 3.1, i.e. by connecting the bisectors of the sides leaving node i. On applying Gauss's theorem, one obtains:

$$\frac{\partial}{\partial t} \int_{\Sigma} f \, \mathrm{d}\sigma + \oint_{\Gamma} \boldsymbol{F} \cdot \hat{n} \, \mathrm{d}\gamma = \int_{\Sigma} s \, \mathrm{d}\sigma, \tag{3.2}$$

where Γ is the boundary of Σ and $\hat{n}$ its external normal versor. Suppose now that all terms are approximated through a first-order approach:

$$\frac{\partial}{\partial t} \int_{\Sigma} f \, \mathrm{d}\sigma \approx \frac{\partial f_i}{\partial t} \Sigma_i, \tag{3.3}$$

$$\oint_{\Gamma} \boldsymbol{F} \cdot \hat{n} \, \mathrm{d}\gamma \approx \sum_j l_{ij} \langle \boldsymbol{F} \cdot \hat{n} \rangle_{ij}, \tag{3.4}$$

$$\int_{\Sigma} s \, \mathrm{d}\sigma \approx s_i \Sigma_i, \tag{3.5}$$

where Σ_i is the area of the box (Fig. 3.1), l_{ij} is the length of the box side lying between node i and node j, $\langle \boldsymbol{F} \cdot \hat{n} \rangle_{ij}$ is a proper average of the normal component (with respect to the box side) of $\boldsymbol{F}$, and s_i is the source term in node i.

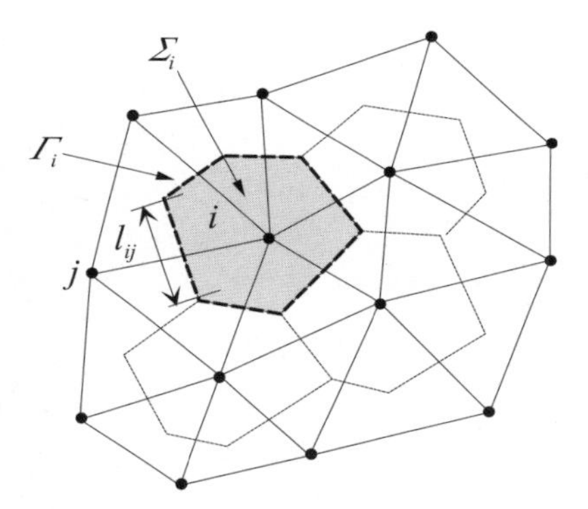

Fig. 3.1. Finite-box discretization of planar domain

The finite-boxes discretization can be directly applied to Poisson's equation (2.2a), which reads, in divergence form:

$$\nabla \cdot \boldsymbol{\mathcal{E}} = \frac{q}{\varepsilon} \left(\sum_k C^+_{\mathrm{D}k} + p - n \right). \tag{3.6}$$

For the sake of simplicity, the material permittivity was assumed to be constant; if this is not the case, the scheme can be easily adapted to accommodate for piecewise constant ϵ. The finite-box first-order approximation is readily obtained as:

$$\sum_j l_{ij} \langle \boldsymbol{\mathcal{E}} \cdot \hat{n} \rangle_{ij} = \frac{q}{\varepsilon} \left(\sum_k C^+_{\mathrm{D}ki} + p_i - n_i \right) \Sigma_i. \tag{3.7}$$

First-order averaging on the box sides yields:

$$\langle \boldsymbol{\mathcal{E}} \cdot \hat{n} \rangle_{ij} \approx \frac{\varphi_i - \varphi_j}{s_{ij}}, \tag{3.8}$$

where s_{ij} is the distance between node i and node j (Fig. 3.2). Notice that the first-order approach implies that the potential varies linearly between neighboring nodes. The resulting discretized Poisson equation is:

$$\sum_j l_{ij} \frac{\varphi_i - \varphi_j}{s_{ij}} = \frac{q}{\varepsilon} \left(\sum_k C^+_{\mathrm{D}ki} + p_i - n_i \right) \Sigma_i. \tag{3.9}$$

In a rectangular grid, the left-hand side approximation reduces to the customary five-point finite-difference formula for the Laplacian of φ.

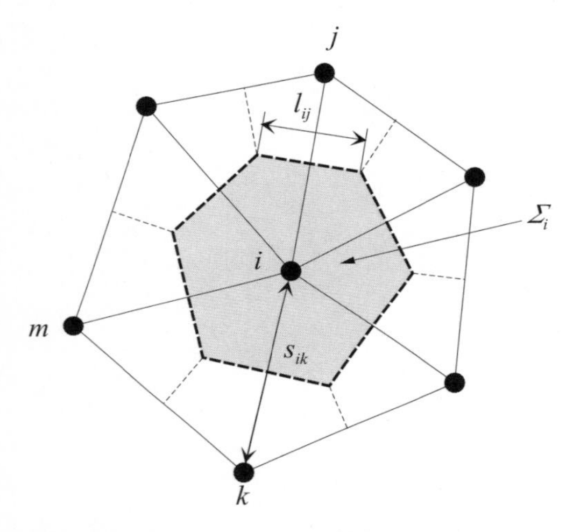

Fig. 3.2. Side lengths and bisectors in finite-box discretization

The application of the finite-box approach to the continuity equation requires some care, since a straightforward first-order finite-difference technique does not lead to a spatially stable scheme. The problem can be addressed either through standard upwind schemes (see [102] and references therein), or by the so-called Scharfetter–Gummel (SG) approach, which is commonly exploited in the field of device simulation. The SG scheme can be derived as follows. To fix the ideas, let us consider the electron continuity equation (2.2b):

$$\frac{\partial n}{\partial t} - \frac{1}{q} \nabla \cdot \boldsymbol{J}_n = -U_n; \tag{3.10}$$

the first-order finite-box discretization immediately yields:

$$\frac{\partial n_i}{\partial t} \Sigma_i - \frac{1}{q} \sum_j l_{ij} \langle \boldsymbol{J}_n \cdot \hat{n} \rangle_{ij} = -U_{ni} \Sigma_i. \tag{3.11}$$

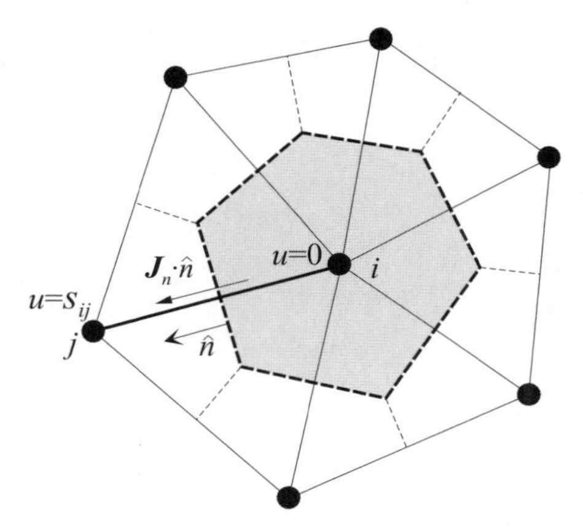

Fig. 3.3. Finite-box discretization of continuity equation

Side averaging can be carried out by taking into account that, according to the drift-diffusion model:

$$\frac{1}{q}\boldsymbol{J}_n \cdot \hat{n} = (-n\mu_n \nabla\varphi + D_n \nabla n) \cdot \hat{n}. \tag{3.12}$$

Now, current conservation suggests that the normal component of current density is almost constant between neighboring nodes, i.e. along the segment of length s_{ij}. On defining a reference frame u running from node i to node j (see Fig. 3.3), and assuming that $\boldsymbol{J}_n \cdot \hat{n}$ is constant between node i and j, (3.12) can be written as an ordinary differential equation:

$$\frac{1}{q}(\boldsymbol{J}_n \cdot \hat{n})_{ij} = -\mu_n n \frac{\mathrm{d}\varphi}{\mathrm{d}u} + D_n \frac{\mathrm{d}n}{\mathrm{d}u} \approx -\mu_n n \frac{\varphi_i - \varphi_j}{s_{ij}} + D_n \frac{\mathrm{d}n}{\mathrm{d}u}. \tag{3.13}$$

Equation (3.13) can be solved in n with boundary conditions $n = n_i$ for $u = 0$ and $n = n_j$ for $u = s_{ij}$. One obtains:

$$n(u) = n_i\,[1 - g(u,\ \Delta_{ij})] + n_j g(u,\ \Delta_{ij}), \tag{3.14}$$

where:

$$g(u,\ \Delta_{ij}) = \frac{1 - \exp(\ \Delta_{ij} u / s_{ij})}{1 - \exp(\ \Delta_{ij})} \tag{3.15}$$

and:

$$\Delta_{ij} = \frac{\varphi_i - \varphi_j}{v_{\mathrm{T}}}. \tag{3.16}$$

From (3.14) it is clear that the behavior of the electron density between nodes i and j is strongly nonlinear, unless the potential difference is as small as v_{T}. Inserting the approximation (3.14) into (3.13) one obtains, since the normal component of the current density is assumed to be constant between neighboring nodes:

$$\frac{1}{q}\langle \boldsymbol{J}_n \cdot \hat{n}\rangle_{ij} \approx \frac{1}{q}(\boldsymbol{J}_n \cdot \hat{n})_{ij} \approx \frac{D_n}{s_{ij}}\left[n_j B(\Delta_{ij}) - n_i B(-\Delta_{ij})\right], \tag{3.17}$$

where B is the Bernoulli function:

$$B(x) = \frac{x}{\exp(x) - 1}. \tag{3.18}$$

The Scharfetter–Gummel approximation for the electron continuity equation in node i finally reads:

$$\frac{\partial n_i}{\partial t}\Sigma_i - \sum_j \frac{D_n l_{ij}}{s_{ij}}\left[n_j B(\Delta_{ij}) - n_i B(-\Delta_{ij})\right] = -U_{ni}\Sigma_i. \tag{3.19}$$

By extending the treatment to the hole continuity equation (2.2b), one finally obtains the discretized bipolar drift-diffusion model (2.2) for node i under the form:

$$\sum_j l_{ij}\frac{\varphi_i - \varphi_j}{s_{ij}} = \frac{q}{\varepsilon}\left(\sum_k C^+_{\mathrm{D}ki} + p_i - n_i\right)\Sigma_i, \tag{3.20a}$$

$$\frac{\partial n_i}{\partial t} - \sum_j \frac{D_n l_{ij}}{s_{ij}\Sigma_i}\left[n_j B(\Delta_{ij}) - n_i B(-\Delta_{ij})\right] + U_{ni} = 0, \tag{3.20b}$$

$$\frac{\partial p_i}{\partial t} + \sum_j \frac{D_p l_{ij}}{s_{ij}\Sigma_i}\left[p_i B(\Delta_{ij}) - p_j B(-\Delta_{ij})\right] + U_{pi} = 0, \tag{3.20c}$$

$$\frac{\partial (C^+_{\mathrm{D}ki})}{\partial t} + \sum_{j\neq k}(p_{jki} - p_{kji}) = 0 \quad \text{for all } k, \tag{3.20d}$$

where the summations on index j refer to the nodes directly connected to node i.

Evaluating the Contact Currents. Within the framework of the finite-boxes discretization scheme, the contact currents can be evaluated as follows. Consider the boundary control volume shown in Fig. 3.4, and define as $\boldsymbol{J}_{\mathrm{t}}$ the total current density, including the displacement current, and as $\boldsymbol{J}_{n\mathrm{t}}$ ($\boldsymbol{J}_{p\mathrm{t}}$) the total electron (hole) current densities, including the displacement currents due to electron (hole) charge. Conservation of the total current density $\boldsymbol{J}_{\mathrm{t}} = \boldsymbol{J}_{n\mathrm{t}} + \boldsymbol{J}_{p\mathrm{t}}$ suggests:

$$\begin{aligned} W\oint_{\Gamma} \boldsymbol{J}_{\mathrm{t}} \cdot \hat{n}\,\mathrm{d}\gamma &= W\oint_{\Gamma_1} \boldsymbol{J}_{\mathrm{t}} \cdot \hat{n}\,\mathrm{d}\gamma + W\oint_{\Gamma_2} \boldsymbol{J}_{\mathrm{t}} \cdot \hat{n}\,\mathrm{d}\gamma \\ &= W\oint_{\Gamma_1} \boldsymbol{J}_{\mathrm{t}} \cdot \hat{n}\,\mathrm{d}\gamma - i_i = 0, \end{aligned} \tag{3.21}$$

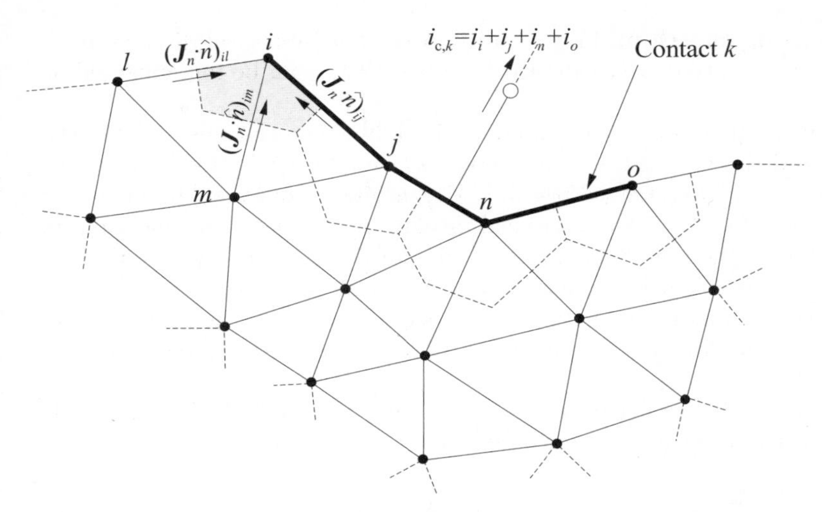

Fig. 3.4. Finite-box evaluation of contact currents

where Γ is the periphery of the box associated with the boundary node i, Γ_1 (Γ_2) the internal (boundary) part of Γ, respectively, W is the contact width, and i_i is the total current *entering* the contact node i. Therefore:

$$i_i = W \oint_{\Gamma_1} \boldsymbol{J}_{\mathrm{t}} \cdot \hat{n}\, \mathrm{d}\gamma = W \oint_{\Gamma_1} \boldsymbol{J}_{n\mathrm{t}} \cdot \hat{n}\, \mathrm{d}\gamma + W \oint_{\Gamma_1} \boldsymbol{J}_{p\mathrm{t}} \cdot \hat{n}\, \mathrm{d}\gamma. \tag{3.22}$$

Application of the finite-box approach, including displacement currents and the effect of recombination–generation terms, yields the following approximation:

$$\begin{aligned} i_i = &\sum_j \left[(\boldsymbol{J}_n \cdot \hat{n})_{ij} + (\boldsymbol{J}_p \cdot \hat{n})_{ij} \right] W \bar{l}_{ij} \\ &+ q \left(-\frac{\partial n_i}{\partial t} + \frac{\partial p_i}{\partial t} - U_{ni} + U_{pi} \right) \Sigma_i W, \end{aligned} \tag{3.23}$$

where $\bar{l}_{ij} = l_{ij}$ if node j is internal, $\bar{l}_{ij} = l_{ij}/2$ if node j lies on the boundary. For the SG discretization technique one has:

$$(\boldsymbol{J}_n \cdot \hat{n})_{ij} \approx \frac{qD_n}{s_{ij}} \left[n_j B(\Delta_{ij}) - n_i B(-\Delta_{ij}) \right], \tag{3.24}$$

$$(\boldsymbol{J}_p \cdot \hat{n})_{ij} \approx \frac{qD_p}{s_{ij}} \left[p_i B(\Delta_{ij}) - p_j B(-\Delta_{ij}) \right]. \tag{3.25}$$

Finally, the total current entering contact k can be expressed as:

$$i_{\mathrm{c},k} = \sum_i i_i, \tag{3.26}$$

where the summation is extended to all nodes belonging to the contact.

Implementing the Boundary Conditions. The numerical discretization of boundary conditions (BC) discussed in Sect. 2.1.1 can be implemented as follows.

Ideal insulating boundaries characterized by homogeneous BC for the potential and charge densities are treated in a simple way; in fact, inspection of the finite-box discretized Poisson and continuity equations immediately suggests that, if node i lies on an ideal insulating boundary, the line integral associated with the boundary must vanish, since the normal components of the electric field or current densities are zero. This amounts to stating that the discretized equation enforcing the proper BC is simply the general form, where the summation on j is extended to all nodes connected to the boundary node.

For a node i belonging to an ohmic contact, the local electron and hole densities satisfy the equilibrium condition:

$$n_i = \frac{1}{2}\left(\sqrt{\sum_k C_{\mathrm{D}ki}^{+2} + 4n_{\mathrm{i}}^2} + \sum_k C_{\mathrm{D}ki}^{+}\right), \tag{3.27}$$

$$p_i = \frac{1}{2}\left(\sqrt{\sum_k C_{\mathrm{D}ki}^{+2} + 4n_{\mathrm{i}}^2} - \sum_k C_{\mathrm{D}ki}^{+}\right), \tag{3.28}$$

where n_{i} is the intrinsic concentration and $\sum_k C_{\mathrm{D}ki}^{+}$ is the local effective ionized donor concentration for level k.

Taking into account that contacts made of ideal conductors (having, in the limit, zero resistivity) are equipotential, i.e. all nodes belonging to a contact have the same potential, we have the following possibilities.

If the ohmic contact k is driven by an ideal voltage source $e_{\mathrm{s},k}$, the Dirichlet BC for the potential simply yields:

$$\varphi_i = e_{\mathrm{s},k} + \varphi_{\mathrm{rif}} \tag{3.29}$$

for any contact node i belonging to contact k. If, on the contrary, the contact is driven by a current generator, one has:

$$\varphi_i = v_k + \varphi_{\mathrm{rif}}, \tag{3.30}$$

where the contact potential v_k is unknown and must be determined by the auxiliary equation:

$$i_{\mathrm{s},k} + G_k v_k + i_{\mathrm{c},k} = 0, \tag{3.31}$$

where G_k is the generator internal conductance, $i_{\mathrm{s},k}$ is the generator short-circuit current, and the total current $i_{\mathrm{c},k}$ entering the contact is evaluated from (3.23)–(3.26).

Finally, rectifying Schottky contacts can be treated as follows. Starting from (2.10) and (2.11) and taking into account (3.23), the equivalent surface recombination BC can be expressed in discretized form, for the contact node i, as follows:

$$\frac{\partial n_i}{\partial t}\Sigma_i - \sum_j \frac{D_n \bar{l}_{ij}}{s_{ij}} \left[n_j B(\Delta_{ij}) - n_i B(-\Delta_{ij})\right] + U_{ni}\Sigma_i = -v_{sn}(n_i - n_\mathrm{s})S_i, \tag{3.32}$$

$$\frac{\partial p_i}{\partial t}\Sigma_i + \sum_j \frac{D_p \bar{l}_{ij}}{s_{ij}} \left[p_i B(\Delta_{ij}) - p_j B(-\Delta_{ij})\right] + U_{pi}\Sigma_i = -v_{sp}(p_i - p_\mathrm{s})S_i, \tag{3.33}$$

where $S_i = (s_{ip} + s_{il})/2$, p and l being the boundary nodes connected to i. The local potential satisfies a simple Dirichlet BC, at least if the contact is driven by an ideal voltage source:

$$\varphi_i = e_{\mathrm{s},k} + v_{\mathrm{bi},k} + \varphi_\mathrm{rif}. \tag{3.34}$$

If the contact is driven by a real source, an additional equation (3.31) must be written as in the ohmic case, see Sect. 2.1.1, (2.13) and the related discussion.

3.1.2 Numerical Solution of the Discretized Model and Small-Signal Analysis

In the previous section, the spatial discretization of the drift-diffusion model and associated boundary conditions has been discussed in detail. Such discretization results in a nonlinear system of ordinary differential equations in time, where the unknowns are arrays of nodal values for the potential, electron and hole densities, while the known (forcing) terms correspond to the generators connected to the device contacts. According to the nature of such generators, several kind of analyses are of practical interest, as already outlined in Sect. 2.1.2 for the continuous problem.

In the most general case, external generators are time-varying and a transient, time-domain analysis must be performed. This can be done by applying suitable, often implicit, time discretization schemes to the differential system, which ultimately is reduced to a nonlinear algebraic system to be solved for each time step [107]. The solution of such a system is carried out through iterative Newton techniques, which are computationally intensive since they require the evaluation of the system Jacobian matrix for each time step. Iterative or mixed approaches have also been successfully applied, see [35,36]. A particular case of time-dependent analysis occurs for a large-signal, (quasi-) periodic excitation, see Sect. 2.1.2; in this case time-domain algorithms can be conveniently replaced by harmonic balance, frequency-domain approaches: see Chap. 5 for a detailed discussion.

In many cases, however, devices operate in small-signal conditions around a DC working point. DC analysis is often the first and simplest step in device simulation. In stationary, time-independent operation time derivatives vanish and the spatially discretized model reduces to a nonlinear algebraic equation system, whose unknowns are the DC values of the nodal potential, electron and hole distributions. The numerical solution of such systems again requires iterative techniques, such as the Newton method and its variations. Faster block iteration schemes have also been applied successfully in many conditions, although a general-purpose device simulation code has to be able to switch to Newton-like iterations whenever other schemes fail to converge, see [35, 36, 108].

Once the DC solution has been evaluated, small-signal analysis can be carried out through linearization. This step leads to a time-domain, linear ordinary differential equation system, which is efficiently solved in the frequency domain through Fourier analysis. Thus, small-signal analysis ultimately amounts to solving a large system of complex linear equations, where the unknowns are the Fourier transforms of the (complex) small-signal potential, electron and hole nodal distributions. It may be remarked that the system matrix is closely related to the Jacobian already exploited in the DC simulation [109].

Since noise analysis can be viewed as a kind of small-signal analysis in the presence of random excitation, a more detailed investigation of the structure of the small-signal system will be presented. We shall denote as $\tilde{\boldsymbol{\varphi}}_{\mathrm{i}}^{\mathrm{ss}}$ the nodal values of the small-signal Fourier transform of potential on the internal N_{i} nodes (including non-contact boundaries), and as $\tilde{\boldsymbol{\varphi}}_{\mathrm{x}}^{\mathrm{ss}}$ the nodal values of potential on the N_{x} nodes lying on the external (ohmic or Schottky) device metallic contacts. Similarly, we shall denote as $\tilde{\boldsymbol{n}}_{\mathrm{i}}^{\mathrm{ss}}$ and $\tilde{\boldsymbol{n}}_{\mathrm{x}}^{\mathrm{ss}}$ ($\tilde{\boldsymbol{p}}_{\mathrm{i}}^{\mathrm{ss}}$ and $\tilde{\boldsymbol{p}}_{\mathrm{x}}^{\mathrm{ss}}$) the Fourier transforms of the electron (hole) density nodal values. Moreover, $\tilde{\boldsymbol{v}}_{\mathrm{c}}^{\mathrm{ss}}$ and $\tilde{\boldsymbol{\imath}}_{\mathrm{c}}^{\mathrm{ss}}$ are the Fourier transforms of the contact potentials and (entering) currents, of size N_{c} corresponding to the number of device terminals minus one (the reference terminal, which is grounded).

Linearization of the drift-diffusion discretized equations and boundary conditions around a DC working point leads to the following set of linear systems. For the sake of simplicity, only band-to-band GR mechanisms have been considered; in the general case additional equations for the linearized trap level rate equations must be either separately added or included in the Poisson equation through back-substitution, see [110]. In all systems, a symbolic notation will be exploited for the relevant matrices, whose detailed expression is omitted for brevity but can be recovered through straightforward linearization of the discretized equations (3.20) and boundary conditions (3.27)–(3.34) and Fourier transformation. Most of the matrices are real, since imaginary components only stem from Fourier transformation of time derivative operator into $\mathrm{i}\omega$. All matrices generally are sparse, but, unless explicitly stated, not diagonal.

The Poisson equation yields the following linear system of N_{i} equations for the internal nodes:

$$\mathsf{A}_{\varphi_{\mathrm{i}}\varphi_{\mathrm{i}}} \cdot \tilde{\boldsymbol{\varphi}}_{\mathrm{i}}^{\mathrm{ss}} + \mathsf{A}_{\varphi_{\mathrm{i}}\varphi_{\mathrm{x}}} \cdot \tilde{\boldsymbol{\varphi}}_{\mathrm{x}}^{\mathrm{ss}} + \mathsf{A}_{\varphi_{\mathrm{i}}n_{\mathrm{i}}} \cdot \tilde{\boldsymbol{n}}_{\mathrm{i}}^{\mathrm{ss}} + \mathsf{A}_{\varphi_{\mathrm{i}}p_{\mathrm{i}}} \cdot \tilde{\boldsymbol{p}}_{\mathrm{i}}^{\mathrm{ss}} = \mathbf{0} \tag{3.35a}$$

and the set of N_{x} Dirichlet BC for the external contact nodes:

$$\mathsf{I}_{N_{\mathrm{x}}} \cdot \tilde{\boldsymbol{\varphi}}_{\mathrm{x}}^{\mathrm{ss}} + \mathsf{A}_{\varphi_{\mathrm{x}} v_{\mathrm{c}}} \cdot \tilde{\boldsymbol{v}}_{\mathrm{c}}^{\mathrm{ss}} = \mathbf{0}, \tag{3.35b}$$

where I_N is the identity matrix of size N, and $\mathsf{A}_{\varphi_{\mathrm{x}} v_{\mathrm{c}}}$ is a placeholder matrix whose elements are either 0 or -1. Notice that the matrices $\mathsf{A}_{\varphi_{\mathrm{i}} n_{\mathrm{i}}}$ and $\mathsf{A}_{\varphi_{\mathrm{i}} p_{\mathrm{i}}}$ are diagonal, see (3.20a).

The electron and hole continuity equations (3.20b) and (3.20c) yield, for internal nodes, two linear systems of N_{i} equations:

$$\mathsf{A}_{n_{\mathrm{i}} \varphi_{\mathrm{i}}} \cdot \tilde{\boldsymbol{\varphi}}_{\mathrm{i}}^{\mathrm{ss}} + \mathsf{A}_{n_{\mathrm{i}} \varphi_{\mathrm{x}}} \cdot \tilde{\boldsymbol{\varphi}}_{\mathrm{x}}^{\mathrm{ss}} + \mathsf{A}_{n_{\mathrm{i}} n_{\mathrm{i}}} \cdot \tilde{\boldsymbol{n}}_{\mathrm{i}}^{\mathrm{ss}} + \mathsf{A}_{n_{\mathrm{i}} n_{\mathrm{x}}} \cdot \tilde{\boldsymbol{n}}_{\mathrm{x}}^{\mathrm{ss}} + \mathsf{A}_{n_{\mathrm{i}} p_{\mathrm{i}}} \cdot \tilde{\boldsymbol{p}}_{\mathrm{i}}^{\mathrm{ss}} = \mathbf{0}, \tag{3.35c}$$

$$\mathsf{A}_{p_{\mathrm{i}} \varphi_{\mathrm{i}}} \cdot \tilde{\boldsymbol{\varphi}}_{\mathrm{i}}^{\mathrm{ss}} + \mathsf{A}_{p_{\mathrm{i}} \varphi_{\mathrm{x}}} \cdot \tilde{\boldsymbol{\varphi}}_{\mathrm{x}}^{\mathrm{ss}} + \mathsf{A}_{p_{\mathrm{i}} n_{\mathrm{i}}} \cdot \tilde{\boldsymbol{n}}_{\mathrm{i}}^{\mathrm{ss}} + \mathsf{A}_{p_{\mathrm{i}} p_{\mathrm{i}}} \cdot \tilde{\boldsymbol{p}}_{\mathrm{i}}^{\mathrm{ss}} + \mathsf{A}_{p_{\mathrm{i}} p_{\mathrm{x}}} \cdot \tilde{\boldsymbol{p}}_{\mathrm{x}}^{\mathrm{ss}} = \mathbf{0}. \tag{3.35d}$$

The related BC can be either Dirichlet, for ohmic contacts, see (3.27) and (3.28), or inhomogeneous Neumann, for Schottky contacts, see (3.32) and (3.33). In both cases, linearization results in a set of N_{x} linear equations:

$$\mathsf{A}_{n_{\mathrm{x}} \varphi_{\mathrm{i}}} \cdot \tilde{\boldsymbol{\varphi}}_{\mathrm{i}}^{\mathrm{ss}} + \mathsf{A}_{n_{\mathrm{x}} \varphi_{\mathrm{x}}} \cdot \tilde{\boldsymbol{\varphi}}_{\mathrm{x}}^{\mathrm{ss}} + \mathsf{A}_{n_{\mathrm{x}} n_{\mathrm{i}}} \cdot \tilde{\boldsymbol{n}}_{\mathrm{i}}^{\mathrm{ss}} + \mathsf{A}_{n_{\mathrm{x}} n_{\mathrm{x}}} \cdot \tilde{\boldsymbol{n}}_{\mathrm{x}}^{\mathrm{ss}} + \mathsf{A}_{n_{\mathrm{x}} p_{\mathrm{x}}} \cdot \tilde{\boldsymbol{p}}_{\mathrm{x}}^{\mathrm{ss}} = \mathbf{0}, \tag{3.35e}$$

$$\mathsf{A}_{p_{\mathrm{x}} \varphi_{\mathrm{i}}} \cdot \tilde{\boldsymbol{\varphi}}_{\mathrm{i}}^{\mathrm{ss}} + \mathsf{A}_{p_{\mathrm{x}} \varphi_{\mathrm{x}}} \cdot \tilde{\boldsymbol{\varphi}}_{\mathrm{x}}^{\mathrm{ss}} + \mathsf{A}_{p_{\mathrm{x}} n_{\mathrm{x}}} \cdot \tilde{\boldsymbol{n}}_{\mathrm{x}}^{\mathrm{ss}} + \mathsf{A}_{p_{\mathrm{x}} p_{\mathrm{i}}} \cdot \tilde{\boldsymbol{p}}_{\mathrm{i}}^{\mathrm{ss}} + \mathsf{A}_{p_{\mathrm{x}} p_{\mathrm{x}}} \cdot \tilde{\boldsymbol{p}}_{\mathrm{x}}^{\mathrm{ss}} = \mathbf{0}. \tag{3.35f}$$

The previous linear systems, obtained through direct linearization of the model physical equations and BC, must be completed by further equations expressing the contact currents and the constraints imposed by the external generators applied to the device. For this reason, we shall denote these as *circuit* equations. Linearization of the N_{c} contact current expressions (3.26) and (3.23) yields:

$$\begin{aligned} \mathsf{A}_{i_{\mathrm{c}} \varphi_{\mathrm{i}}} \cdot \tilde{\boldsymbol{\varphi}}_{\mathrm{i}}^{\mathrm{ss}} + \mathsf{A}_{i_{\mathrm{c}} \varphi_{\mathrm{x}}} \cdot \tilde{\boldsymbol{\varphi}}_{\mathrm{x}}^{\mathrm{ss}} + \mathsf{A}_{i_{\mathrm{c}} n_{\mathrm{i}}} \cdot \tilde{\boldsymbol{n}}_{\mathrm{i}}^{\mathrm{ss}} + \mathsf{A}_{i_{\mathrm{c}} n_{\mathrm{x}}} \cdot \tilde{\boldsymbol{n}}_{\mathrm{x}}^{\mathrm{ss}} & \\ + \mathsf{A}_{i_{\mathrm{c}} p_{\mathrm{i}}} \cdot \tilde{\boldsymbol{p}}_{\mathrm{i}}^{\mathrm{ss}} + \mathsf{A}_{i_{\mathrm{c}} p_{\mathrm{x}}} \cdot \tilde{\boldsymbol{p}}_{\mathrm{x}}^{\mathrm{ss}} + \tilde{\boldsymbol{\imath}}_{\mathrm{c}}^{\mathrm{ss}} = \mathbf{0}. & \end{aligned} \tag{3.35g}$$

In order to close the system, N_{c} additional equations are needed. These are given by the (already linear) constraints imposed by the external generators applied to the linearized device, see (3.30), (3.31) and (3.34). The resulting equations can be expressed in a unified form as:

$$\mathsf{A}_{c v_{\mathrm{c}}} \cdot \tilde{\boldsymbol{v}}_{\mathrm{c}}^{\mathrm{ss}} + \mathsf{A}_{c i_{\mathrm{c}}} \cdot \tilde{\boldsymbol{\imath}}_{\mathrm{c}}^{\mathrm{ss}} = \tilde{\boldsymbol{s}}_{\mathrm{s}}^{\mathrm{ss}}, \tag{3.35h}$$

where $\tilde{\boldsymbol{s}}_{\mathrm{s}}^{\mathrm{ss}}$ is the Fourier transform of the set of small-signal (voltage or current) generators, applied to the device. Notice that, according to the nature of the generators, the elements of matrices A_{c} can be dimensionless, impedances or admittances. In limiting cases (ideal voltage or current generators) some of the matrices vanish.

In total, the small-signal system (3.35) includes $3(N_{\mathrm{i}} + N_{\mathrm{x}}) + 2N_{\mathrm{c}}$ linear equations in the $3(N_{\mathrm{i}} + N_{\mathrm{x}})$ physical unknowns ($\tilde{\boldsymbol{\varphi}}_{\mathrm{i}}^{\mathrm{ss}}, \tilde{\boldsymbol{\varphi}}_{\mathrm{x}}^{\mathrm{ss}}, \tilde{\boldsymbol{n}}_{\mathrm{i}}^{\mathrm{ss}}, \tilde{\boldsymbol{n}}_{\mathrm{x}}^{\mathrm{ss}}, \tilde{\boldsymbol{p}}_{\mathrm{i}}^{\mathrm{ss}}, \tilde{\boldsymbol{p}}_{\mathrm{x}}^{\mathrm{ss}}$) and $2N_{\mathrm{c}}$ circuit unknowns ($\tilde{\boldsymbol{v}}_{\mathrm{c}}^{\mathrm{ss}}, \tilde{\boldsymbol{\imath}}_{\mathrm{c}}^{\mathrm{ss}}$).

3.2 Numerical Noise Simulation: Formulation

Although any kind of Green's function, as defined in Sect. 2.2.1, can be numerically evaluated through finite-boxes discretization of the physical model, we shall confine our treatment to the classical case of the impedance field method (IFM), wherein the Green's function of interest is the open-circuit potential induced at the device contacts. The extension of the discretized formulation to the general case is straightforward.

In order to evaluate the open-circuit Green's functions or the impedance fields, see (2.42), system (2.31) with the open-circuit BC (2.28) must be discretized according to the finite-box strategy. The resulting equation system basically is the same as the small-signal system (3.35), but all small-signal values have to be interpreted as fluctuations, and the forcing terms have to be properly modified. Basically, two cases must be considered: in the first, current injection occurs in an internal node. In this case, all non-reference contacts are connected to open circuits, which can be directly expressed by discarding system (3.35h) and setting the terminal current (fluctuation) $\delta\tilde{\boldsymbol{\imath}}_{\mathrm{c}} = \mathbf{0}$ in system (3.35g). In the second case, current injection takes place on a node belonging to contact k; in this case, all contacts but contact k are open-circuited, and contact k becomes connected to a current source, i.e. the condition is the same as in small-signal analysis when the impedance matrix elements are sought. This corresponds to the well-known result that the limiting value of the impedance field on the contacts are the elements of the impedance matrix. In the following analysis, a unified treatment of internal and contact injection will be provided.

As a last remark, consistent with the notation followed in Sect. 2.2.1, the potential fluctuation will be directly referred to as the discretized (nodal) Green's function, so that the physical unknown set will be:

$$\tilde{\boldsymbol{G}}_{\varphi,\alpha\mathrm{i}}(k), \tilde{\boldsymbol{G}}_{\varphi,\alpha\mathrm{x}}(k), \delta\tilde{\boldsymbol{n}}_{\mathrm{i}}(k), \delta\tilde{\boldsymbol{n}}_{\mathrm{x}}(k), \delta\tilde{\boldsymbol{p}}_{\mathrm{i}}(k), \delta\tilde{\boldsymbol{p}}_{\mathrm{x}}(k), \tag{3.36}$$

where $\alpha = n, p$ denotes that the internal source term appears in the electron or hole continuity equation, respectively, and the argument k refers to injection in node k. Notice that, for the sake of simplicity, $k = 1, \ldots, (N_{\mathrm{i}} + N_{\mathrm{c}})$, meaning that injection either occurs in the internal nodes, or into a contact (rather than into each individual node belonging to a contact); this simplification is allowed by the fact that all contacts are equipotential, i.e. all contact nodes belonging to the same contact are electrically connected through a short-circuit, thus making injection in any of them equivalent.

Concerning the circuit unknowns, since contact current fluctuations are zero, only contact voltage fluctuations $\delta\tilde{\boldsymbol{e}}(k) \equiv \delta\tilde{\boldsymbol{v}}_{\mathrm{c}}(k)$ remain, which for the sake of consistency, will be denoted as $\tilde{\boldsymbol{G}}_{e,\alpha}(k)$. Notice, however, that this contact Green's function adds no further information with respect to $\tilde{\boldsymbol{G}}_{\varphi,\alpha\mathrm{x}}(k)$, since, from (3.35b), the potential fluctuations of all nodes pertaining to the same contact are equal to the corresponding contact potential fluctuation.

As a further remark, it should be noted that electron or hole injection into a contact is equivalent, apart from the sign. This is more easily seen in terms of impedance fields, since an electron or hole scalar current injection into a contact leads to the same contact potentials. The connection with the Green's function is obtained from (2.42), which becomes, in discretized form:

$$\tilde{\boldsymbol{Z}}_\alpha(k) = \frac{\kappa_\alpha}{q}\tilde{\boldsymbol{G}}_{e,\alpha}(k) \qquad \alpha = n, p; \tag{3.37}$$

but, for $k = (N_\mathrm{i} + 1), \ldots, (N_\mathrm{i} + N_\mathrm{c})$, $\tilde{\boldsymbol{Z}}_n(k) = \tilde{\boldsymbol{Z}}_p(k)$, and therefore:

$$\tilde{\boldsymbol{G}}_{e,n}(k) = -\tilde{\boldsymbol{G}}_{e,p}(k), \qquad k = (N_\mathrm{i} + 1), \ldots, (N_\mathrm{i} + N_\mathrm{c}). \tag{3.38}$$

Finally, the Green's function elements arising from contact injection are related to the elements of the small-signal impedance matrix Z. Since:

$$\tilde{Z}_{nj}(k) = \tilde{Z}_{pj}(k) = Z_{jk},$$

where j denotes that the observation point is contact j, one has:

$$\tilde{G}_{e,nj}(k) = -\frac{1}{q}Z_{jk}, \tag{3.39}$$

$$\tilde{G}_{e,pj}(k) = \frac{1}{q}Z_{jk}. \tag{3.40}$$

We now discuss the discretization of the forcing term arising in system (2.31), which is a spatial delta function appearing either in the electron or in the hole continuity equation. Application of the finite-box discretization leads directly to a discretized forcing term (appearing in the proper continuity equation) which is zero for all nodes but one, where it equals unity. In other words, in evaluating the Green's function relative to an electron or hole excitation injected in node k, the discretized forcing term must be zero for all equations, and unity for the electron or hole discretized continuity equation relative to node k. Rather than setting up a separate forcing term for each node k, and then evaluating the related nodal Green's function, it is helpful to write (at least formally) a matrix forcing term that collects injection for every node k and for both continuity equations (electron and hole). Correspondingly, the resulting unknowns (and, in particular, the discretized Green's functions) become matrices properly collecting all solutions pertaining to different electron or hole injection points. Namely, the structure of the matrix discretized Green's functions $\tilde{\mathsf{G}}_{\varphi\mathrm{i}}$ and $\tilde{\mathsf{G}}_{\varphi\mathrm{x}}$ (of dimensions $N_\mathrm{i} \times 2\,(N_\mathrm{i} + N_\mathrm{c})$ and $N_\mathrm{x} \times 2\,(N_\mathrm{i} + N_\mathrm{c})$, respectively) becomes:

$$\tilde{\mathsf{G}}_{\varphi\mathrm{i}} = \left[\tilde{\boldsymbol{G}}_{\varphi,n\mathrm{i}}(1), \ldots, \tilde{\boldsymbol{G}}_{\varphi,n\mathrm{i}}(N_\mathrm{i} + N_\mathrm{c}), \tilde{\boldsymbol{G}}_{\varphi,p\mathrm{i}}(1), \ldots, \tilde{\boldsymbol{G}}_{\varphi,p\mathrm{i}}(N_\mathrm{i} + N_\mathrm{c})\right], \tag{3.41}$$

$$\tilde{\mathsf{G}}_{\varphi\mathrm{x}} = \left[\tilde{\boldsymbol{G}}_{\varphi,n\mathrm{x}}(1), \ldots, \tilde{\boldsymbol{G}}_{\varphi,n\mathrm{x}}(N_\mathrm{i} + N_\mathrm{c}), \tilde{\boldsymbol{G}}_{\varphi,p\mathrm{x}}(1), \ldots, \tilde{\boldsymbol{G}}_{\varphi,p\mathrm{x}}(N_\mathrm{i} + N_\mathrm{c})\right]. \tag{3.42}$$

Concerning the contact Green's functions $\tilde{\boldsymbol{G}}_{e,n}(k)$ and $\tilde{\boldsymbol{G}}_{e,p}(k)$, these are collected in the matrix $\tilde{\mathsf{G}}_e$ of size $N_c \times 2(N_i + N_c)$ such as:

$$\tilde{\mathsf{G}}_e = \left[\tilde{\boldsymbol{G}}_{e,n}(1), \ldots, \tilde{\boldsymbol{G}}_{e,n}(N_i + N_c), \tilde{\boldsymbol{G}}_{e,p}(1), \ldots, \tilde{\boldsymbol{G}}_{e,p}(N_i + N_c)\right]. \quad (3.43)$$

3.2.1 Evaluating the Discretized Green's Functions

Having introduced the relevant matrix unknowns, we now explicitly formulate the related complex equation system; the coefficient matrices are the same exploited in the small-signal analysis, see (3.35). In the Poisson equation, no forcing terms arise, thus yielding the following linear system of N_i equations for the internal nodes:

$$\mathsf{A}_{\varphi_i\varphi_i} \cdot \tilde{\mathsf{G}}_{\varphi i} + \mathsf{A}_{\varphi_i\varphi_x} \cdot \tilde{\mathsf{G}}_{\varphi x} + \mathsf{A}_{\varphi_i n_i} \cdot \delta\tilde{\mathsf{n}}_i + \mathsf{A}_{\varphi_i p_i} \cdot \delta\tilde{\mathsf{p}}_i = \mathsf{0}_{N_i,2(N_i+N_c)}, \quad (3.44a)$$

and the following set of N_x equations which impose that contacts are equipotential:

$$\mathsf{I}_{N_x} \cdot \tilde{\mathsf{G}}_{\varphi x} + \mathsf{A}_{\varphi_x v_c} \cdot \tilde{\mathsf{G}}_e = \mathsf{0}_{N_x,2(N_i+N_c)}, \quad (3.44b)$$

where $\mathsf{0}_{N,M}$ is a null matrix of size $N \times M$. The electron and hole continuity equations for the internal nodes yield the following two linear systems of N_i equations which contain the discretized delta function matrix forcing term:

$$\mathsf{A}_{n_i\varphi_i} \cdot \tilde{\mathsf{G}}_{\varphi i} + \mathsf{A}_{n_i\varphi_x} \cdot \tilde{\mathsf{G}}_{\varphi x} + \mathsf{A}_{n_i n_i} \cdot \delta\tilde{\mathsf{n}}_i + \mathsf{A}_{n_i n_x} \cdot \delta\tilde{\mathsf{n}}_x + \mathsf{A}_{n_i p_i} \cdot \delta\tilde{\mathsf{p}}_i = \mathsf{S}_n, \quad (3.44c)$$

$$\mathsf{A}_{p_i\varphi_i} \cdot \tilde{\mathsf{G}}_{\varphi i} + \mathsf{A}_{p_i\varphi_x} \cdot \tilde{\mathsf{G}}_{\varphi x} + \mathsf{A}_{p_i n_i} \cdot \delta\tilde{\mathsf{n}}_i + \mathsf{A}_{p_i p_i} \cdot \delta\tilde{\mathsf{p}}_i + \mathsf{A}_{p_i p_x} \cdot \delta\tilde{\mathsf{p}}_x = \mathsf{S}_p. \quad (3.44d)$$

Matrices S_n and S_p have size $N_i \times 2(N_i + N_c)$ and are defined as follows:

$$\mathsf{S}_n = \left[\mathsf{I}_{N_i}, \mathsf{0}_{N_i,N_i+2N_c}\right], \quad (3.44e)$$

$$\mathsf{S}_p = \left[\mathsf{0}_{N_i,N_i}, \mathsf{I}_{N_i}, \mathsf{0}_{N_i,2N_c}\right]. \quad (3.44f)$$

The related BC lead to the two sets of N_x equations:

$$\mathsf{A}_{n_x\varphi_i} \cdot \tilde{\mathsf{G}}_{\varphi i} + \mathsf{A}_{n_x\varphi_x} \cdot \tilde{\mathsf{G}}_{\varphi x} + \mathsf{A}_{n_x n_i} \cdot \delta\tilde{\mathsf{n}}_i + \mathsf{A}_{n_x n_x} \cdot \delta\tilde{\mathsf{n}}_x + \mathsf{A}_{n_x p_x} \cdot \delta\tilde{\mathsf{p}}_x = \mathsf{0}_{N_x,2(N_i+N_c)}, \quad (3.44g)$$

$$\mathsf{A}_{p_x\varphi_i} \cdot \tilde{\mathsf{G}}_{\varphi i} + \mathsf{A}_{p_x\varphi_x} \cdot \tilde{\mathsf{G}}_{\varphi x} + \mathsf{A}_{p_x n_x} \cdot \delta\tilde{\mathsf{n}}_x + \mathsf{A}_{p_x p_i} \cdot \delta\tilde{\mathsf{p}}_i + \mathsf{A}_{p_x p_x} \cdot \delta\tilde{\mathsf{p}}_x = \mathsf{0}_{N_x,2(N_i+N_c)}. \quad (3.44h)$$

For open-circuit BC, the contact current equations yield the following system of size N_c:

$$\mathsf{A}_{i_c\varphi_i} \cdot \tilde{\mathsf{G}}_{\varphi i} + \mathsf{A}_{i_c\varphi_x} \cdot \tilde{\mathsf{G}}_{\varphi x} + \mathsf{A}_{i_c n_i} \cdot \delta\tilde{\mathsf{n}}_i + \mathsf{A}_{i_c n_x} \cdot \delta\tilde{\mathsf{n}}_x + \mathsf{A}_{i_c p_i} \cdot \delta\tilde{\mathsf{p}}_i + \mathsf{A}_{i_c p_x} \cdot \delta\tilde{\mathsf{p}}_x = \mathsf{S}_c, \quad (3.44i)$$

where the matrix $\mathsf{S}_{\rm c}$ is defined as:

$$\mathsf{S}_{\rm c} = \left[\mathsf{0}_{N_{\rm c},2N_{\rm i}}, -\mathsf{I}_{N_{\rm c}}, \mathsf{I}_{N_{\rm c}} \right]. \tag{3.44j}$$

From a numerical standpoint, the solution of (3.44) amounts to solving $2(N_{\rm i} + N_{\rm c})$ complex linear systems of size $3(N_{\rm i} + N_{\rm x}) + N_{\rm c}$ for each operating frequency of interest. Although techniques like the LU factorization allow the solution process to be divided into two steps, factorization and back-substitution, in which a new solution with a different r.h.s. only requires a comparatively fast substitution, the extremely high number of systems to be solved makes this approach (which shall be referred to hereafter as the *direct* approach) computationally intensive. Time-domain techniques for response evaluation through Fourier transformation lead to the same problem [111]. Moreover, most of the resulting information is actually redundant, since the only elements of the solution which really matter for noise evaluation are the contact Green's function elements $\tilde{\boldsymbol{G}}_{e,n}(k)$ and $\tilde{\boldsymbol{G}}_{e,p}(k)$, i.e. the open-circuit potential fluctuations induced by a unit injection (hole or electron) into each internal node k of the device. This information only amounts to $2N_{\rm c}$ unknowns for each injection node.

3.2.2 Evaluating the Correlation Spectra

Once the scalar discretized Green's functions $\tilde{\mathsf{G}}_e$ have been evaluated, the correlation matrix of the open-circuit voltage fluctuations $\mathsf{S}_{\delta \boldsymbol{e},\delta \boldsymbol{e}}$ can be estimated by straightforward discretization of (2.36). In order to apply this equation, the vector Green's functions must be recovered by computing the numerical gradient of $\tilde{\mathsf{G}}_e$ with respect to the injection point. Since the scalar Green's functions are known as nodal quantities, a suitable approximation for the components of the gradient can be derived from standard approaches. In particular, within the framework of a two-dimensional triangular discretization, the Green's functions can be linearly approximated on each element, thus leading to a constant element-by-element gradient $\tilde{\boldsymbol{G}}_{\varphi,\alpha i}(l;\omega)$ where l refers to element l. A nodal value for the discretized vector Green's function can be derived by an average procedure carried out on the control subregions Ω_{kl} (of surface Σ_{kl}) pertaining to node k and element l as follows:

$$\left\langle \tilde{\boldsymbol{G}}_{\varphi,\alpha i}(\omega) \right\rangle (k) = \frac{1}{\Sigma_k} \sum_l \tilde{\boldsymbol{G}}_{\varphi,\alpha i}(l;\omega) \Sigma_{kl}, \tag{3.45}$$

where the summation is extended to all elements having a common vertex in k. Taking into account (3.45), the element i,j of the correlation matrix $\mathsf{S}_{\delta \boldsymbol{e},\delta \boldsymbol{e}}$ can be approximated as:

$$S_{\delta e_i,\delta e_j}(\omega) \approx \sum_{\alpha,\beta=n,p} \sum_k \tilde{G}_{\varphi,\alpha i}(k;\omega) K_{\gamma_\alpha,\gamma_\beta}(k;\omega) \tilde{G}^*_{\varphi,\beta j}(k;\omega) \Sigma_k$$

$$+ \sum_{\alpha,\beta=n,p} \kappa_\alpha \kappa_\beta \sum_k \left\langle \tilde{\boldsymbol{G}}_{\varphi,\alpha i}(\omega) \right\rangle (k) \cdot \mathsf{K}_{\delta \boldsymbol{J}_\alpha, \delta \boldsymbol{J}_\beta}(k;\omega) \cdot \left\langle \tilde{\boldsymbol{G}}^\dagger_{\varphi,\beta j}(\omega) \right\rangle (k) \Sigma_k . \tag{3.46}$$

Care must be exerted in numerically estimating the vector Green's function in regions (like, for example, the depletion region of *pn* junctions) in which the corresponding scalar quantity exhibits sharp gradients; in this case it may be advisable to exploit higher-order numerical differentiation schemes.

3.3 Efficient Evaluation of the Green's Functions

In order to greatly reduce the computational burden of the evaluation of the Green's functions through the direct approach, two strategies have been proposed so far.The first technique, proposed in 1989 by Ghione et al. [3] was derived from the so-called *adjoint* approach to efficient noise evaluation in electrical networks [112,113]. This method is based on Tellegen's theorem and its practical application to a discretized semiconductor device model is confined to the monopolar, drift-diffusion case. The adjoint approach, however, is physically appealing, and will be described in more detail in Sect. 3.3.1. A second technique, which readily applies to any form of discretized PDE-based model, and in particular to the drift-diffusion bipolar case discussed here, was proposed in 1995 by Bonani et al. [5, 6, 114]. This method, which shall be referred to as the *generalized adjoint* approach, ultimately derives from the application of Branin's method [115], originally developed for network small-change sensitivity and noise analysis, to PDE-based discretized device models. The generalized adjoint approach is discussed in Sect. 3.3.2.

3.3.1 The Adjoint Approach

Consider an electrical network including many noisy elements and suppose that we are interested in the noise equivalent circuit at a limited number of external ports. Assume that the noisy elements are characterized through a parallel equivalent circuit and that all current noise sources are uncorrelated. The open-circuit voltage induced on the external port k by a generator $\tilde{\imath}_{ij}$ connected between nodes i and j is then :

$$\tilde{v}_k = [Z_{ki}(\omega) - Z_{kj}(\omega)]\, \tilde{\imath}_{ij}, \tag{3.47}$$

where Z_{kl} is the transimpedance between ports k and l (i.e. the voltage response at port k to a unit current generator injected in port l; port l is identified by node l and the reference or ground node, and the current generator enters node l, see Fig. 3.5). From (3.47), and taking into account that noise sources are uncorrelated, the power and correlation spectra of the open-circuit voltages at the external ports k and l can be expressed as:

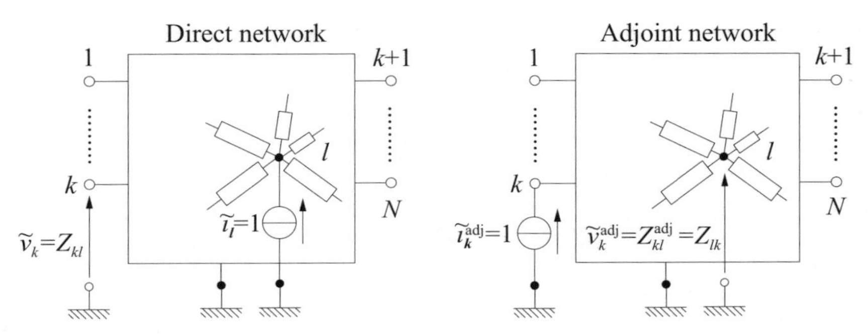

Fig. 3.5. Properties of the direct/adjoint network pair

$$S_{v_k,v_l} = \sum_{i,j} \left[Z_{ki}(\omega) - Z_{kj}(\omega)\right] \left[Z_{li}(\omega) - Z_{lj}(\omega)\right]^* S_{i_{ij},i_{ij}}, \tag{3.48}$$

where the summation is extended to all node pairs (i, j) to which noisy sides are connected. Thus, the basic step in network noise analysis is the evaluation of the transimpedance Z_{ij} between the "internal" port j and the "external" or "contact" port i. Typically, a low number of external ports is present, whereas the number of internal nodes (and ports) can be high. Thus, evaluating Z_{ij} requires a large number of analyses carried out by successively placing a current source in all of the internal ports and evaluating the open-circuit voltage induced on the external ports.

Suppose now we identify a network (called the adjoint network) having the same topology as the original (or direct) one, but with elements such as to lead to the following *interreciprocity* property:

$$Z_{ij} = Z_{ji}^{\text{adj}}, \tag{3.49}$$

where Z_{ji}^{adj} is the transimpedance between the external port i and the internal node (port) j of the adjoint network. In order to evaluate Z_{ji}^{adj}, only as many analyses as the number of external ports have to be made, in which the (open-circuit) potential induced on the internal nodes is computed. The impedance matrix (defined by taking as the reference ports the internal and external ones) of the adjoint network can be directly identified from the impedance matrix of the direct network by application of Tellegen's theorem (e.g. see [63]). Tellegen's theorem in the frequency domain states that, given two networks a and b having the same topology, but different elements:

$$\sum_k \tilde{v}_k^a \tilde{\imath}_k^b = \sum_k \tilde{v}_k^b \tilde{\imath}_k^a = 0, \tag{3.50}$$

where $\tilde{v}_k^\alpha$ $(\tilde{\imath}_k^\alpha)$, $\alpha = a, b$ are the branch voltages (currents) of network α. Since the direct and adjoint networks have the same topology, Tellegen's theorem immediately yields:

$$\sum_k \tilde{v}_k \tilde{\imath}_k^{\text{adj}} = \sum_k \sum_l Z_{kl} \tilde{\imath}_l \tilde{\imath}_k^{\text{adj}} = \sum_k \tilde{v}_k^{\text{adj}} \tilde{\imath}_k = \sum_k \sum_l Z_{kl}^{\text{adj}} \tilde{\imath}_l^{\text{adj}} \tilde{\imath}_k = 0. \quad (3.51)$$

Suppose now that in the direct network all impressed currents are zero but into port k, where $\tilde{\imath}_k = 1$ while in the adjoint network all impressed currents are zero but into port l, where $\tilde{\imath}_l^{\text{adj}} = 1$; one has from (3.51):

$$Z_{kl}^{\text{adj}} = Z_{lk}, \quad (3.52)$$

or, in other words, the impedance matrix of the adjoint network simply is the transpose of that of the direct network:

$$\mathsf{Z}^{\text{adj}} = \mathsf{Z}^{\text{T}}. \quad (3.53)$$

This immediately ensures that property (3.49) is verified. As a further step, discussed in [112, 113] the elements of the adjoint network can be identified from the ones of the direct network, and the adjoint network can be introduced also in the time domain. However, these steps are not needed in our case, in which property (3.53) suffices.

To summarize, the adjoint approach to noise analysis requires first to identify, from the direct network, the adjoint one; second, to perform as many analyses on the adjoint network as the number of contacts (besides the reference contact) [112, 113] in order to evaluate a set of transimpedances. There is a clear similarity between the concept of transimpedance between internal and external ports and the concept of impedance field. The question is therefore whether the adjoint approach can be exploited within the framework of a physics-based model.

As a matter of principle, the answer is yes, since the small-signal model exploited for noise analysis is, either in continuous or discretized form, a linear system. Directly operating on the continuous system (i.e. on the frequency-domain small-signal drift-diffusion model written as a set of partial differential equations) requires us to identify the adjoint of a linear operator [4]. The interreciprocity property of the direct/adjoint device pair is outlined in Fig. 3.6. This approach is interesting from a theoretical standpoint, and the adjoint operator has been used [76] to establish general properties in noise analysis. Nevertheless, identification of the continuous adjoint system is not needed, if the direct system has already been discretized, provided that the direct discretized equation can be cast in a form which allows property (3.53) to be exploited, i.e. in impedance or admittance form.

This is indeed always possible by properly eliminating from system (3.44) the electron and hole concentrations. Unfortunately, in the bipolar case this process requires the explicit inversion of large (though sparse) matrices, thus making the approach numerically inconvenient. However, in the monopolar case only diagonal matrices have to be inverted, and the system can be easily cast in admittance form [4] as follows.

In the monopolar case (to fix the ideas, electrons are the majority carriers) system (3.44) becomes:

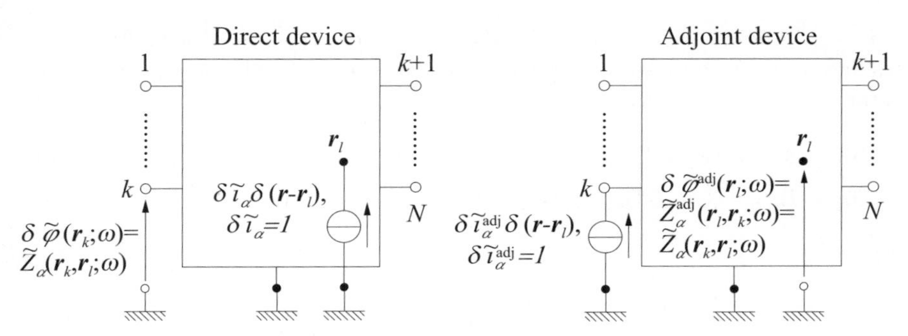

Fig. 3.6. Properties of the direct/adjoint device pair

$$\mathsf{A}_{\varphi_\mathrm{i}\varphi_\mathrm{i}} \cdot \tilde{\mathsf{G}}_{\varphi\mathrm{i}} + \mathsf{A}_{\varphi_\mathrm{i}\varphi_\mathrm{x}} \cdot \tilde{\mathsf{G}}_{\varphi\mathrm{x}} + \mathsf{A}_{\varphi_\mathrm{i}n_\mathrm{i}} \cdot \delta\tilde{\mathsf{n}}_\mathrm{i} = \mathsf{0}_{N_\mathrm{i},N_\mathrm{i}+N_\mathrm{c}}, \tag{3.54a}$$

$$\mathsf{I} \cdot \tilde{\mathsf{G}}_{\varphi\mathrm{x}} + \mathsf{A}_{\varphi_\mathrm{x}v_\mathrm{c}} \cdot \tilde{\mathsf{G}}_e = \mathsf{0}_{N_\mathrm{x},N_\mathrm{i}+N_\mathrm{c}}, \tag{3.54b}$$

$$\mathsf{A}_{n_\mathrm{i}\varphi_\mathrm{i}} \cdot \tilde{\mathsf{G}}_{\varphi\mathrm{i}} + \mathsf{A}_{n_\mathrm{i}\varphi_\mathrm{x}} \cdot \tilde{\mathsf{G}}_{\varphi\mathrm{x}} + \mathsf{A}_{n_\mathrm{i}n_\mathrm{i}} \cdot \delta\tilde{\mathsf{n}}_\mathrm{i} + \mathsf{A}_{n_\mathrm{i}n_\mathrm{x}} \cdot \delta\tilde{\mathsf{n}}_\mathrm{x} = \mathsf{S}'_n, \tag{3.54c}$$

$$\mathsf{A}_{n_\mathrm{x}\varphi_\mathrm{i}} \cdot \tilde{\mathsf{G}}_{\varphi\mathrm{i}} + \mathsf{A}_{n_\mathrm{x}\varphi_\mathrm{x}} \cdot \tilde{\mathsf{G}}_{\varphi\mathrm{x}} + \mathsf{A}_{n_\mathrm{x}n_\mathrm{i}} \cdot \delta\tilde{\mathsf{n}}_\mathrm{i} + \mathsf{A}_{n_\mathrm{x}n_\mathrm{x}} \cdot \delta\tilde{\mathsf{n}}_\mathrm{x} = \mathsf{0}_{N_\mathrm{x},N_\mathrm{i}+N_\mathrm{c}}, \tag{3.54d}$$

$$\mathsf{A}_{i_\mathrm{c}\varphi_\mathrm{i}} \cdot \tilde{\mathsf{G}}_{\varphi\mathrm{i}} + \mathsf{A}_{i_\mathrm{c}\varphi_\mathrm{x}} \cdot \tilde{\mathsf{G}}_{\varphi\mathrm{x}} + \mathsf{A}_{i_\mathrm{c}n_\mathrm{i}} \cdot \delta\tilde{\mathsf{n}}_\mathrm{i} + \mathsf{A}_{i_\mathrm{c}n_\mathrm{x}} \cdot \delta\tilde{\mathsf{n}}_\mathrm{x} = \mathsf{S}'_\mathrm{c}. \tag{3.54e}$$

In system (3.54) the source matrices have been modified as follows:

$$\mathsf{S}'_n = \left[\mathsf{I}_{N_\mathrm{i}}, \mathsf{0}_{N_\mathrm{i},N_\mathrm{c}}\right], \tag{3.55}$$

$$\mathsf{S}'_\mathrm{c} = \left[\mathsf{0}_{N_\mathrm{c},N_\mathrm{i}}, \mathsf{I}_{N_\mathrm{c}}\right]. \tag{3.56}$$

In order to cast system (3.54) into an admittance form and directly apply the adjoint strategy, we first eliminate the electron density fluctuations by expressing them in terms of the potential fluctuations via the Poisson equation (3.54a). This approach is numerically feasible since matrix $\mathsf{A}_{\varphi_\mathrm{i}n_\mathrm{i}}$ is *diagonal* within the finite-boxes discretization scheme or whenever charge lumping is applied. One has, for the electron-density fluctuations in the internal nodes:

$$\delta\tilde{\mathsf{n}}_\mathrm{i} = -\mathsf{A}^{-1}_{\varphi_\mathrm{i}n_\mathrm{i}} \cdot \left[\mathsf{A}_{\varphi_\mathrm{i}\varphi_\mathrm{i}} \cdot \tilde{\mathsf{G}}_{\varphi\mathrm{i}} + \mathsf{A}_{\varphi_\mathrm{i}\varphi_\mathrm{x}} \cdot \tilde{\mathsf{G}}_{\varphi\mathrm{x}}\right]; \tag{3.57}$$

concerning the external contact nodes one has from (3.54d):

$$\delta\tilde{\mathsf{n}}_\mathrm{x} = -\mathsf{A}^{-1}_{n_\mathrm{x}n_\mathrm{x}} \cdot \left[\mathsf{A}_{n_\mathrm{x}\varphi_\mathrm{i}} \cdot \tilde{\mathsf{G}}_{\varphi\mathrm{i}} + \mathsf{A}_{n_\mathrm{x}\varphi_\mathrm{x}} \cdot \tilde{\mathsf{G}}_{\varphi\mathrm{x}} + \mathsf{A}_{n_\mathrm{x}n_\mathrm{i}} \cdot \delta\tilde{\mathsf{n}}_\mathrm{i}\right], \tag{3.58}$$

so that, taking into account (3.57), the electron-density fluctuations on the contact nodes can be expressed as a function of potential fluctuations. Although the matrix $\mathsf{A}_{n_\mathrm{x}n_\mathrm{x}}$ is generally not diagonal, its size $N_\mathrm{x} \times N_\mathrm{x}$ is small enough to enable inversion with negligible computational burden. Moreover, if the device contacts are ohmic, $\delta\tilde{\mathsf{n}}_\mathrm{x}$ is simply zero.

Using (3.54b), one can formally express the electron-density fluctuations as a function of the potential fluctuations in the internal nodes and of the contact voltage fluctuations:

$$\delta\tilde{\mathsf{n}}_{\mathrm{i}} = \mathsf{B}_{n_{\mathrm{i}}\varphi_{\mathrm{i}}} \cdot \tilde{\mathsf{G}}_{\varphi\mathrm{i}} + \mathsf{B}_{n_{\mathrm{i}}v_{\mathrm{c}}} \cdot \tilde{\mathsf{G}}_{e}, \tag{3.59a}$$

$$\delta\tilde{\mathsf{n}}_{\mathrm{x}} = \mathsf{B}_{n_{\mathrm{x}}\varphi_{\mathrm{i}}} \cdot \tilde{\mathsf{G}}_{\varphi\mathrm{i}} + \mathsf{B}_{n_{\mathrm{x}}v_{\mathrm{c}}} \cdot \tilde{\mathsf{G}}_{e}, \tag{3.59b}$$

where:

$$\mathsf{B}_{n_{\mathrm{i}}\varphi_{\mathrm{i}}} = -\mathsf{A}^{-1}_{\varphi_{\mathrm{i}}n_{\mathrm{i}}} \cdot \mathsf{A}_{\varphi_{\mathrm{i}}\varphi_{\mathrm{i}}}, \tag{3.60a}$$

$$\mathsf{B}_{n_{\mathrm{i}}v_{\mathrm{c}}} = \mathsf{A}^{-1}_{\varphi_{\mathrm{i}}n_{\mathrm{i}}} \cdot \mathsf{A}_{\varphi_{\mathrm{i}}\varphi_{\mathrm{x}}} \cdot \mathsf{A}_{\varphi_{\mathrm{x}}v_{\mathrm{c}}}, \tag{3.60b}$$

$$\mathsf{B}_{n_{\mathrm{x}}\varphi_{\mathrm{i}}} = -\mathsf{A}^{-1}_{n_{\mathrm{x}}n_{\mathrm{x}}} \cdot \left[\mathsf{A}_{n_{\mathrm{x}}\varphi_{\mathrm{i}}} + \mathsf{A}_{n_{\mathrm{x}}n_{\mathrm{i}}} \cdot \mathsf{B}_{n_{\mathrm{i}}\varphi_{\mathrm{i}}}\right], \tag{3.60c}$$

$$\mathsf{B}_{n_{\mathrm{x}}v_{\mathrm{c}}} = \mathsf{A}^{-1}_{n_{\mathrm{x}}n_{\mathrm{x}}} \cdot \left[\mathsf{A}_{n_{\mathrm{x}}\varphi_{\mathrm{x}}} \cdot \mathsf{A}_{\varphi_{\mathrm{x}}v_{\mathrm{c}}} - \mathsf{A}_{n_{\mathrm{x}}n_{\mathrm{i}}} \cdot \mathsf{B}_{n_{\mathrm{i}}v_{\mathrm{c}}}\right]. \tag{3.60d}$$

On substituting (3.59) into (3.54c) and (3.54e) we directly obtain an admittance formulation which reads:

$$\begin{bmatrix} \mathsf{Y}_{n_{\mathrm{i}}\varphi_{\mathrm{i}}} & \mathsf{Y}_{n_{\mathrm{i}}v_{\mathrm{c}}} \\ \mathsf{Y}_{i_{\mathrm{c}}\varphi_{\mathrm{i}}} & \mathsf{Y}_{i_{\mathrm{c}}v_{\mathrm{c}}} \end{bmatrix} \begin{bmatrix} \tilde{\mathsf{G}}_{\varphi\mathrm{i}} \\ \tilde{\mathsf{G}}_{e} \end{bmatrix} = \begin{bmatrix} \mathsf{S}'_{n} \\ \mathsf{S}'_{\mathrm{c}} \end{bmatrix}, \tag{3.61}$$

where the admittance matrix elements are:

$$\mathsf{Y}_{n_{\mathrm{i}}\varphi_{\mathrm{i}}} = \mathsf{A}_{n_{\mathrm{i}}\varphi_{\mathrm{i}}} + \mathsf{A}_{n_{\mathrm{i}}n_{\mathrm{i}}} \cdot \mathsf{B}_{n_{\mathrm{i}}\varphi_{\mathrm{i}}} + \mathsf{A}_{n_{\mathrm{i}}n_{\mathrm{x}}} \cdot \mathsf{B}_{n_{\mathrm{x}}\varphi_{\mathrm{i}}}, \tag{3.62a}$$

$$\mathsf{Y}_{n_{\mathrm{i}}v_{\mathrm{c}}} = -\mathsf{A}_{n_{\mathrm{i}}\varphi_{\mathrm{x}}} \cdot \mathsf{A}_{\varphi_{\mathrm{x}}v_{\mathrm{c}}} + \mathsf{A}_{n_{\mathrm{i}}n_{\mathrm{i}}} \cdot \mathsf{B}_{n_{\mathrm{i}}v_{\mathrm{c}}} + \mathsf{A}_{n_{\mathrm{i}}n_{\mathrm{x}}} \cdot \mathsf{B}_{n_{\mathrm{x}}v_{\mathrm{c}}}, \tag{3.62b}$$

$$\mathsf{Y}_{i_{\mathrm{c}}\varphi_{\mathrm{i}}} = \mathsf{A}_{i_{\mathrm{c}}\varphi_{\mathrm{i}}} + \mathsf{A}_{i_{\mathrm{c}}n_{\mathrm{i}}} \cdot \mathsf{B}_{n_{\mathrm{i}}\varphi_{\mathrm{i}}} + \mathsf{A}_{i_{\mathrm{c}}n_{\mathrm{x}}} \cdot \mathsf{B}_{n_{\mathrm{x}}\varphi_{\mathrm{i}}}, \tag{3.62c}$$

$$\mathsf{Y}_{i_{\mathrm{c}}v_{\mathrm{c}}} = -\mathsf{A}_{i_{\mathrm{c}}\varphi_{\mathrm{x}}} \cdot \mathsf{A}_{\varphi_{\mathrm{x}}v_{\mathrm{c}}} + \mathsf{A}_{i_{\mathrm{c}}n_{\mathrm{i}}} \cdot \mathsf{B}_{n_{\mathrm{i}}v_{\mathrm{c}}} + \mathsf{A}_{i_{\mathrm{c}}n_{\mathrm{x}}} \cdot \mathsf{B}_{n_{\mathrm{x}}v_{\mathrm{c}}}. \tag{3.62d}$$

In order to follow the adjoint strategy, it is better to write the system (3.61) with a vector unknown and forcing term:

$$\mathsf{Y} \cdot \tilde{\boldsymbol{G}}_{n}(k) = \boldsymbol{S}'(k), \tag{3.63}$$

where Y is the device nodal admittance matrix:

$$\mathsf{Y} = \begin{bmatrix} \mathsf{Y}_{n_{\mathrm{i}}\varphi_{\mathrm{i}}} & \mathsf{Y}_{n_{\mathrm{i}}v_{\mathrm{c}}} \\ \mathsf{Y}_{i_{\mathrm{c}}\varphi_{\mathrm{i}}} & \mathsf{Y}_{i_{\mathrm{c}}v_{\mathrm{c}}} \end{bmatrix}, \tag{3.64}$$

while:

$$\tilde{\boldsymbol{G}}_{n}(k) = \begin{bmatrix} \tilde{\boldsymbol{G}}_{\varphi,n\mathrm{i}}(k) \\ \tilde{\boldsymbol{G}}_{e,n}(k) \end{bmatrix}. \tag{3.65}$$

In system (3.63) k refers to the injection node or contact, i.e. $1 \leq k \leq (N_{\mathrm{i}} + N_{\mathrm{c}})$; the vector forcing term is zero apart from element k, which is unity.

In order to evaluate $\tilde{\boldsymbol{G}}_{e,n}(k)$ for all k, we need to perform $N_\mathrm{i}+N_\mathrm{c}$ solutions of (3.63). We now introduce the adjoint of (3.63); this is simply obtained from rule (3.53), applied to the admittance rather than to the impedance representation. The adjoint system then reads:

$$\mathsf{Y}^\mathrm{T} \cdot \tilde{\boldsymbol{G}}_n^\mathrm{adj}(k) = \boldsymbol{S}'(k), \tag{3.66}$$

where:

$$\mathsf{Y}^\mathrm{T} = \begin{bmatrix} \mathsf{Y}^\mathrm{T}_{n_\mathrm{i}\varphi_\mathrm{i}} & \mathsf{Y}^\mathrm{T}_{i_\mathrm{c}\varphi_\mathrm{i}} \\ \mathsf{Y}^\mathrm{T}_{n_\mathrm{i}v_\mathrm{c}} & \mathsf{Y}^\mathrm{T}_{i_\mathrm{c}v_\mathrm{c}} \end{bmatrix}, \tag{3.67}$$

and:

$$\tilde{\boldsymbol{G}}_n^\mathrm{adj}(k) = \begin{bmatrix} \tilde{\boldsymbol{G}}^\mathrm{adj}_{\varphi,n\mathrm{i}}(k) \\ \tilde{\boldsymbol{G}}^\mathrm{adj}_{e,n}(k) \end{bmatrix} \tag{3.68}$$

are the adjoint potential fluctuations induced (in the internal and contact nodes, respectively), by a unit current injection in node k (internal or contact). The forcing term is the same as in the direct system.

We now exploit the basic property of the adjoint system, i.e. interreciprocity, which reads, in this case:

$$\tilde{\boldsymbol{G}}_{n,l}(k) = \tilde{\boldsymbol{G}}^\mathrm{adj}_{n,k}(l), \tag{3.69}$$

where $k, l = 1, \ldots, N_\mathrm{i} + N_\mathrm{c}$. Therefore, by injecting a unit current into a contact of the adjoint system, i.e. for $l = N_\mathrm{i} + 1, \ldots, N_\mathrm{i} + N_\mathrm{c}$ in the r.h.s. of (3.69) and evaluating the induced adjoint potential $\tilde{\boldsymbol{G}}_n^\mathrm{adj}$ in all nodes and contacts, i.e. for $k = 1, \ldots, N_\mathrm{i} + N_\mathrm{c}$, one immediately recovers the potential $\tilde{\boldsymbol{G}}_n$ induced on the contacts by a unit electron current injection taking place in one of the internal nodes or contacts, see Fig. 3.6. This means that the evaluation of the scalar Green's function on the contacts only amounts to a number of solutions of the adjoint system equal to the (low) number of (non-reference) contacts, N_c. Note that the elements $\tilde{\boldsymbol{G}}^\mathrm{adj}_{n,k}(l)$ with $k = l = N_\mathrm{i}+1, \ldots, N_\mathrm{i}+N_\mathrm{c}$ describe the open-circuit voltage response on the contacts to a unit injection into the contacts, i.e. the element Z^adj_{ji} (where $j = k - N_\mathrm{i}$, $i = l - N_\mathrm{i}$) of the small-signal impedance matrix of the adjoint device, or equivalently, the element Z_{ij} of the small-signal impedance matrix of the direct device. Therefore, the small-signal parameters can be obtained as a by-product of the Green's function evaluation.

3.3.2 The Generalized Adjoint Approach

Despite its appealing physical interpretation, the extension of the adjoint approach beyond the monopolar drift-diffusion case is impractical owing to the need to obtain a network equivalent for the physics-based model.

A more general technique, applying to any kind of discretized PDE-based physical model, was proposed in [5,114]. The method, which shall be referred to here as the *generalized adjoint* method, is similar to Branin's technique [115] for the noise and sensitivity analysis of electrical networks.

To discuss the method in a convenient way, we introduce the following notation. System (3.44) will be written in compact form as:

$$\mathsf{A} \cdot \mathsf{x} = \mathsf{b}, \tag{3.70}$$

where A is a square matrix of size $3(N_\mathrm{i} + N_\mathrm{x}) + N_\mathrm{c}$ explicitly given as:

$$\mathsf{A} = \begin{bmatrix} \mathsf{A}_{\varphi_\mathrm{i}\varphi_\mathrm{i}} & \mathsf{A}_{\varphi_\mathrm{i}\varphi_\mathrm{x}} & \mathsf{A}_{\varphi_\mathrm{i}n_\mathrm{i}} & 0_{N_\mathrm{i},N_\mathrm{x}} & \mathsf{A}_{\varphi_\mathrm{i}p_\mathrm{i}} & 0_{N_\mathrm{i},N_\mathrm{x}} & 0_{N_\mathrm{i},N_\mathrm{c}} \\ 0_{N_\mathrm{x},N_\mathrm{i}} & \mathsf{I}_{N_\mathrm{x}} & 0_{N_\mathrm{x},N_\mathrm{i}} & 0_{N_\mathrm{x},N_\mathrm{x}} & 0_{N_\mathrm{x},N_\mathrm{i}} & 0_{N_\mathrm{x},N_\mathrm{x}} & \mathsf{A}_{\varphi_\mathrm{x}v_\mathrm{c}} \\ \mathsf{A}_{n_\mathrm{i}\varphi_\mathrm{i}} & \mathsf{A}_{n_\mathrm{i}\varphi_\mathrm{x}} & \mathsf{A}_{n_\mathrm{i}n_\mathrm{i}} & \mathsf{A}_{n_\mathrm{i}n_\mathrm{x}} & \mathsf{A}_{n_\mathrm{i}p_\mathrm{i}} & 0_{N_\mathrm{i},N_\mathrm{x}} & 0_{N_\mathrm{i},N_\mathrm{c}} \\ \mathsf{A}_{p_\mathrm{i}\varphi_\mathrm{i}} & \mathsf{A}_{p_\mathrm{i}\varphi_\mathrm{x}} & \mathsf{A}_{p_\mathrm{i}n_\mathrm{i}} & 0_{N_\mathrm{i},N_\mathrm{x}} & \mathsf{A}_{p_\mathrm{i}p_\mathrm{i}} & \mathsf{A}_{p_\mathrm{i}p_\mathrm{x}} & 0_{N_\mathrm{i},N_\mathrm{c}} \\ \mathsf{A}_{n_\mathrm{x}\varphi_\mathrm{i}} & \mathsf{A}_{n_\mathrm{x}\varphi_\mathrm{x}} & \mathsf{A}_{n_\mathrm{x}n_\mathrm{i}} & \mathsf{A}_{n_\mathrm{x}n_\mathrm{x}} & 0_{N_\mathrm{x},N_\mathrm{i}} & \mathsf{A}_{n_\mathrm{x}p_\mathrm{x}} & 0_{N_\mathrm{x},N_\mathrm{c}} \\ \mathsf{A}_{p_\mathrm{x}\varphi_\mathrm{i}} & \mathsf{A}_{p_\mathrm{x}\varphi_\mathrm{x}} & 0_{N_\mathrm{x},N_\mathrm{i}} & \mathsf{A}_{p_\mathrm{x}n_\mathrm{x}} & \mathsf{A}_{p_\mathrm{x}p_\mathrm{i}} & \mathsf{A}_{p_\mathrm{x}p_\mathrm{x}} & 0_{N_\mathrm{x},N_\mathrm{c}} \\ \mathsf{A}_{i_\mathrm{c}\varphi_\mathrm{i}} & \mathsf{A}_{i_\mathrm{c}\varphi_\mathrm{x}} & \mathsf{A}_{i_\mathrm{c}n_\mathrm{i}} & \mathsf{A}_{i_\mathrm{c}n_\mathrm{x}} & \mathsf{A}_{i_\mathrm{c}p_\mathrm{i}} & \mathsf{A}_{i_\mathrm{c}p_\mathrm{x}} & 0_{N_\mathrm{c},N_\mathrm{c}} \end{bmatrix}, \tag{3.71}$$

while the unknown is:

$$\mathsf{x} = \begin{bmatrix} \tilde{\mathsf{G}}_{\varphi\mathrm{i}} \\ \tilde{\mathsf{G}}_{\varphi\mathrm{x}} \\ \delta\tilde{\mathsf{n}}_\mathrm{i} \\ \delta\tilde{\mathsf{n}}_\mathrm{x} \\ \delta\tilde{\mathsf{p}}_\mathrm{i} \\ \delta\tilde{\mathsf{p}}_\mathrm{x} \\ \tilde{\mathsf{G}}_e \end{bmatrix}, \tag{3.72}$$

and the forcing term:

$$\mathsf{b} = \begin{bmatrix} 0_{N_\mathrm{i},2(N_\mathrm{i}+N_\mathrm{c})} \\ 0_{N_\mathrm{x},2(N_\mathrm{i}+N_\mathrm{c})} \\ \mathsf{S}_n \\ \mathsf{S}_p \\ 0_{N_\mathrm{x},2(N_\mathrm{i}+N_\mathrm{c})} \\ 0_{N_\mathrm{x},2(N_\mathrm{i}+N_\mathrm{c})} \\ \mathsf{S}_\mathrm{c} \end{bmatrix}. \tag{3.73}$$

In order to extract from the unknown x the significant part, i.e. $\tilde{\mathsf{G}}_e$, we introduce the matrix e, of size $[3(N_\mathrm{i} + N_\mathrm{x}) + N_\mathrm{c}] \times N_\mathrm{c}$, defined as:

$$\mathsf{e} = \left[0_{N_\mathrm{c},3(N_\mathrm{i}+N_\mathrm{x})}, \mathsf{I}_{N_\mathrm{c}}\right]^\mathrm{T}. \tag{3.74}$$

The following relationship can be easily proven by inspection:

$$\tilde{\mathsf{G}}_e = \mathsf{e}^\mathrm{T} \cdot \mathsf{x}. \tag{3.75}$$

However, system (3.70) can be formally solved to yield:

$$\mathsf{x} = \mathsf{A}^{-1} \cdot \mathsf{b}\,, \tag{3.76}$$

thus giving, after substitution in (3.75):

$$\tilde{\mathsf{G}}_e = \mathsf{e}^{\mathrm{T}} \cdot \mathsf{A}^{-1} \cdot \mathsf{b} = \mathsf{y}^{\mathrm{T}} \cdot \mathsf{b}, \tag{3.77}$$

where:

$$\mathsf{y}^{\mathrm{T}} = \mathsf{e}^{\mathrm{T}} \cdot \mathsf{A}^{-1} \tag{3.78}$$

can be obtained from the solution of the transposed linear system:

$$\mathsf{A}^{\mathrm{T}} \cdot \mathsf{y} = \mathsf{e}\,. \tag{3.79}$$

Therefore, in order to recover y, only N_{c} transposed linear systems have to be solved, as opposed to the $2(N_{\mathrm{i}} + N_{\mathrm{c}})$ systems whose solution is needed in the direct approach. The numerical advantage is dramatic, since, already in a 2D simulation, $N_{\mathrm{i}} \approx 10^3 \div 10^4$ while $N_{\mathrm{c}} \approx 2 \div 3$ in most cases. Thus, the same computational efficiency is obtained as in the adjoint approach. Notice that (3.77), which is finally exploited to evaluate the contact Green's functions, only implies the product of two matrices, and has therefore negligible computational intensity.

According to the actual numerical implementation of the physics-based model, slightly different formulations for the generalized adjoint approach can turn out to be convenient, instead of the canonical procedure described here; for an example of alternative implementation, see [6].

3.3.3 Extensions of the Generalized Adjoint Approach

Taking into account that the Green's function technique applies to any linear or linearized PDE system, this approach can be directly exploited not only in the case discussed in the previous section, but also in the treatment of more complex physical models including a larger number of equations. A first, straightforward example is provided by the drift-diffusion approach wherein trap-assisted transitions are included by adding the trap-level rate equations (2.2d); other examples are higher-order transport models such as the energy balance and full hydrodynamic model.

From a numerical standpoint, the generalized adjoint approach directly applies to any of the aforementioned cases, since (3.70) still holds provided that the system matrix is properly defined according to the model under consideration, and the purpose of the method is the same, i.e. evaluating the solution only in correspondence of contacts, rather than on the whole simulation domain. In a similar way, the generalized adjoint approach can also be exploited, as discussed in Chap. 5, in the efficient implementation of the large-signal noise analysis in (quasi-) periodic operation through the drift-diffusion approach. In this case, the nodal unknowns are still the potential and carrier density distributions, but, owing to multi-frequency operation, more than one frequency component must be included for each node, thus leading to an augmented system which, however, can be formally cast in the form (3.70).

4. Results and Case Studies

The purpose of this chapter is to present a few results and case studies on the physics-based numerical simulation of noise in small-signal conditions for some selected classes of devices: resistors, *pn* diodes, field-effect and bipolar transistors. As a general remark, numerical noise modeling yields results which are in agreement with analytical approaches in all simple cases, but also enables us to cast light on more complex behaviors. Examples of the latter instance are the noise modeling of intrinsic or weakly doped short semiconductor samples, whose noise properties can significantly deviate from the Nyquist law already at comparatively low field, and the excess noise due to GR effects, whose frequency behavior in short samples is not strictly Lorentzian, with cutoff angular frequency simply equal to the inverse of the minority-carrier lifetime. Finally, numerical noise modeling can handle situations wherein analytical methods fail to provide a simple answer, as in bipolar devices under high-injection conditions, or in short-channel FETs.

In many examples shown the discussion will start from the physical structure and will then proceed to discussing the behavior of the Green's functions (scalar and vector impedance fields, or other, more convenient, Green's functions) first in equilibrium and then out of equilibrium. Then, the distribution of the spatial noise densities, whose integral on the device volume provides the noise current or voltage correlation matrix, is investigated. This allows the spatial origin of noise to be understood in detail. Finally, the overall behavior of the correlation spectrum is discussed and compared, where possible, to analytical approaches.

To conclude these introductory remarks, it should be noticed that comparison between physics-based noise models and experimental results measured on real-world devices (either on-wafer or in package) is heavily influenced by the presence of parasitics or, in some instances, of technological or process features which are intrinsically difficult to detect and model. While noise in bipolar devices (with the notable exception of heterojunction bipolar transistors, whose behavior is still heavily influenced by surface effects) more or less exhibits, at least in the active region, a shot-like behavior which is consistent with a simple theoretical framework (with some corrections arising from parasitic resistances), the behavior of field-effect transistors is more involved. In such devices the DC and, above all, the small-signal behavior is

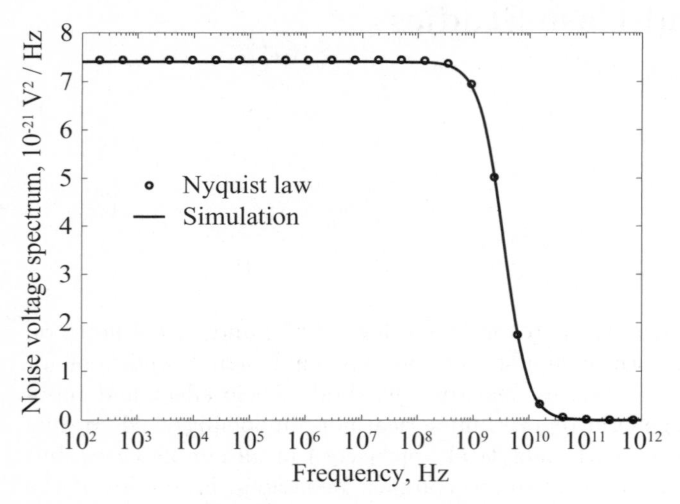

Fig. 4.1. Frequency behavior of voltage fluctuation spectrum for a doped semiconductor sample; the applied voltage is $v = 1$ V

more or less heavily influenced by resistive and reactive parasitics which are difficult to account for within the framework of physics-based simulation; for this reason, meaningful comparisons often require to be carried out on intrinsic parameters obtained from de-embedding techniques. As wisely noted by Nougier [88], noise simulation can hardly be expected to provide results in agreement with the experiment whenever DC and small-signal simulation fails to do so accurately; comparisons between measured results and physics-based simulation therefore have to be carried out with some care, and, above all in III-V devices, model parameters have to be properly tuned around textbook values in order to provide good agreement with the experiment. From this standpoint, many successes have been obtained in the domain of FET noise modeling through the so-called quasi-2D approach, whereby Green's function techniques are applied to a simplified, quasi-monodimensional physics-based model; for a detailed treatment, see [98].

4.1 Resistors

4.1.1 Noise in a Doped Semiconductor Resistor

As a first example, we examine the noise behavior of a simple, uniformly doped, semiconductor sample, i.e. a semiconductor resistor. The sample length is L=100 μm, and the doping concentration is $N_D = 10^{14}$ cm^{-3}. GR processes have been made negligible by assigning minority carriers a suitably long lifetime. The resulting open-circuit voltage fluctuation spectrum, shown in Fig. 4.1, exhibits a cutoff frequency which is related to the RC constant arising from the sample resistance and the sample geometrical capacitance. The fluctuation spectrum is virtually bias independent and exactly follows the

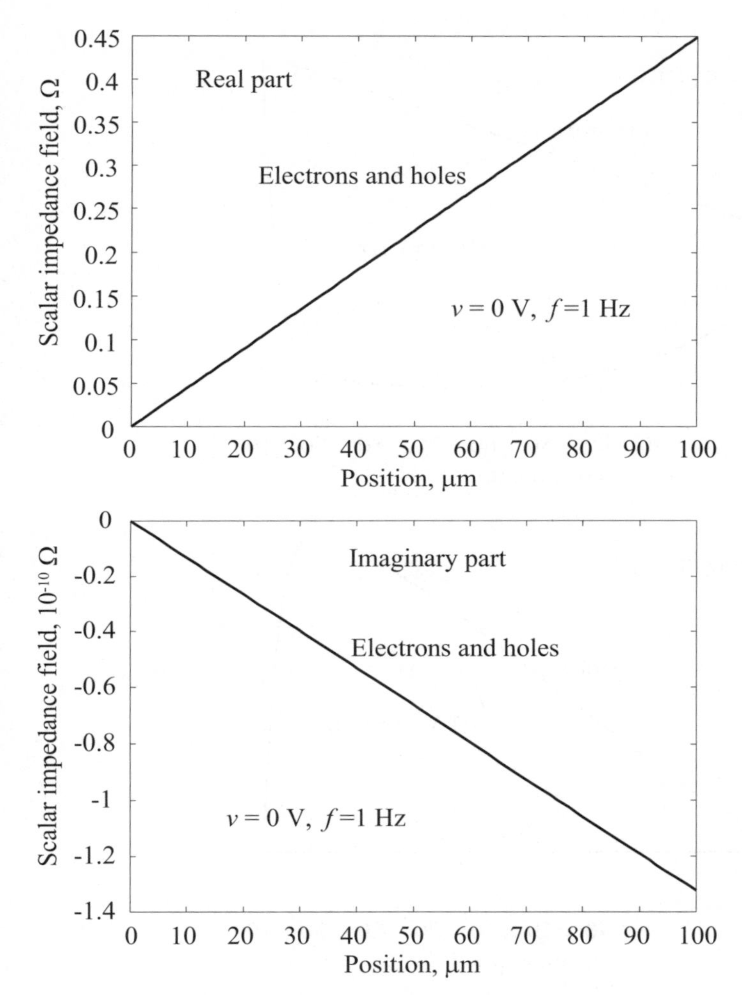

Fig. 4.2. Electron and hole scalar impedance fields at equilibrium, $f = 1$ Hz

Nyquist law at all frequencies. The ideal, Nyquist-law behavior of this simple example should, however, be investigated in greater detail. The Nyquist law has been shown to hold, for a linear resistor, also in nonequilibrium conditions, as the result of a simple, analytical Green's function approach (see Sect. 2.3.1), wherein minority carriers were neglected. In particular, the analytical theory suggests that the majority-carrier impedance field is linear. However, strictly speaking, we expect the Nyquist law to hold exactly only in quilibrium; thus, some further complexities must occur out of equilibrium. This is indeed the case: in equilibrium conditions the electron and hole scalar impedance fields are linear and equal, as shown in Fig. 4.2.

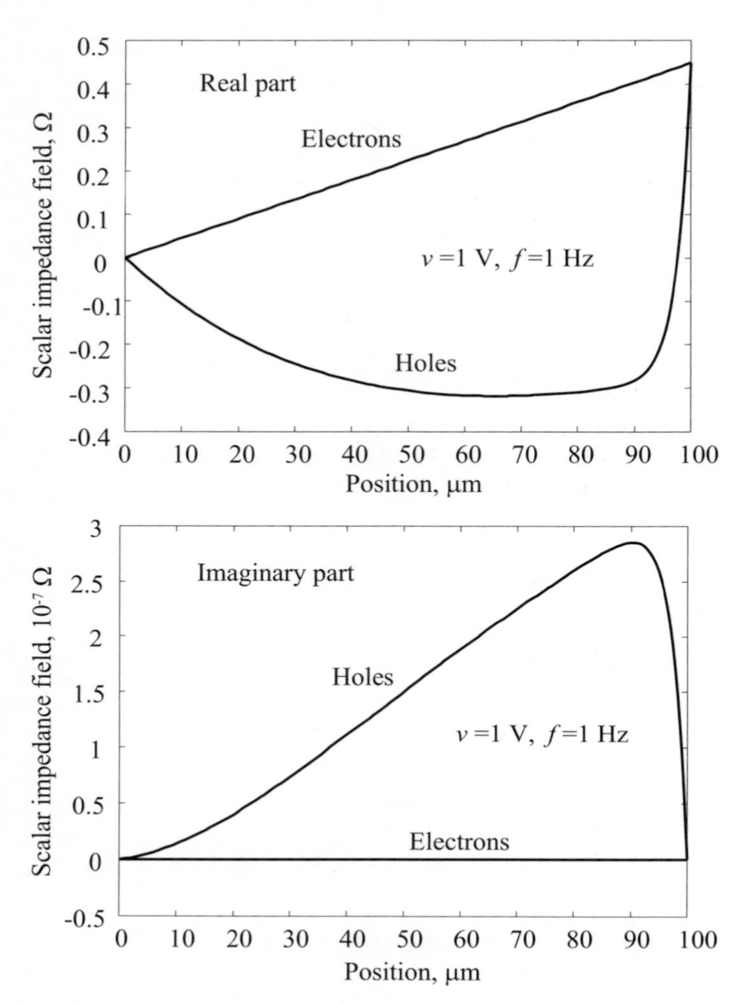

Fig. 4.3. Electron and hole scalar impedance fields out of equilibrium, $v = 1$ V, $f = 1$ Hz

Of course, the minority-carrier contribution is negligible here, since the microscopic noise source is much smaller than that of the majority carriers. Nevertheless, if we increase the applied bias, while the majority-carrier impedance field still has a linear behavior, a strong distortion takes place in the minority-carrier impedance field, see Fig. 4.3. Therefore, the minority-carrier noise contribution deviates greatly from the equilibrium value, albeit remaining at a much lower level than the majority-carrier contribution, see Fig. 4.4. Notice that, owing to the distorted shape of the minority-carrier impedance field near the contacts, the vector impedance field increases there and so does

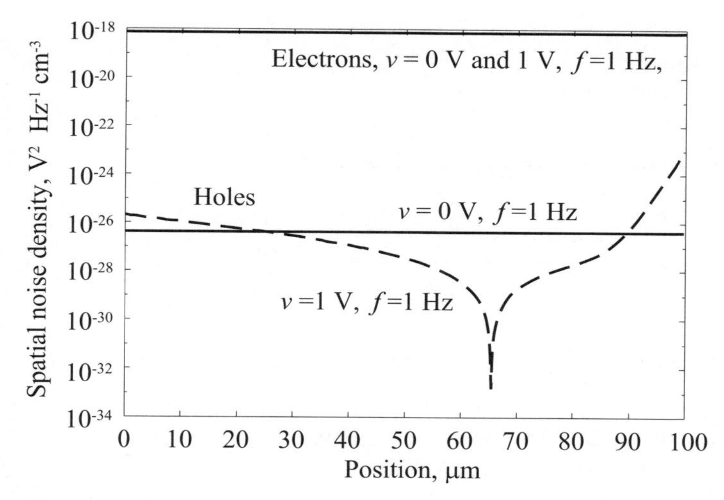

Fig. 4.4. Electron and hole spatial noise density at equilibrium and for $v = 1$ V

the minority-carrier noise density when compared to the equilibrium value, thus leading to an increase of minority-carrier noise. The total noise remains unaffected, however, due to the negligibly small value of the minority-carrier microscopic source.

We therefore conclude that in a doped semiconductor sample the Nyquist-like behavior occurs, out of equilibrium, only because minority-carrier noise is altogether negligible. This is not expected to be true if the doping level is decreased, as will be discussed in the next section.

4.1.2 Noise in a Quasi-Intrinsic Resistor

The noise behavior of weakly doped samples is made more complex by the fact that the contribution of minority carriers is not always negligible with respect to the majority contribution. For the sake of simplicity, suppose that GR effects are negligible and that the sample under consideration is n-type. Figure 4.5 shows the current fluctuation spectrum at low frequency (1 Hz) for a weakly doped sample having different doping levels; the sample length is 100 μm. Although at equilibrium (zero bias) the Nyquist law is satisfied in any case, increasing the bias (expressed here in terms of the DC field, which actually turns out to be uniform within the sample as numerically simulated) current fluctuations increase. The excess noise is inversely proportional, at fixed bias, to the doping level, i.e. vanishes, as expected, in a high-doped sample.

The excess noise can be shown to be related not only to the sample doping, but also to the sample length. As shown in Fig. 4.6, if we keep the doping level constant and compare the excess noise of samples having different lengths at

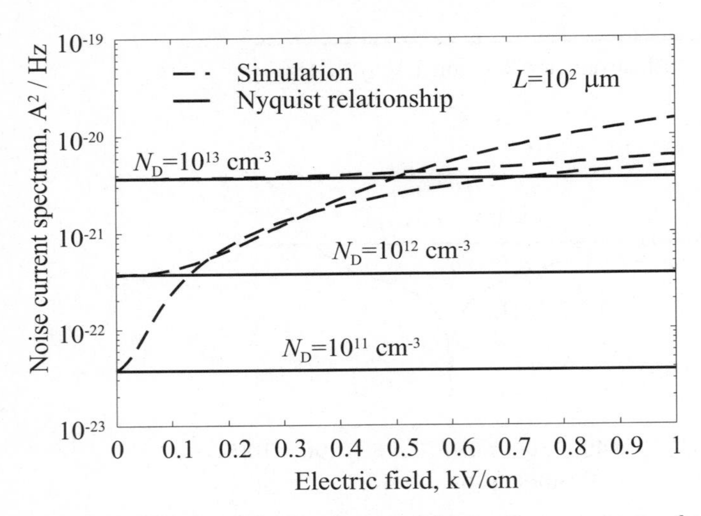

Fig. 4.5. Bias dependence of current fluctuation spectrum of a low doping sample ($f = 1$ Hz) for different doping levels

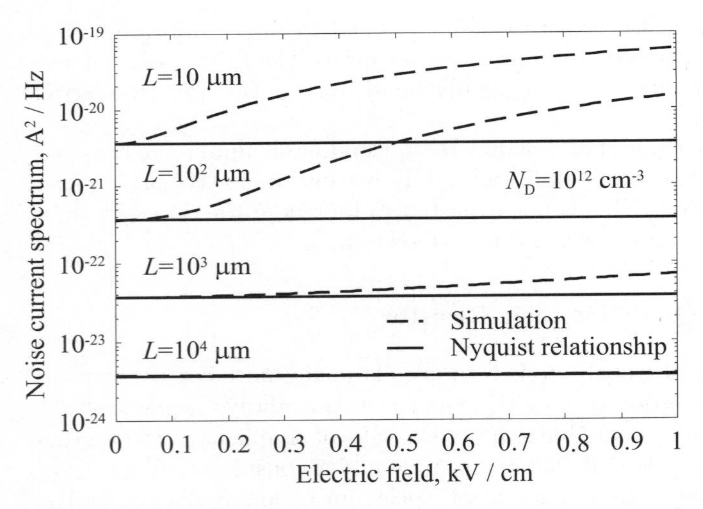

Fig. 4.6. Bias dependence of current fluctuation spectrum of a low doping sample ($f = 1$ Hz) for different sample lengths

constant electric field, we see that the excess noise increases when the sample length decreases.

From a microscopic standpoint, this behavior can be investigated by comparing the short-circuit current scalar Green's functions in and out of equilibrium, shown in Fig. 4.7. The sample length is $L = 100$ μm and the sample doping is $N_D = 10^{11}$ cm^{-3}. While the behavior is, as expected, linear for both minority and majority carriers at equilibrium, out of equilibrium both

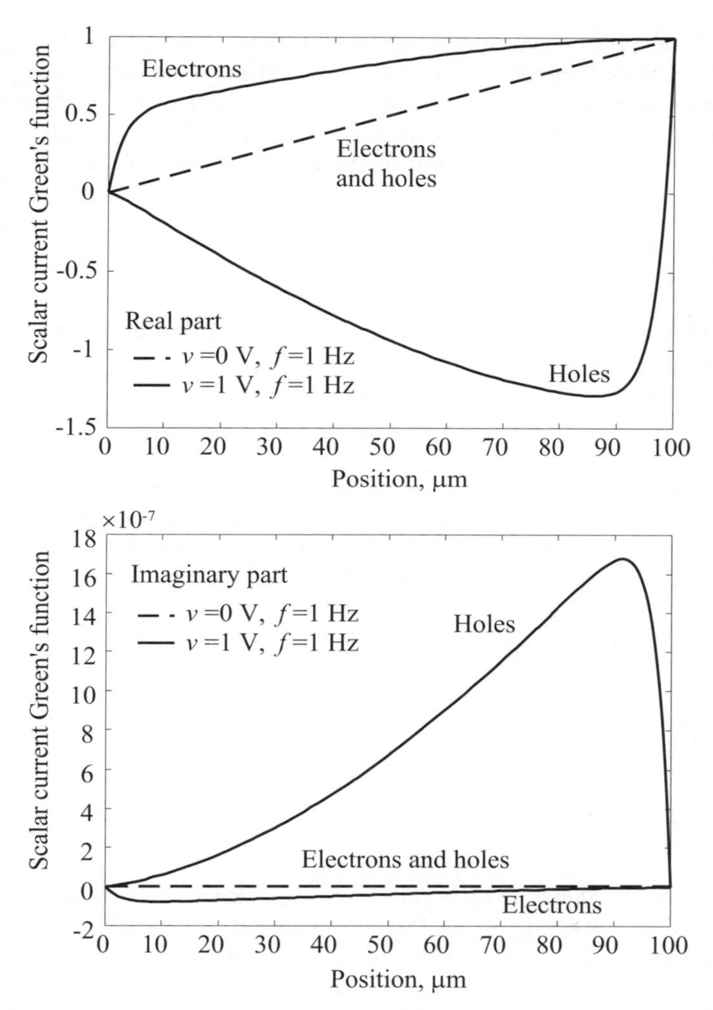

Fig. 4.7. Short-circuit current scalar Green's functions at equilibrium and for $v =$ 1 V for a quasi-intrinsic sample

the minority- and the majority-carrier Green's functions undergo a strong distortion near the sample ends, which in turn causes a large gradient to arise in the corresponding vector Green's function. As a consequence, both the minority- and the majority-carrier noise densities increase. Notice that the high-noise region arising at the sample ends for minority carriers has a fixed extent, depending mainly on bias, not on the doping level (compare with Fig. 4.3). This remark is confirmed by an analytical investigation carried out on a low-injection model at zero frequency. This simplified model assumes that the bias point minority-carrier density is negligible with respect to the excess (small-signal) one, and that the excess (small-signal) minority-carrier drift current is again negligible with respect to diffusion currents, thus decoupling the minority-carrier continuity equation from the rest of the drift-diffusion system. To fix the ideas, consider an n-type sample where minority carriers are holes; for hole current injection in point x' the hole continuity equation reduces to:

$$D_h \frac{\mathrm{d}^2 \delta\tilde{p}}{\mathrm{d}x^2} - \mu_h \mathcal{E}_0 \frac{\mathrm{d}\delta\tilde{p}}{\mathrm{d}x} - \frac{\delta\tilde{p}}{\tau_h} - \delta\left(x - x'\right) = 0. \tag{4.1}$$

On solving (4.1) the excess hole distribution profile becomes dominated by the characteristic lengths $\pm L_E$, where:

$$L_E = \frac{2v_{\mathrm{T}}}{\mathcal{E}_0}, \tag{4.2}$$

where $\mathcal{E}_0$ is the DC electric field in the sample; if $L_E \gg L$ the minority charge profile is piecewise linear. Once $\delta\tilde{p}(x)$ is known, by combining the linearized electron continuity equation and Poisson equation one obtains the following equation for the small-signal electron distribution:

$$\frac{\mathrm{d}^2 \delta\tilde{n}}{\mathrm{d}x^2} + \frac{\mu_n \mathcal{E}_0}{D_n} \frac{\mathrm{d}\delta\tilde{n}}{\mathrm{d}x} + \frac{q\mu_n N_D}{D_n \varepsilon} \left(\delta\tilde{p} - \delta\tilde{n}\right) - \frac{\delta\tilde{p}}{D_n \tau_h} = 0. \tag{4.3}$$

The majority-carrier excess distribution turns out to be dominated by the characteristic lengths:

$$\frac{1}{L_\pm} = \frac{1}{L_E} \pm \sqrt{\frac{1}{L_D^2} + \frac{1}{L_E^2}}, \tag{4.4}$$

where L_D is the extrinsic Debye length. The excess potential distribution can be finally derived by integrating the Poisson equation. Concerning the short-circuit Green's function, this can be obtained by evaluating the current density distribution derived from the excess minority and majority carriers. The behavior of the short-circuit Green's function with respect to the injection point turns out to be linear, almost independent of the doping level, whenever the condition $|L_E| \gg L$ is observed; this, of course, requires the applied bias to be very low, of the order of a few v_{T} units. For larger applied voltages the minority-carrier Green's scalar function is strongly nonlinear, thus leading to an increase of the vector Green's function and of minority-carrier noise.

For majority-carrier injection, the simplified model leads to negligible induced excess minority-carrier density, while the induced majority-carrier profile is again dominated by the characteristic lengths defined in (4.4). If $L_D \ll L_E$ and $L_D \ll L$ the injected majority-carrier profile is symmetrical (and almost zero, apart from a small region near to the injection point). This leads to constant induced excess field on most of the sample and, ultimately, to a linear Green's function with respect to the injection point. At low bias and moderately high-doped conditions $L_D \ll L_E$ and $L_D \ll L$ are well verified, and, increasing the resistor bias, $L_D \ll L_E$ is still verified if the doping level is high, since this amounts to decreasing the extrinsic Debye length. At a certain bias value, however, $L_D \approx L_E$ and the two characteristic lengths become different, leading to a strongly nonlinear Green's function and therefore to an excess majority-carrier noise.

In summary, the insurgence of excess noise in minority carriers is practically independent of the sample doping, and only depends on the sample bias and length L; conversely, the onset of excess noise in majority carriers is favored by high bias, low doping, and a short sample length. In a moderately doped sample the minority-carrier noise is negligible, and therefore the external behavior still follows the Nyquist law even if minority carriers exhibit excess noise, while in an intrinsic or almost intrinsic sample the onset of excess noise takes place at an applied voltage which is of the order of a few v_{T} units.

4.1.3 GR Noise in Semiconductor Samples

In the analysis of uniform, finite-length samples we have up to now neglected GR noise. GR noise can be considered as a typical example of excess noise, i.e. noise arising out of thermodynamic equilibrium. This nature is clearly visible by considering the so-called equivalent noise source in terms of current fluctuations defined in (1.104), see the discussion in Sect. 1.2.2; since the noise source is proportional to the mean carrier velocity, it vanishes in equilibrium and is, to a first approximation, proportional to the square of the current (and therefore to the square of the applied voltage) out of equilibrium. A further well-known feature of GR noise is its Lorentzian frequency dependence, with a cutoff angular frequency of the order of the inverse of the majority-carrier lifetime. The aforementioned dependencies on frequency and bias explicitly appear in the so-called equivalent monopolar noise source, as expressed in [28] for an n-type sample:

$$\mathsf{K}_{\delta \boldsymbol{J}_n, \delta \boldsymbol{J}_n}(\boldsymbol{r}; \omega) = \frac{\boldsymbol{J}_{n0} \boldsymbol{J}_{n0}}{n_0} \frac{4\alpha\tau_{\mathrm{eq}}}{1 + \omega^2 \tau_{\mathrm{eq}}^2}, \tag{4.5}$$

where both α and τ_{eq} are field-dependent parameters [28, 116], the latter being often approximated to the minority-carrier lifetime.

In order to introduce the following discussion, which is mainly based on [39], we recall that GR noise can be modelled, as discussed in Sect. 1.2.2, according either to a fundamental (exact) or to an equivalent (approximate) approach. In the exact approach white population fluctuations appear as Langevin sources applied to a bipolar model. In the equivalent approach, (nonwhite) current fluctuation sources are derived from (white) population fluctuations in homogeneous conditions, and then inserted as Langevin sources in a model which can be either bipolar or monopolar. In the simplest, but also widely exploited, equivalent monopolar model, the local noise source (4.5) is the Langevin forcing term of a majority-carrier model.

A main difference can be immediately detected between the fundamental and the equivalent approaches: in the former, the local noise sources are white and (at least for a uniformly doped sample) bias independent, and the Lorentzian-like and bias dependence arises, in a rather complex way, from the combined effect of the Green's functions and of fluctuation correlation, see [39]; in the latter case, the frequency and bias dependence is already embedded into the equivalent local noise source. We will investigate here to what extent the fundamental and equivalent (bipolar and monopolar) approaches can be expected to provide comparable results.

We shall consider first the dependence of the GR noise spectrum on the applied bias. The low-frequency open-circuit GR noise voltage spectrum is shown as a function of the (uniform) applied electric field in Fig. 4.8 for a sample with length $L = 100$ μm and for several values of the doping level. A band-to-band GR process with $\tau_r = 100$ ns was considered and $\tau_{\mathrm{eq}} \approx \tau_r$ was assumed in the equivalent approach. The continuous curve corresponds to the fundamental approach, the dotted (dashed-dotted) line to the equivalent bipolar (monopolar) cases. In the monopolar case, no correction was introduced, i.e. $\alpha = 1$ in (4.5). As expected, in all approaches the noise spectrum is proportional, at low field, to the square of the DC bias; however, the low-field values differ and in the high-field region the fundamental theory yields a saturation of the noise spectrum, the equivalent approach a monotonically increasing spectrum.

This behavior, which is almost independent of the doping level, is found to be influenced, for the bipolar model, by the sample length, whereas in the monopolar equivalent model the bias dependence is not sensitive to the sample length, as shown in Fig. 4.9; here the sample length is 1 cm for the long sample, 100 μm for the short one (for reference, the minority-carrier diffusion length is 10 μm for $N_D = 10^{16}$ cm^{-3}).

Figures 4.8 and 4.9 suggest that the bipolar equivalent model overestimates low-field noise in the short sample. This behavior, as discussed in [39], is related to the ohmic boundary conditions and becomes negligible, at low field, if the sample length is much larger than the diffusion length. A similar argument applies for the noise saturation exhibited by the fundamental approach at increasing electric field, where the field-dependent diffusion length is anisotropic, with the effect shown in Fig. 4.10 on charge spreading. At high electric fields the sample length ultimately becomes comparable with the (high-field) diffusion length, making the effect of ohmic boundary conditions prevail, with saturation of the noise spectrum. Such an effect is not accounted for in the equivalent noise source. Finally, the monopolar model disagrees with the fundamental approach independent of the sample length,

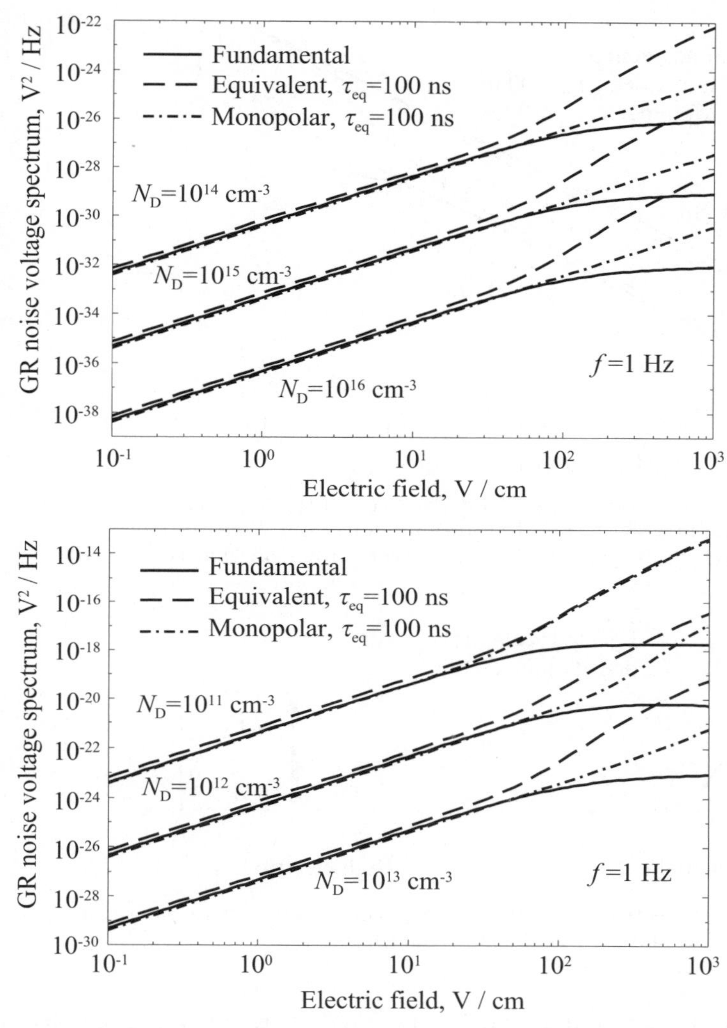

Fig. 4.8. Field dependence of the low-frequency GR voltage noise spectrum in a short ($L = 100$ μm) n-doped sample according to the fundamental and the equivalent bipolar and monopolar approaches. The doping levels are $N_D = 10^{16}$ cm^{-3}, $N_D = 10^{15}$ cm^{-3}, $N_D = 10^{14}$ cm^{-3} (*above*), $N_D = 10^{13}$ cm^{-3}, $N_D = 10^{12}$ cm^{-3}, $N_D = 10^{11}$ cm^{-3} (*below*). The recombination lifetime is $\tau_r = 100$ ns (from [39])

mainly because in the direct-like GR process considered minority-carrier noise is not negligible, see [39].

The frequency dependence from the fundamental and the equivalent (bipolar and monopolar) approaches is shown in Fig. 4.11 for a sample with $L = 100$ μm, $N_D = 10^{14}$ cm^{-3}, $\tau_r = 100$ ns and an applied electric field of 10 V/cm; in the monopolar equivalent model, the local noise source (4.5)

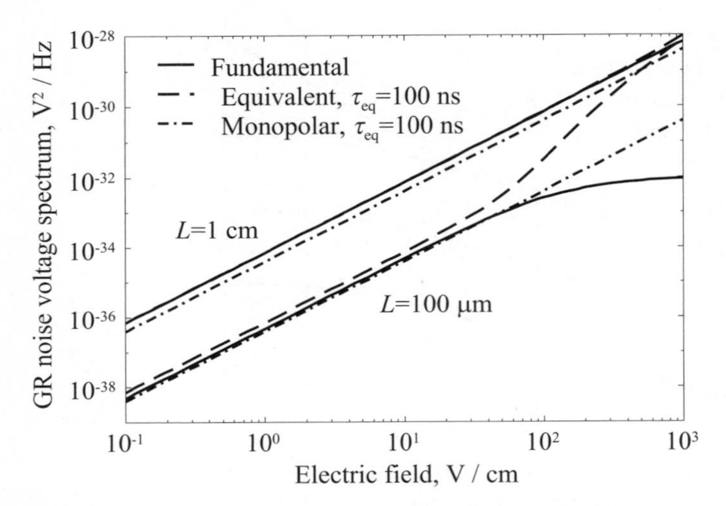

Fig. 4.9. Comparison between the field dependence of the low-frequency GR voltage noise spectrum in a short ($L = 100$ μm) and long ($L = 1$ cm) n-doped sample. The doping is $N_D = 10^{16}$ cm^{-3} and the recombination lifetime is $\tau_r = 100$ ns (from [39])

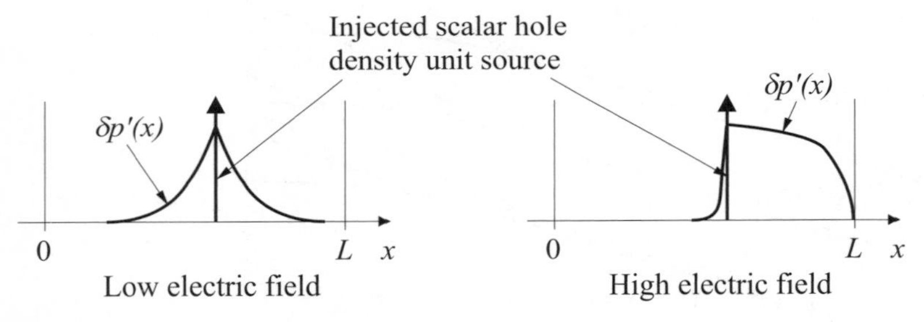

Fig. 4.10. Effect of electric field and boundary conditions on an excess minority-carrier density in low and high field conditions (from [39])

has been used with $\tau_{\text{eq}} = 100$ ns and $\alpha = p_0/N_D$. The results point out that the actual corner frequency is directly related to the carrier lifetime only for low fields, long samples; in general, it is field- and boundary-conditions-dependent. For the equivalent model to match the exact frequency dependence of GR noise, τ_{eq} and α, see (4.5), must be independently corrected; this clearly results from Fig. 4.11. The same disagreement occurs from the equivalent monopolar approach with nominal parameters; by proper adjustment of τ_{eq} and α, however ($\tau_{\text{eq}} = 85$ ns and $\alpha = 0.7637$, see [39]), excellent agreement is obtained. This confirms that this simplified model is able to accurately reproduce experimental data, as already suggested in [28].

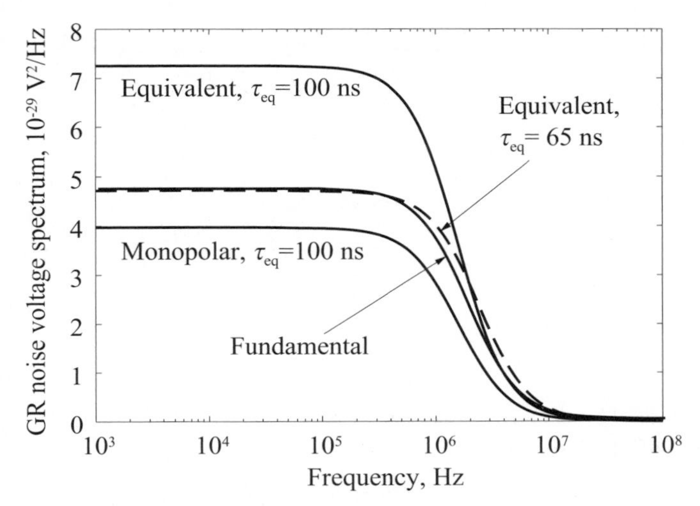

Fig. 4.11. GR voltage noise spectrum in a short sample (L = 100 µm) at E = 10 V/cm, evaluated according to the fundamental and equivalent approaches. The doping is $N_D = 10^{14}$ cm^{-3}, the physical lifetime is 100 ns, and the lifetime fitted to reproduce the low-frequency value (with $\alpha = 1$) is 65 ns (from [39])

4.2 Semiconductor Diodes

4.2.1 Noise Modeling of *pn* junctions

In order to introduce junction modeling, let us discuss a simple, monodimensional case: a symmetric silicon diode with doping $N_D = N_A = 10^{16}$ cm^{-3}, with side lengths $W_n = W_p = 5$ µm and cross section $A = 1$ µm^2. The doping profile is abrupt and constant mobilities have been assumed for electrons (μ_n = 1390 cm^2 V^{-1} s^{-1}) and holes (μ_h = 470 cm^2 V^{-1} s^{-1}). The minority-carrier lifetime is 1 ns, and the diode is long (i.e. much longer than the diffusion length) in both sides.[1]

The resulting DC curve for the diode is shown in Fig. 4.12. This figure clearly shows that the ideality factor is also one in low injection, meaning that GR currents are negligible with respect to diffusion currents; deviations arise beyond 0.7 V owing to the effect of the side resistance (small in this case) and, above all, of high-injection effects. According to the analytical Green's function theory in Sect. 2.3.4, the diode noise originates from the minority-carrier injection regions. This is confirmed by the electron and hole impedance fields simulated at low frequency (f = 10 kHz).

In thermodynamic equilibrium, the short-circuit Green's functions have the appearance shown in Fig. 4.13. The behavior is linear (almost constant because of the high-doped level) in the regions where the relevant carriers

[1] In the following figures, the n side will be shown on the left-hand side with respect to the junction plane.

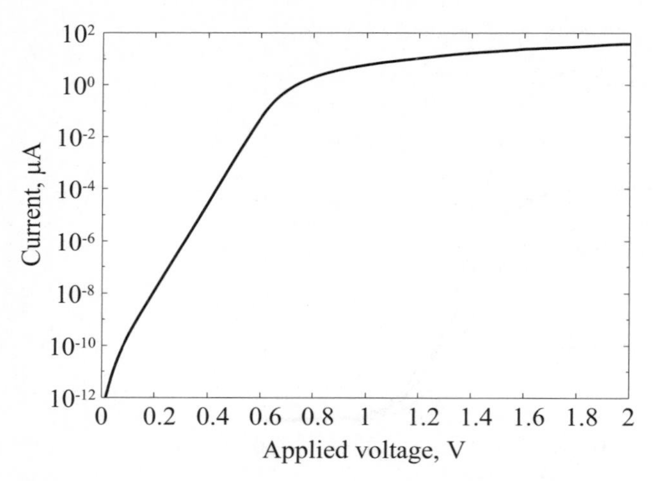

Fig. 4.12. Voltage–current characteristic of symmetric abrupt *pn* diode

are majority carriers; in the minority-carrier injection regions the impedance fields are monotonic, with a large slope, see Fig. 4.13. In such conditions, the diode noise exactly satisfies the Nyquist law. Increasing the diode bias drives the junction from equilibrium to the low-injection regime, without any appreciable change in the short-circuit Green's functions with respect to the equilibrium; for example at 0.5 V bias the Green's functions can be hardly distinguished from the equilibrium case. In order to justify the spatial origin of noise within the diode volume, we notice that the vector Green's function for electrons is neglibile in the n side, large in the p side, whereas the opposite occurs for holes. As a consequence, the vector Green's functions are negligible where the local noise source is large (i.e. in those regions where electrons and holes are majority carriers, respectively) whereas it is large in the injection regions. This implies that the noise generated by majority carriers is weakly propagated to the external circuit, while injected minority-carrier noise is strongly coupled. This picture is confirmed by Fig. 4.14, which shows the local noise density for electrons and holes in low-injection conditions. The electron noise is almost entirely originated by the electron-injection region in the p side (on the right) while, correspondingly, hole noise comes from the hole injection noise in the n side (on the left). The relative magnitude of the two noise sources originates, in this symmetrical structure, from the different carrier mobilities. The shape of the minority-carrier noise source closely follows the injected excess minority charge profile, which is, in a long diode, exponential. Finally, notice that negligible noise is generated in the depletion region across the junction.

This simple picture is modified in high injection, where the slope of the majority-carrier scalar Green's function is no longer negligible. In fact, injected minority carriers induce, for quasi-neutrality, an equal amount of majority carriers, which is no longer negligible with respect to the background

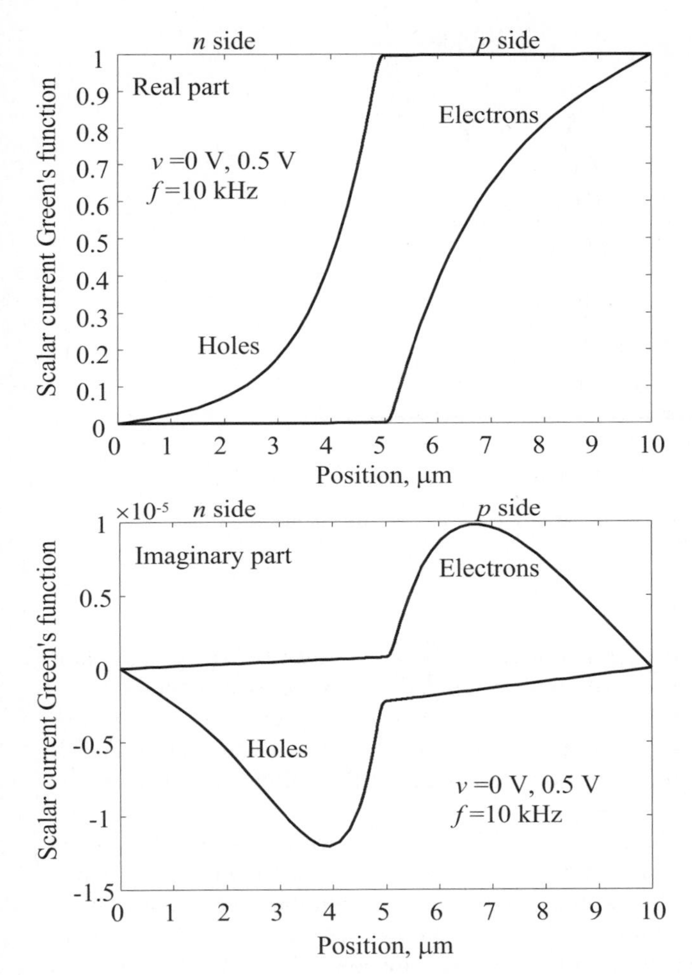

Fig. 4.13. Diode hole and electron short-circuit Green's functions, real and imaginary parts in equilibrium and in low injection, $v = 0.5$ V

doping. Such excess majority carriers yield an increasingly large noise source. These remarks are supported by Fig. 4.15, showing the scalar short-circuit current Green's function for electrons and holes, and by Fig. 4.16, where the electron and hole noise densities are shown. In high injection, noise is no longer originated from minority carriers only.

The overall behavior of the diode low-frequency spectrum of current fluctuations is finally shown in Fig. 4.17, where this is compared to the analytical approach in Sect. 2.3.4. Basically, at low injection the shot-noise pattern is closely followed; deviations arise in high injection, where the simple shot-noise model underestimates noise. At equilibrium the spectrum tends to the value foreseen by the Nyquist law.

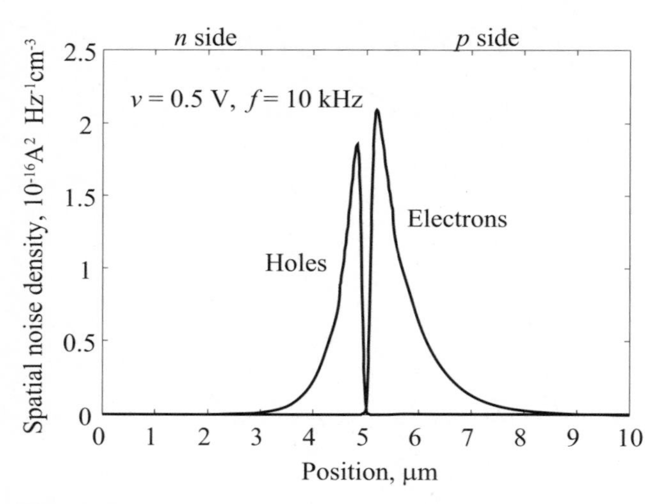

Fig. 4.14. Electron and hole noise densities in low injection, $v = 0.5$ V

4.2.2 A 2D *pn* Diode

The second simulation example is a 2D *pn* diode obtained by a *p*-type implant (peak concentration of 5×10^{17} cm^{-3}) on a uniformly doped *n*-type ($N_D = 2 \times 10^{15}$ cm^{-3}) substrate so as to obtain a 0.5-μm deep junction. The device is 3 μm wide and 3 μm deep, discretized with 744 grid points. Only half of the device has been simulated so as to reduce the computational intensity, see Fig. 4.18.

As in the 1D case, the simulated noise current spectrum at low frequency ($f = 10$ kHz, see Fig. 4.19) is in excellent agreement with the theory, while discrepancies arise at the onset of high injection. Similar remarks apply for the noise density, whose section (along a line located in the middle of the *p*-implanted region) is shown in Fig. 4.20 both for low (above, $V = 0.5$ V) and for high (below, $V = 0.8$ V) injection. Finally, Fig. 4.21 shows the real part of the short-circuit current scalar electron Green's function in low (above) and high (below) injection. In low injection, the scalar Green's function is flat in the *n* region of the device, and therefore the vector Green's function is not zero only in the *p*-type region. This picture no longer holds in high injection (see Fig. 4.21, below), thus explaining the spreading of the noise density in the majority-carrier regions shown in Fig. 4.20, below. It may be further noticed that the local noise source due to injected minority carriers is about the same order of magnitude in both sides of the junction, despite the large difference in doping; this suggests that the higher injection efficiency of holes into the *n* substrate is compensated by the larger magnitude of the electron vector Green's function in the *p* region.

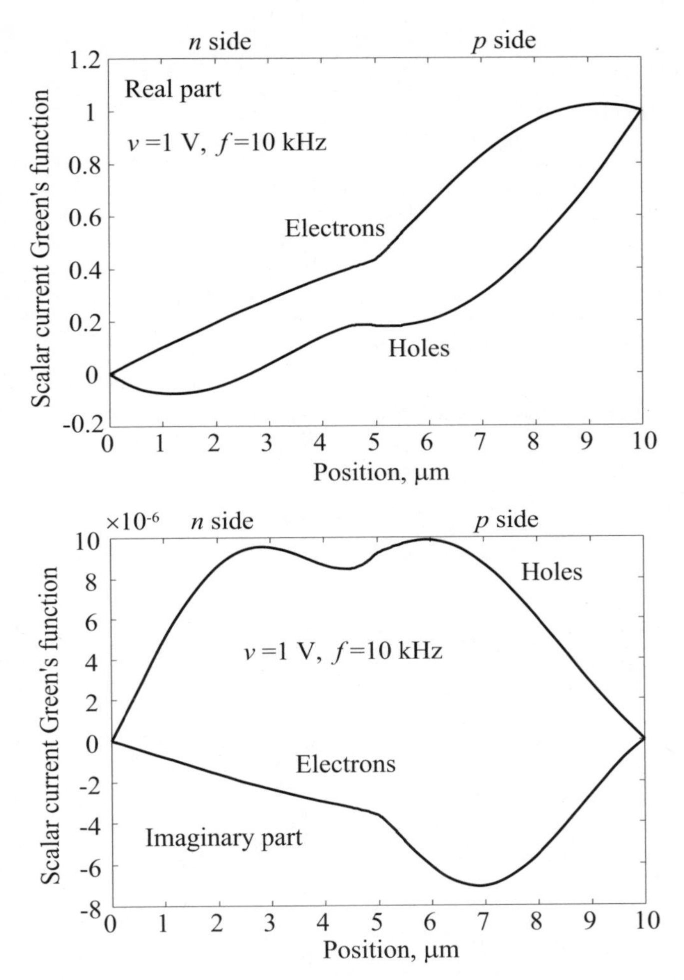

Fig. 4.15. Diode hole end electron short-circuit Green's functions, real and imaginary parts in high injection, $v = 1$ V

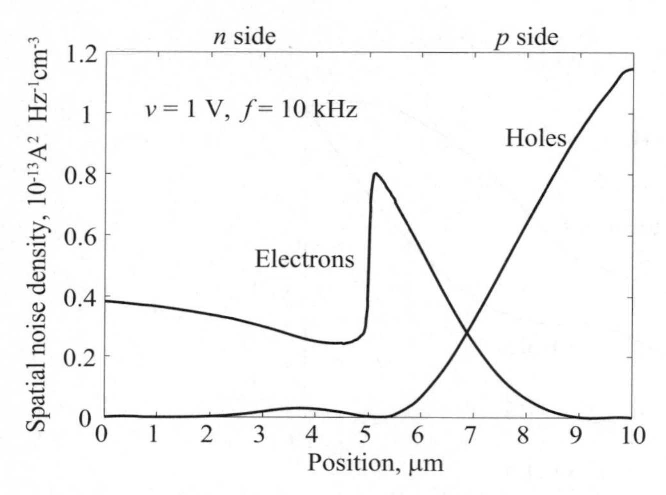

Fig. 4.16. Electron and hole noise density in high injection, $v = 1$ V

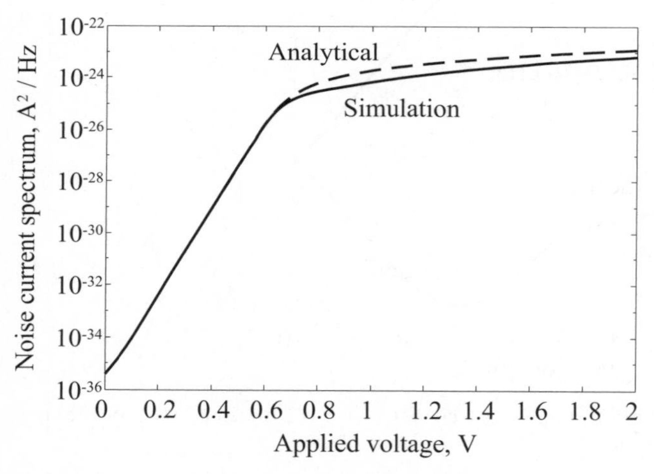

Fig. 4.17. Low-frequency short-circuit current fluctuation spectrum for 1D abrupt *pn* junction: comparison between the analytical shot-like approach and the physics-based simulation

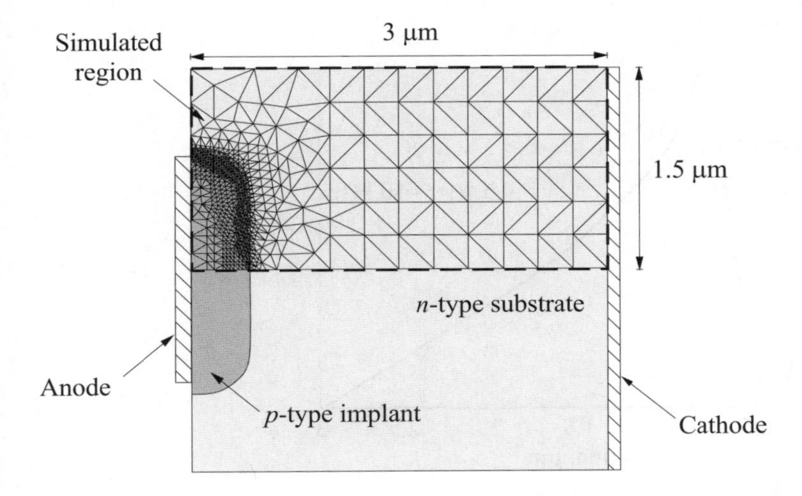

Fig. 4.18. Simulated 2D diode structure, showing the discretization mesh

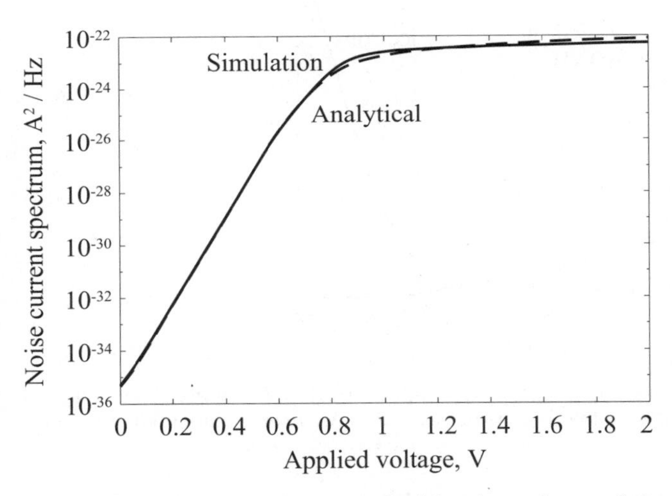

Fig. 4.19. Comparison between the bias dependence of the theoretical and simulated noise current spectrum of the 2D diode ($f = 10$ kHz, from [6])

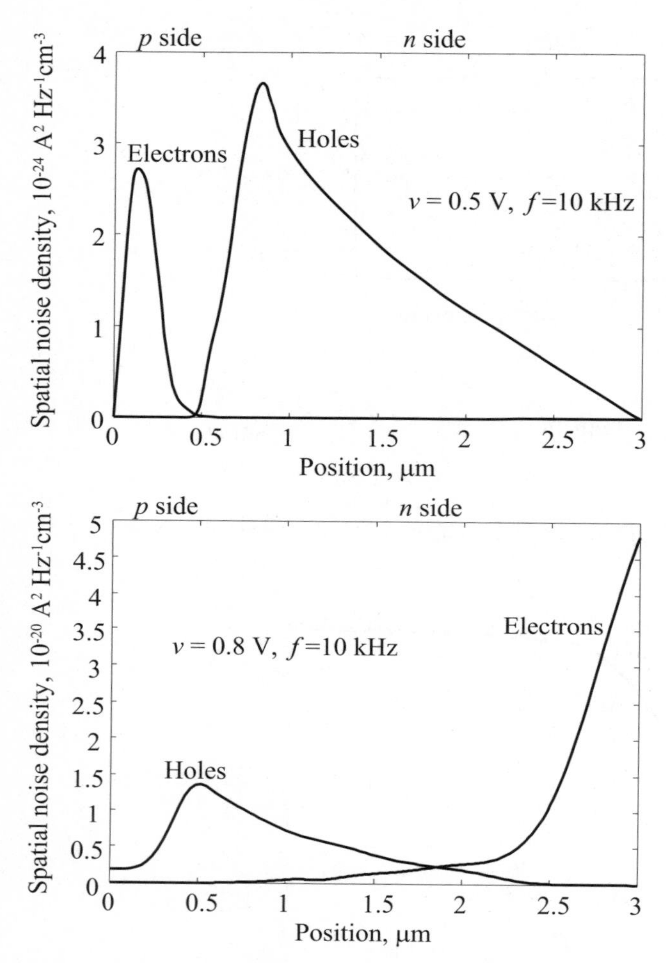

Fig. 4.20. Depth dependence (section for $x = 0.25$ μm) of the noise density ($f =$ 10 kHz) for the 2D diode in low injection ($v_d = 0.5$ V, *above*) and high injection ($v_d = 0.8$ V, *below*)

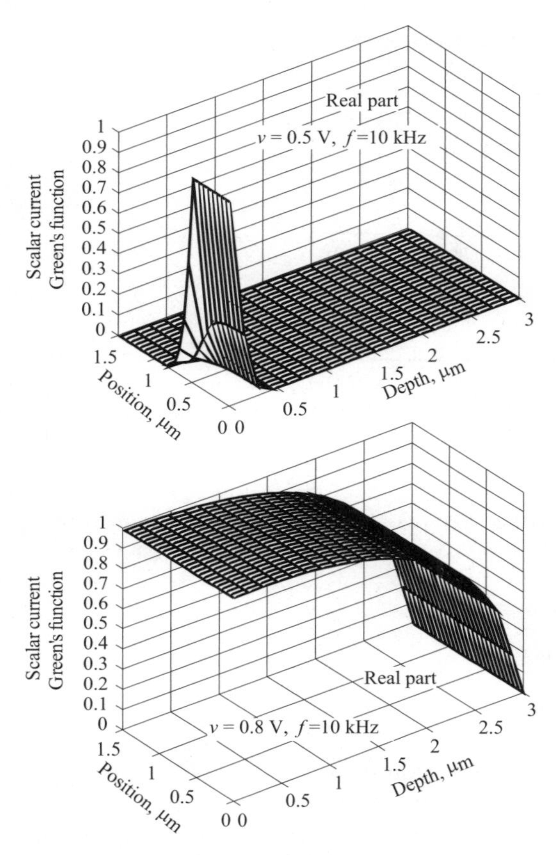

Fig. 4.21. Real part (f = 10 kHz) of short-circuit current electron scalar Green's function for a 2D diode in low injection (v_d = 0.5 V, *above*) and high injection (v_d = 0.8 V, *below*)

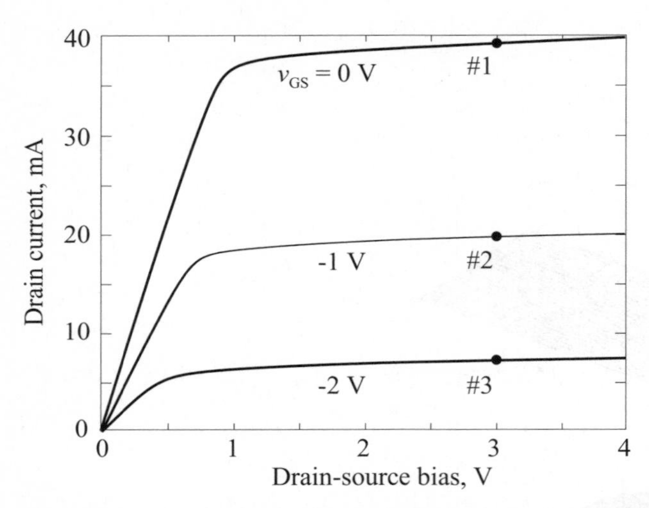

Fig. 4.22. Simulated DC characteristics of epitaxial 1 μm MESFET. The operating points are: #1, $v_{GS} = 0$ V, $v_{DS} = 3$ V; #2, $v_{GS} = -1$ V, $v_{DS} = 3$ V; #3, $v_{GS} = -2$ V, $v_{DS} = 3$ V (from [4])

4.3 Field-Effect Transistors

4.3.1 MESFET Noise Modeling

In the present section, a simple MESFET case study is presented, concerning an epitaxial MESFET with 1 μm gate length, 300 μm gate periphery, active layer thickness of 0.2 μm, epilayer doping $N_D = 10^{17}$ cm^{-3} on an ideal semi-insulating buffer layer. The analysis was two-dimensional, using a non-uniform triangular mesh having about 1800 nodes. Only diffusion noise has been included in the simulation. The analysis was carried out through a monopolar model, as is usually appropriate to this kind of device; further details can be found in [4]. The simulated DC curves are shown in Fig. 4.22; a small signal and noise analysis was performed up to 14 GHz on the operating points marked in Fig. 4.22. Notice that only the top layer of the substrate was actually included in the simulation.

The scalar Green's functions for the gate and the drain short-circuit currents are shown in Fig. 4.23 and Fig. 4.24, respectively. The frequency is $f = 4$ GHz and the bias point is #1 of Fig. 4.22. Concerning the gate Green's function, Fig. 4.23 suggests that the coupling between a scalar current source impressed into the device and the gate short-circuit current is mainly capacitive, as already postulated in the analytical model presented in Sect. 2.3.2; in fact, the in-phase (real) component of the gate short-circuit current Green's function is almost zero everywhere, while the quadrature (imaginary) component is different from zero, with a maximum near the drain edge of the channel. On the other hand, as seen from Fig. 4.24, the behavior of the drain

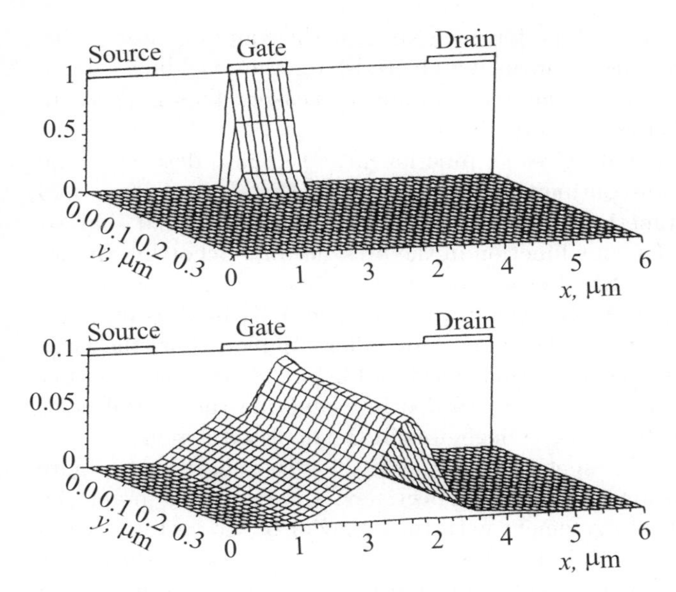

Fig. 4.23. Short-circuit gate current scalar Green's function: real part (above), imaginary part (*below*). The frequency is $f = 4$ GHz; the bias point is #1 of Fig. 4.22 (from [4])

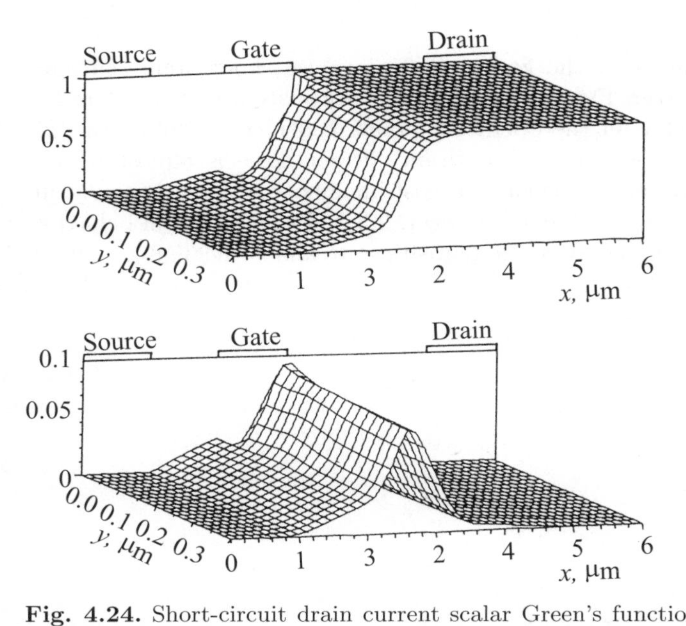

Fig. 4.24. Short-circuit drain current scalar Green's function: real part (above), imaginary part (*below*). The frequency is $f = 4$ GHz; the bias point is #1 of Fig. 4.22 (from [4])

short-circuit current Green's function suggests that the coupling between internal fluctuations and the drain in mainly resistive, thus yielding a large in-phase component, while capacitive coupling is weaker. This is again in agreement with the analytical model in Sect. 2.3.2.

The behavior of the gate Green's function near the gate deserves some comment. In fact, the discontinuity in the real part, as is shown in Fig. 4.23 is an artifact of the numerical discretization. In fact, a more refined analysis of the behavior of the Green's function in the presence of a Schottky contact made on the basis of the Sze–Bethe model [72], which is not reported here for brevity, suggests that the Green's function can actually be discontinuous across a Schottky barrier, and therefore, not differentiable. Thus, the large gradient of the scalar Green's function shown in Fig. 4.23 below the Schottky contact is unphysical and must be discarded in evaluating the spatial noise density, which can be shown to be negligible in the depleted region.

Figure 4.25 shows the spatial noise density of the short-circuit gate (above) and drain (below) currents, respectively; the operating frequency is 4 GHz. The result is in agreement with the behavior of the vector Green's function, which is large, both in the drain and in the gate case, in the device channel, and the fact that the local noise source is large in the active region and small in the weakly doped buffer. Therefore, the major noise contributions are seen to originate from the ohmic and saturated part of the channel, in agreement with the analytical model. Notice that, at the simulation frequency, the gate noise density is orders of magnitude smaller than the drain noise density.

The overall behavior of the gate and drain noise short-circuit current power spectra are shown Fig. 4.26, while their correlation coefficient is reported in Fig. 4.27, for all the operating points considered. Notice the f^2 behavior of the gate spectrum, while drain current noise is almost white; moreover, the correlation coefficient is almost imaginary at low frequency, in agreement with the analytical model in Sect. 2.3.2. The quasi-ideal behavior suggests that in the simulated device parasitics are almost negligible.

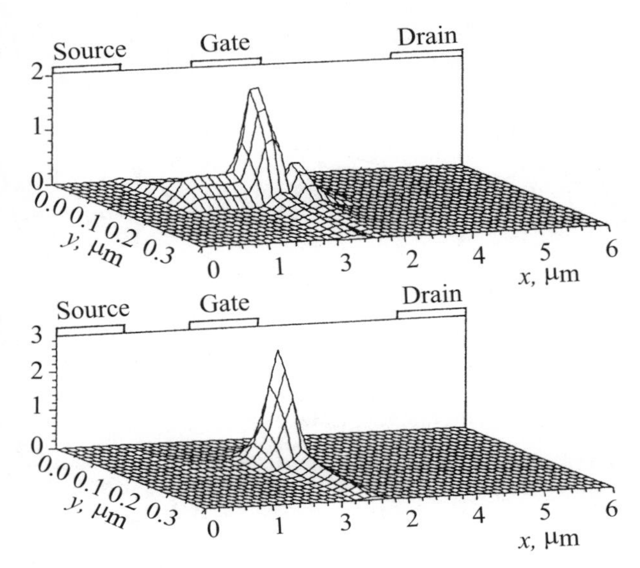

Fig. 4.25. Spatial density of power spectra of gate (*above*) and drain (*below*) short-circuit current noise generators; the bias point is #1 of Fig. 4.22 and the frequency is 4 GHz (from [4])

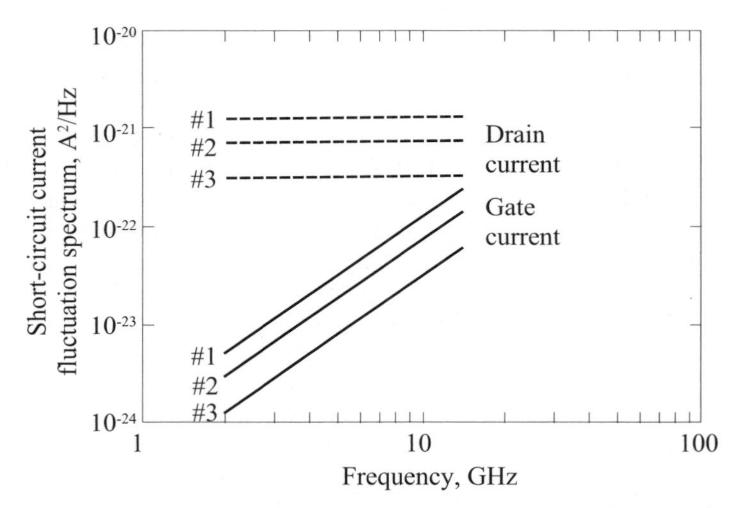

Fig. 4.26. Power spectra of gate and drain short-circuit noise generators, as a function of frequency; the bias points are those in Fig. 4.22 (from [4])

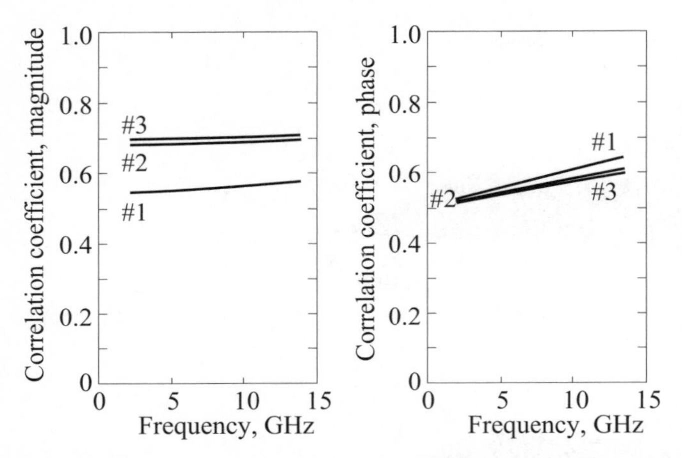

Fig. 4.27. Amplitude (*left*) and phase (*right*, expressed in radians) of the correlation coefficient of gate and drain short-circuit noise currents; the bias points are those in Fig. 4.22 (from [4])

4.3.2 MOSFET Noise Modeling

In this section we analyze two n-channel MOSFETs as case studies: an idealized, long-channel device (gate length 2 μm) and a more realistic short-channel 0.5-μm gate length device. In both cases the oxide thickness is 90 Å, and the substrate p-type doping is $N_A = 5 \times 10^{17}$ cm^{-3}; the gate width is 10 μm. Conventional n^+ drain and source ohmic contacts are modeled as implanted regions. In both devices the threshold voltage is around 1 V.

The purpose of the first example is to show that the numerical noise model is in fair agreement with simple, conventional analytical approaches, such as van der Ziel's, see Sect. 2.3.2. Within this framework, simulation of the long-channel device was carried out with a constant-mobility model. The DC characteristics of the device are shown in Fig. 4.28, continuous line. Despite the large gate length, the van der Ziel model with linear charge control and constant threshold voltage fails, with nominal parameters, to quantitatively reproduce the simulated characteristics. However, the parameters can be fitted around their nominal values on the simulated characteristics in the linear region, thus providing the result shown in Fig. 4.28 which is in excellent agreement with the simulation, apart from a slight discrepancy in the saturation region. In the fitting process, the oxide thickness was kept at the nominal value, together with the gate width and length; the mobility, threshold voltage, body-effect factor and inversion surface potential $2\varphi_p$ were fitted. It should be noted, however, that more refined charge-control models (e.g. see [90]) allow the DC current of long-gate devices to be accurately simulated with nominal parameters.

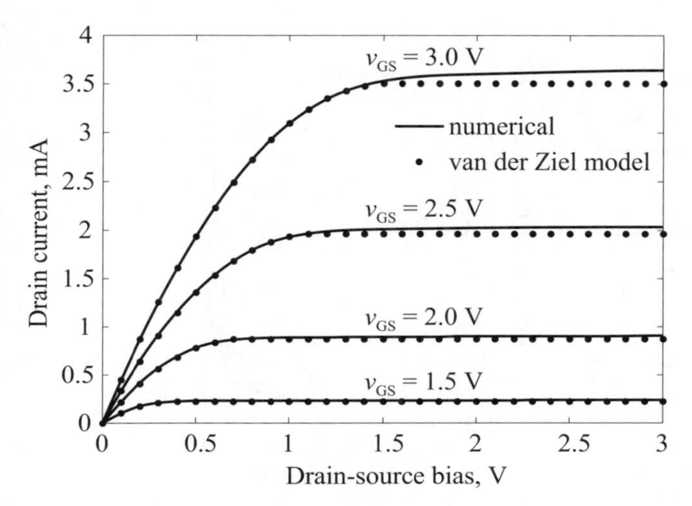

Fig. 4.28. DC characteristics of the 2-μm gate length MOS: comparison between the numerical simulation and the (fitted) van der Ziel model

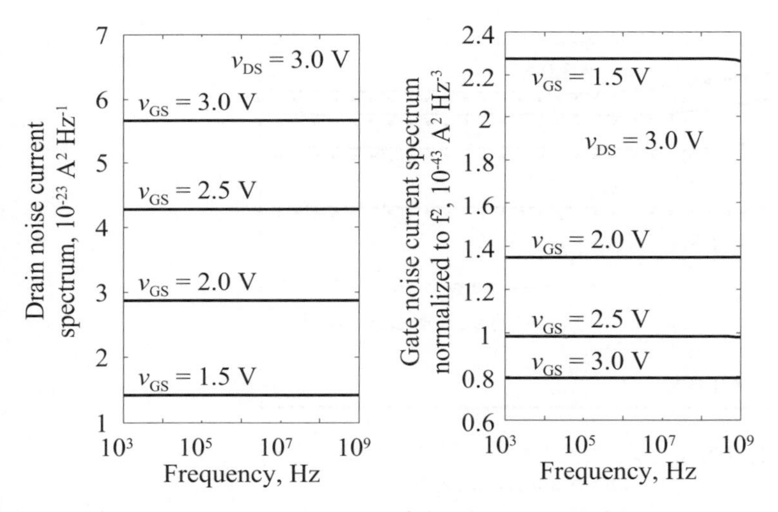

Fig. 4.29. Frequency dependence of the short-circuit drain noise current spectrum (*left*) and of the short-circuit gate noise current spectrum (*right*, divided by f^2) for the 2-μm gate length MOS

Noise simulation was then carried out as a function of frequency for several bias points. As expected from the theory, the short-circuit drain current noise spectrum is white, while the short-circuit gate current noise spectrum is proportional to f^2, see Fig. 4.29. The correlation coefficient of the gate and drain noise currents is again almost frequency independent and close to the theoretical value of 0.4 in magnitude and to $\pi/2$ in phase, see Fig. 4.30.

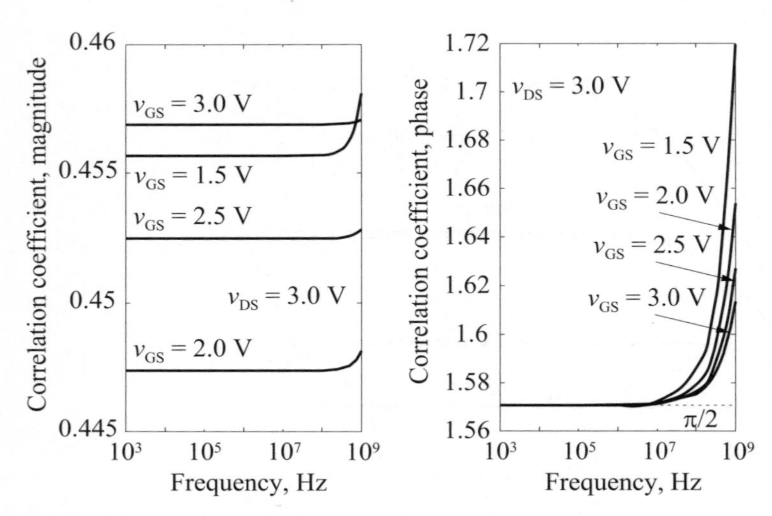

Fig. 4.30. Frequency dependence of the correlation coefficient of short-circuit drain and gate noise currents: amplitude (*left*) and phase (*right*, expressed in radians) for the 2-μm gate length MOS

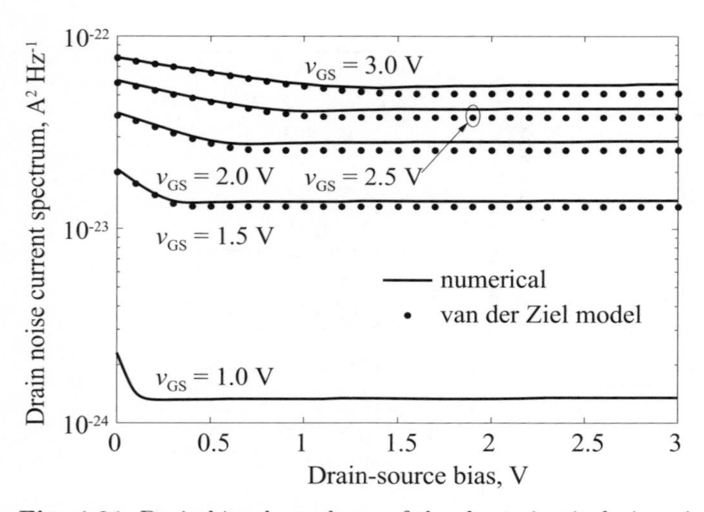

Fig. 4.31. Drain bias dependence of the short-circuit drain noise current spectrum for the 2-μm gate length MOS: comparison between the numerical simulation and the (fitted) van der Ziel model. The operating frequency is 1 kHz

Although a bipolar noise model was used in the numerical simulation, hole noise turns out to be negligible altogether, in agreement with the theoretical analyses only considering electron noise in the channel.

Figure 4.31 shows the drain bias dependence of the short-circuit drain noise current spectrum for several gate bias values at low frequency (f = 1 kHz). The results from the numerical model (continuous line) are in good

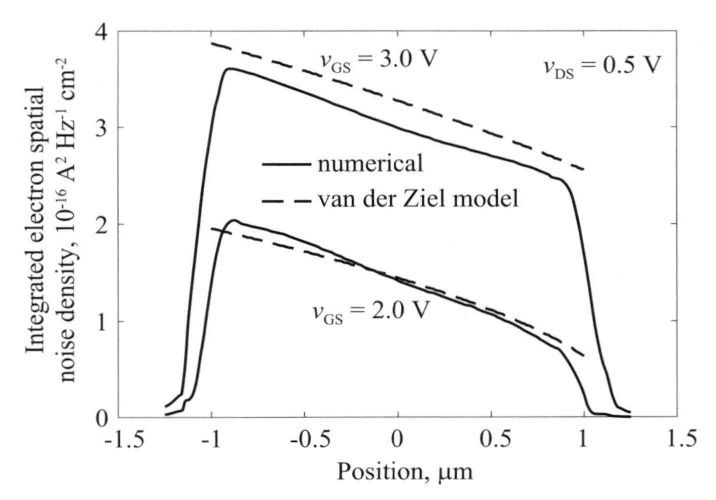

Fig. 4.32. Integrated electron spatial noise density for the 2-μm gate length MOS in the linear region: comparison between the numerical simulation and the (fitted) van der Ziel model. The operating frequency is 1 kHz

agreement with van der Ziel's analytical approach exploiting the same parameters as derived from the DC fitting. A slight disagreement is found in saturation, while the linear region behavior is correctly reproduced. No results are shown for van der Ziel's model near threshold, since in these conditions the linear charge control model is very inaccurate.

In order to gain a better insight into the origin of such a good agreement, the spatial noise densities derived from the numerical and van der Ziel's model were compared. In the numerical model, we integrated the noise density along the channel depth in order to make the comparison possible. Figure 4.32 and Fig. 4.33 show the resulting spatial noise densities in the linear region and in saturation, respectively. It can be noticed that in the linear region the agreement is good, while some discrepancy arises in saturation, where the numerical simulation clearly exhibits some channel length modulation effect towards the drain end of the channel, which is not accounted for by van der Ziel's simple model.

Figures 4.34 and 4.35 show the bias behavior of the short-circuit gate noise current spectrum and of the magnitude of the correlation coefficient between the gate and drain noise currents, respectively, at low frequency. The phase of the correlation coefficient is omitted since it is constant and equal to $\pi/2$ in all bias conditions.

Finally, the standard noise parameters γ and β introduced by van der Ziel according to (2.91) and (2.99), respectively, were evaluated from the numerical model as a function of the drain and gate bias, see Fig. 4.36. In saturation γ is almost constant and weakly dependent on the gate bias, at least well above threshold; values in saturation are close to the theoretical

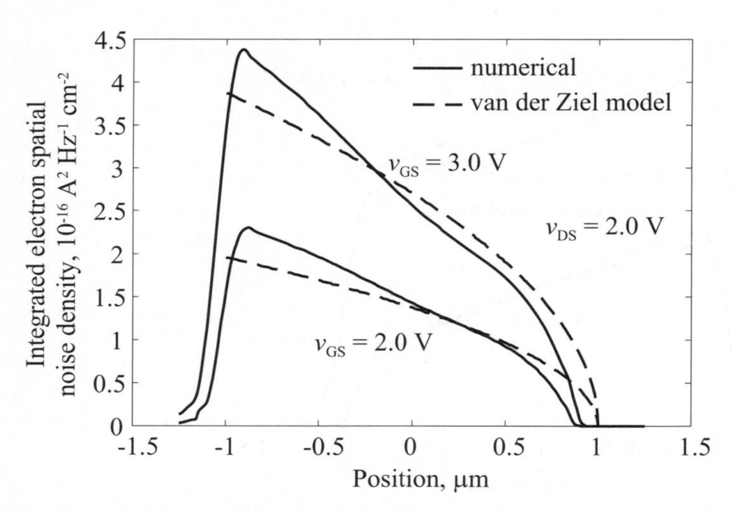

Fig. 4.33. Integrated electron spatial noise density for the 2-μm gate length MOS in saturation: comparison between the numerical simulation and the (fitted) van der Ziel model. The operating frequency is 1 kHz

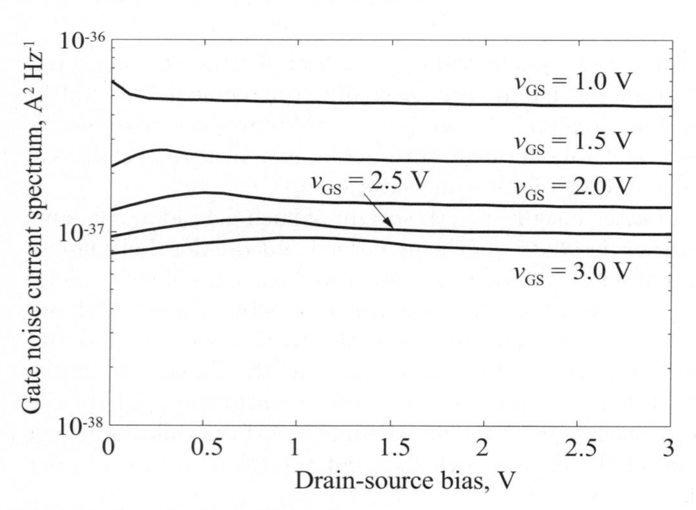

Fig. 4.34. Drain bias dependence of the short-circuit gate noise current spectrum for the 2-μm gate length MOS. The operating frequency is 1 kHz

estimate $\gamma = 2/3$. On the other hand β, while again being constant in saturation, shows a larger spread around the theoretical estimate $\beta = 4/3$.

A second, short gate MOSFET was then simulated in order to gain insight into possible effects connected with device scaling and with the use of more realistic and complex models for the transport parameters, such as the mo-

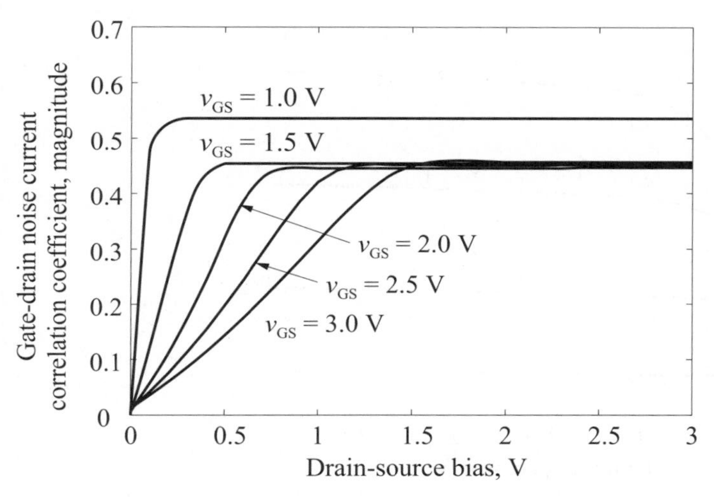

Fig. 4.35. Drain bias dependence of the magnitude of the correlation coefficient between the gate and drain noise currents for the 2-μm gate length MOS. The operating frequency is 1 kHz

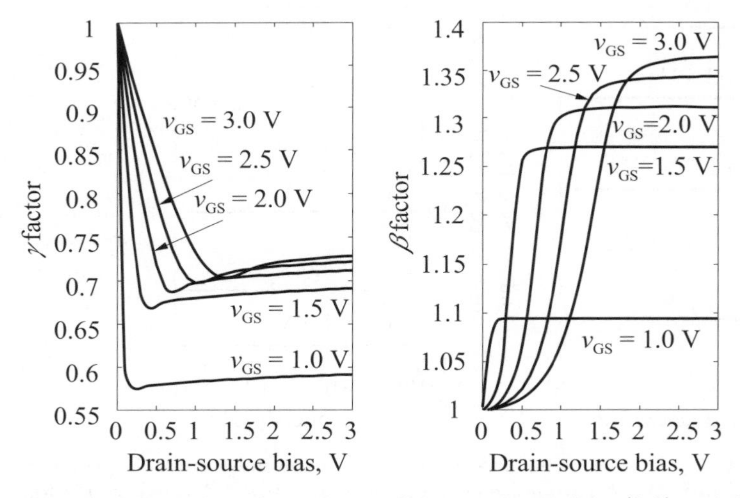

Fig. 4.36. Drain bias dependence of the van der Ziel's γ (*left*) and β (*right*) factors for the 2-μm gate length MOS. The operating frequency is 1 kHz

bility. In particular, an advanced mobility model accounting for the channel and lateral electric field dependence was exploited [117, 118].

The resulting DC curves are shown in Fig. 4.37 while the frequency dependencies of the short-circuit noise current spectra and their correlation coefficient are reported in Figs. 4.38 and 4.39, respectively. The qualitative frequency behavior is the same as in the long-gate device, and also in this

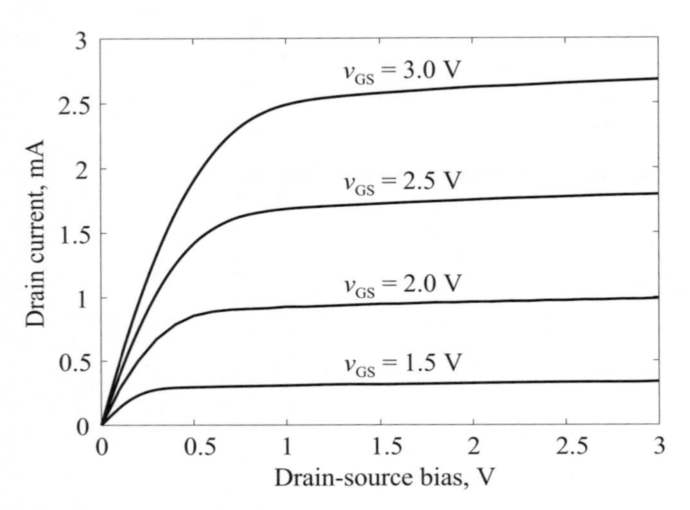

Fig. 4.37. DC characteristics of the 0.5-μm gate length MOS

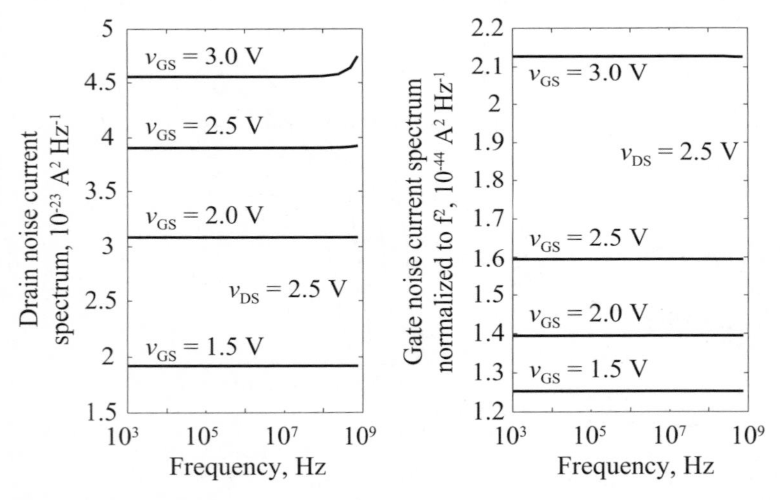

Fig. 4.38. Frequency dependence of the short-circuit drain noise current spectrum (*left*) and of the short-circuit gate noise current spectrum (*right*, divided by f^2) for the 0.5-μm gate length MOS

case the hole contribution to noise is negligible. As expected, the simple analytical van der Ziel's model was found to be unable to satisfactorily reproduce even the DC behavior of the short-gate device considered; however, more accurate compact models can be successfully exploited to simulate the DC [119, 120] and (to a certain extent) the noise behavior of short-gate devices, e.g. see [91, 120]. Compact noise models for short-gate MOSFETs have been proposed that account for several effects, such as velocity satura-

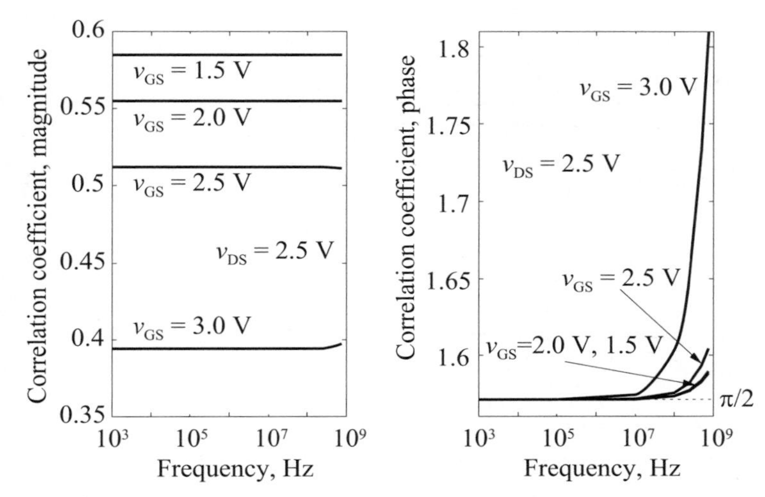

Fig. 4.39. Frequency dependence of the correlation coefficient of short-circuit drain and gate noise currents: amplitude (*left*) and phase (*right*, expressed in radians) for the 0.5-μm gate length MOS

tion [121–123] and electron heating in the channel [83, 123, 124]; nevertheless, despite the ability shown by these models to accurately simulate the experimental noise behavior of specific devices, the physical interpretation of some mechanisms (in particular, electron heating) is still controversial [91] and requires further investigation.

The possibility of successfully simulating the noise behavior of short-gate MOSFETs through compact models is supported by the fact that no strong qualitative difference exists between long-gate and short-gate devices with respect to the noise dependence on bias and frequency, see Fig. 4.40 for the short-circuit drain noise current spectrum as a function of the drain bias. However, a closer look into the spatial noise densities reveals the possible appearance of new features in the shape of the noise distribution. For example, a secondary peak can be present toward the drain end of the channel (Fig. 4.41); this peak becomes dominant in saturation, above all for high channel current, see Fig. 4.42. The presence of a secondary peak is related to the vector Green's function, rather than to the microscopic noise source, which is approximately constant (or even decreasing) towards the drain end of the channel. Although the underlying behavior of the vector Green's function is generally related to the onset of velocity saturation, further investigations suggest that this feature is also influenced by the detailed choice of the mobility model. Similar remarks can be made for the bias behavior of the short-circuit gate noise current spectrum and for the magnitude of the correlation coefficient, see Figs. 4.43 and 4.44. It can be noticed that the gate noise spectrum decreases with respect to the long-gate case and becomes almost independent of the gate bias, at least in saturation. On the other hand,

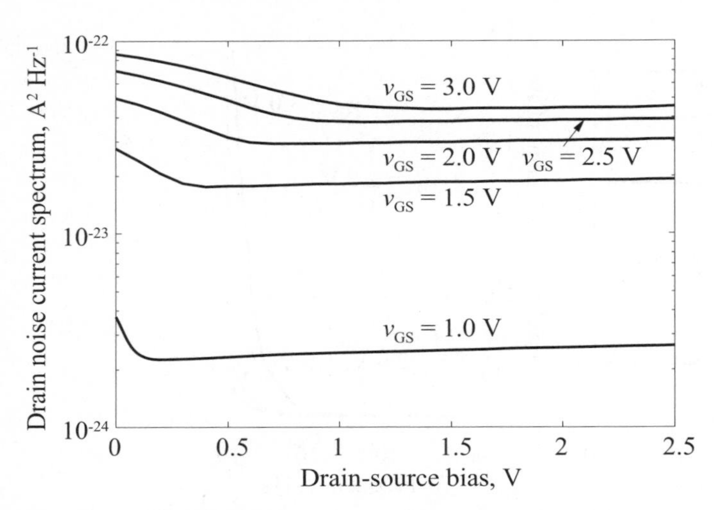

Fig. 4.40. Drain bias dependence of the short-circuit drain noise current spectrum for the 0.5-μm gate length MOS. The operating frequency is 1 kHz

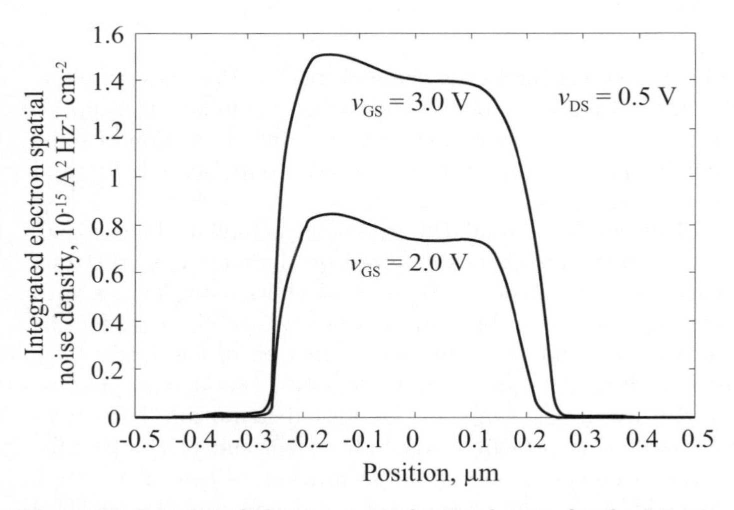

Fig. 4.41. Integrated electron spatial noise density for the 0.5-μm gate length MOS in the linear region. The operating frequency is 1 kHz

a greater spreading of the correlation coefficient takes place in the short-gate case. As a final remark on short-channel MOSFET noise modeling, it can be concluded that the increased complexity in the device physics, and the wide variety of available models for the transport parameters (including the driving force), suggest great care is needed in the interpretation of physics-based simulations. In particular, the present models for the transport parameters in the MOS channel are somewhat unsatisfactory for noise analysis, since they

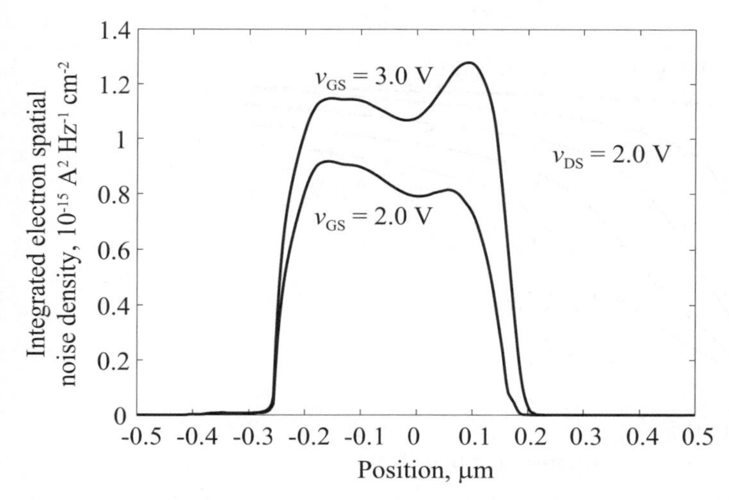

Fig. 4.42. Integrated electron spatial noise density for the 0.5-μm gate length MOS in saturation. The operating frequency is 1 kHz

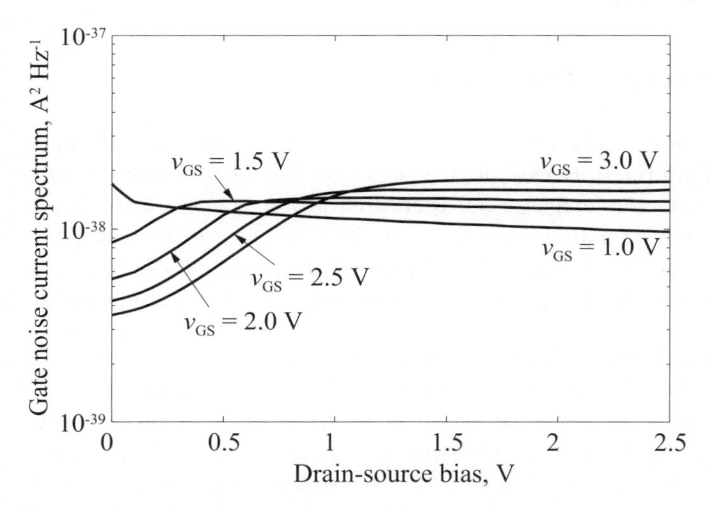

Fig. 4.43. Drain bias dependence of the short-circuit gate noise current spectrum for the 0.5-μm gate length MOS. The operating frequency is 1 kHz

are optimized, often on an empirical basis, on the DC characteristics only; improved models could be obtained, for example, through more fundamental simulations, like the full-band Monte Carlo technique.

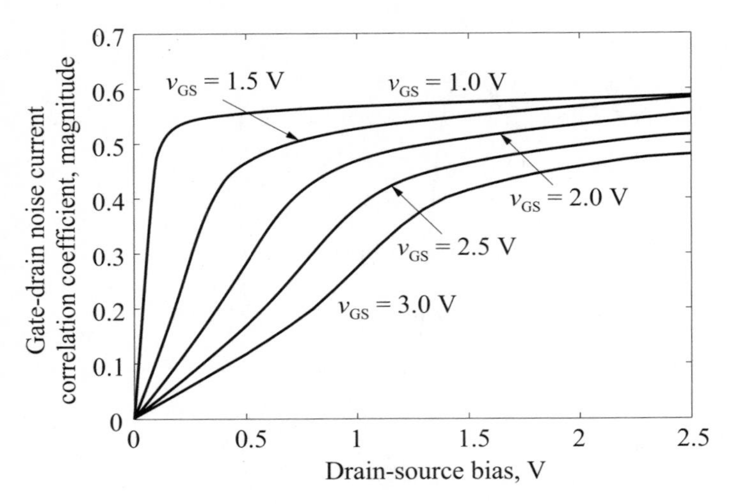

Fig. 4.44. Drain bias dependence of the magnitude of the correlation coefficient between the gate and drain noise currents for the 0.5-μm gate length MOS. The operating frequency is 1 kHz

4.4 Bipolar Junction Transistors

This section addresses the noise modeling of bipolar junction transistors. As a case study, a simple, intrinsic device structure has been analyzed and the short-circuit base and collector current fluctuation spectra have been computed both in low and high-injection conditions. The results obtained confirm the validity of the shot-like model presented in Sect. 2.3.4, but also point out the impact of high-injection effects on the collector current noise spectrum.

The structure analyzed is an intrinsic *npn* transistor with emitter doping $N_{DE} = 10^{18}$ cm^{-3}, base doping $N_{AB} = 10^{16}$ cm^{-3}, collector doping $N_{DC} = 10^{15}$ cm^{-3}. The emitter and collector regions are 2 μm long, whereas the base length is 1 μm. The doping profile is abrupt and constant mobility was used in all simulations. The GR model is direct, with minority-carrier lifetime of 20 μs in all regions.

The short-circuit base and collector current fluctuation spectra at 1 kHz and $v_{CE} = 3$ V are shown in Figs. 4.45 and 4.46, respectively, as a function of the base-emitter bias. Both spectra are almost white up to the device cutoff frequency, since the GR noise contribution is negligible with respect to diffusion noise in all operating conditions. Concerning the bias behavior of the base-current fluctuation spectrum, this is compared in Fig. 4.45 to the shot noise formula (2.136), where the base current is evaluated from the simulation. The agreement found is excellent both in low-and in high-injection conditions. Concerning the collector current fluctuation spectrum, Fig. 4.46 shows that the simple shot noise approach holds well in low injection,

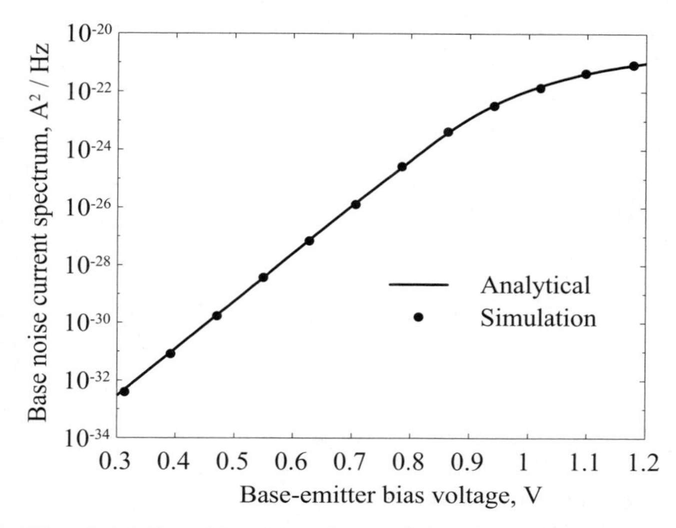

Fig. 4.45. Base bias dependence of the short-circuit base current noise spectrum at 1 kHz and $v_{\mathrm{CE}} = 3$ V

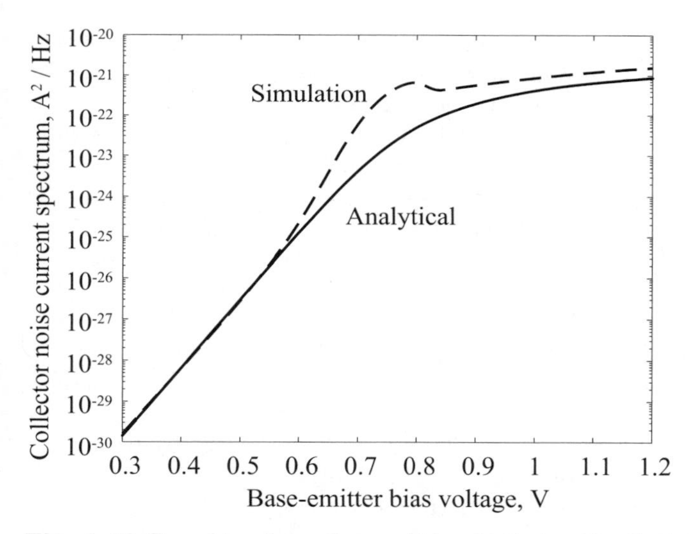

Fig. 4.46. Base bias dependence of the short-circuit collector current noise spectrum at 1 kHz and $v_{\mathrm{CE}} = 3$ V

but fails to yield the correct behavior in high-injection conditions. The local maximum that is clearly visible in the spectrum bias behavior is correlated to the onset of base pushout in the transistor. This is confirmed by the current gain β, shown in Fig. 4.47, which suddenly drops (around 0.75 V base-emitter bias) from a low-injection value of around 600, which is consistent with the analytical estimate of the parameter according to Shockley's diffusion theory.

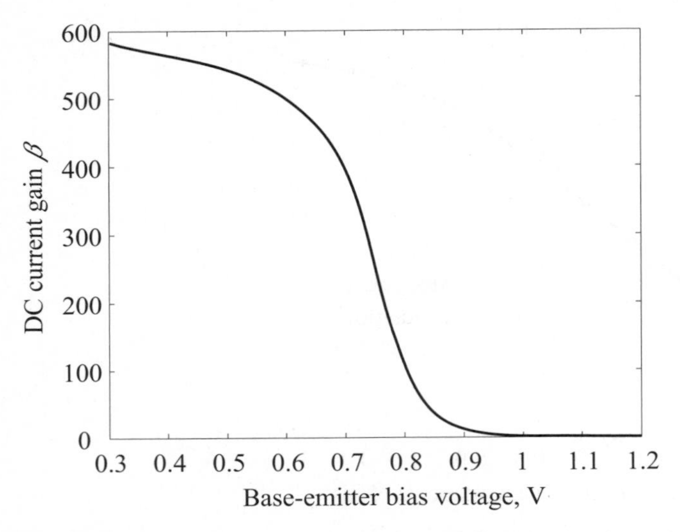

Fig. 4.47. Base bias dependence of the DC current gain

In order to achieve a better insight into the bias behavior of noise, the base and collector short-circuit current spatial noise densities (integrated across the device depth) have been computed. Figure 4.48 shows the electron (above) and the hole (below) contribution to the base noise density for different values of the base-emitter bias. Electron noise is dominant in the base region, while hole noise prevails in the emitter, as suggested by the interpretation of junction noise as originated by injected minority carriers in direct bias [76]. In high injection, the hole contribution in the base raises, again as expected from the junction noise behavior; moreover, a significant noise contribution is originated by minority and majority carriers injected into the collector because of the base pushout effects, clearly visible for $v_{\mathrm{BE}} = 0.84$ V. Nevertheless, this additional noise contribution caused by high injection remains small, thus preserving the overall shot-like behavior.

Figure 4.49 shows the electron (above) and the hole (below) contribution to the collector noise density for different values of the base-emitter bias. At low bias, electron noise is dominated by the electrons injected into the base region, whereas the hole noise is again maximum in the base, as expected from the quasi-neutral base ohmic behavior. In high injection, a large amount of base pushout is again visible, this time with a considerable impact on the total noise spectrum, which finally deviates, as already noted, from the simple shot-like law.

Figure 4.50 finally shows the total noise densities for the base (above) and collector (below) short-circuit noise currents. The base current noise clearly originates from both the emitter and the base regions; in high injection, the collector region contribution increases, but not enough to outweigh the

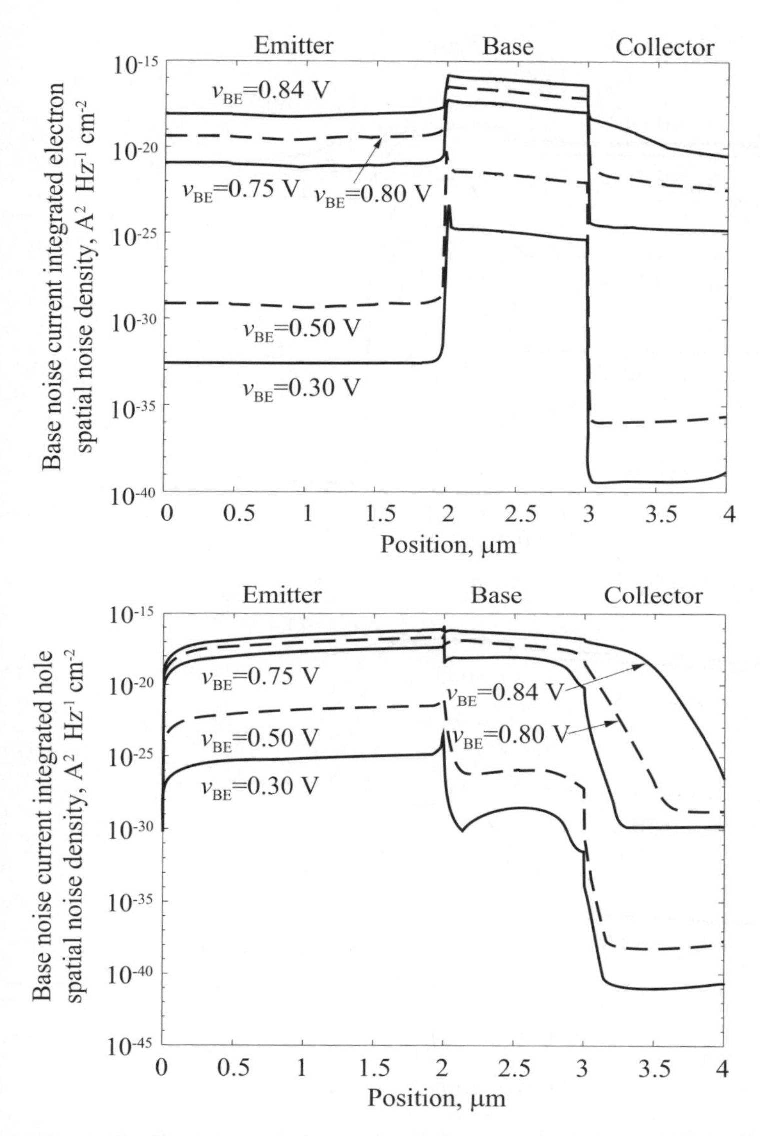

Fig. 4.48. Spatial dependence of the electron (*above*) and hole (*below*) base short-circuit noise current spatial noise density at 1 kHz and $v_{CE} = 3$ V

base region contribution. The collector current noise is dominated by the base region contribution in low injection, but the collector region becomes significant after the onset of base pushout.

Similar results can be obtained from the analysis of a more realistic implanted device, as discussed in [6]. It should be noted that the bipolar noise

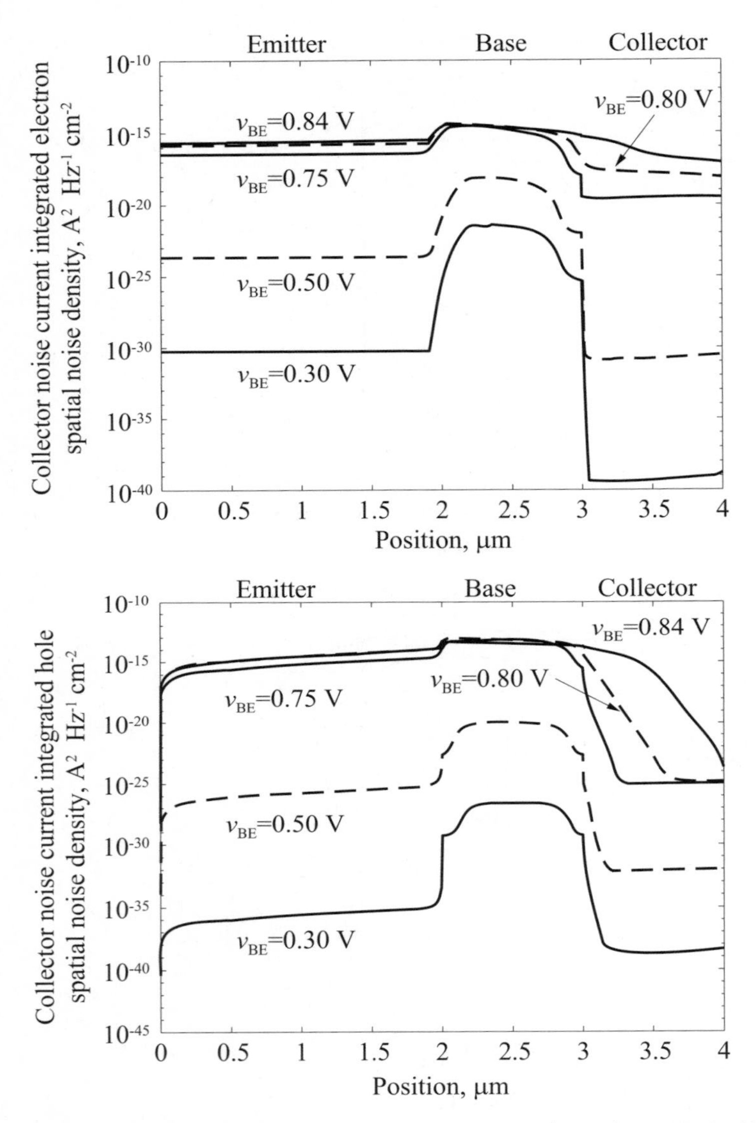

Fig. 4.49. Spatial dependence of the electron (*above*) and hole (*below*) collector short-circuit noise current spatial noise density at 1 kHz and $v_{CE} = 3$ V

simulation requires considerable attention in designing a proper discretization grid, which must be dense enough in the depletion regions, where a strong gradient is observed for both the Green's functions and the microscopic noise source, thus leading to a critical behavior for the spatial noise density. This point becomes even more significant near equilibrium, where an inadequate discretization can lead to results in disagreement with the Nyquist theorem. It

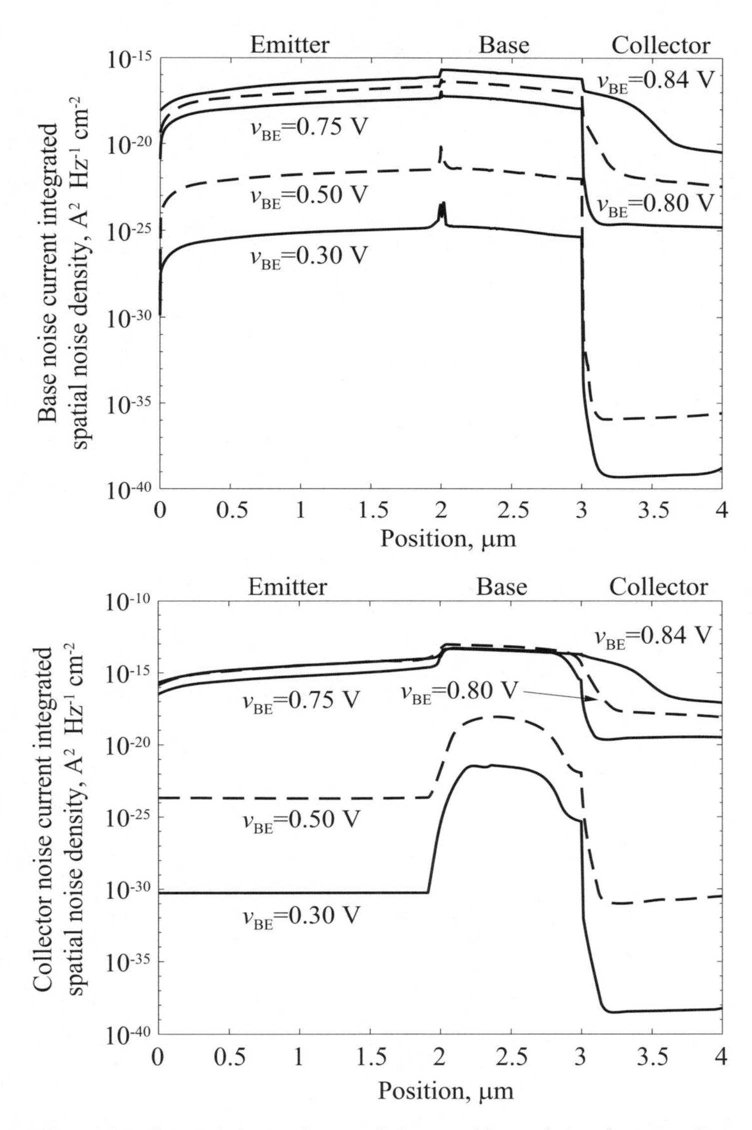

Fig. 4.50. Spatial dependence of the total base (*above*) and collector (*below*) short-circuit noise current spatial noise density at 1 kHz and $v_{\mathrm{CE}} = 3$ V

can be suggested that a first check on the quality of the discretization can be made by verifying that the small-signal device parameters satisfy reciprocity in or around equilibrium.

5. Noise in Large-Signal Operation

5.1 A System-Oriented Introduction

Large-signal (LS) periodic or quasi-periodic device operation is customarily defined as the operation regime characterized by electrical signals made of a superposition of *harmonic components*, often named *tones* as well. Such signals are large enough to require a complete, time-varying device analysis, meaning that the basic assumption of small-signal operation, i.e. that the time-dependent signals are a linear perturbation of the DC operating point, is no longer valid.

The large-signal regime can be either *forced* or *autonomous*. In autonomous operation, the time-varying signals are obtained as a result of an unstable device operating point, and therefore this case occurs for oscillators. If the time-varying signals represent the device response to an external large excitation, the operating regime is termed forced. From the standpoint of noise analysis, forced operation can be treated by means of linear perturbation analysis, in much the same way as discussed for noise in small-signal conditions, although the statistical properties of the stochastic processes describing noise are heavily influenced by the LS regime. On the other hand, autonomous systems are perturbed by noise sources in a very complex way, whose analysis is still a matter of open research; for details, see, for example, [125, 126]. For the sake of simplicity, we shall consider hereafter forced operation only.

In order to fix the ideas, let us consider first a simple, first-order dynamic nonlinear system with memoryless nonlinearity (see Fig. 5.1), meaning that

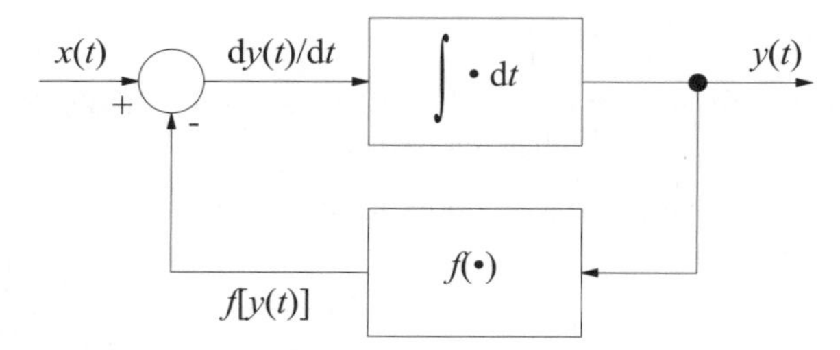

Fig. 5.1. Representation of a nonlinear system with memoryless nonlinearity

the nonlinear part of the system is an instantaneous function of its control variable $y(t)$:

$$\frac{\mathrm{d}y}{\mathrm{d}t} + f\left[y(t)\right] = x(t), \tag{5.1}$$

where $f(\cdot)$ is the nonlinear function and $x(t)$ is the large-signal forcing term, made of a superposition of harmonic components:

$$x(t) = \sum_{k=-N_{\mathrm{F}}}^{N_{\mathrm{F}}} X_k \mathrm{e}^{\mathrm{i}\omega_{\mathrm{i},k} t}, \tag{5.2}$$

where X_k is the *complex amplitude* of tone $\omega_{\mathrm{i},k}$. Notice that, hereafter, capital variables will always denote the frequency components of the corresponding time-domain quantities. The $2N_{\mathrm{F}} + 1$ complex spectral amplitudes correspond to $2N_{\mathrm{F}} + 1$ independent real numbers, since signal $x(t)$ is real, and therefore $X_{-k} = X_k^*$ for any k. This complex representation is used in the present formulation because it is well suited to the discussion, although from a practical standpoint, a direct (real) Fourier representation is more convenient [127], and the numerical implementation has to be made accordingly. According to the number of input tones N_{F}, several operating conditions are possible. In all cases, nonlinearity results in the generation of harmonics and intermodulation products, meaning that the output signal $y(t)$ includes spectral components at the input signal frequencies and at their harmonics, i.e. at their linear combinations with integer coefficients. Practically important cases are:

- the *single-tone* input ($N_{\mathrm{F}} = 1$), when the input signal is strictly periodic with (angular) frequency $\omega_{\mathrm{i},1}$. Device nonlinearity produces signals with spectrum:

$$\omega_k = k\omega_{\mathrm{i},1} \qquad k \text{ integer}; \tag{5.3}$$

- the *two-tone* input ($N_{\mathrm{F}} = 2$), when the input frequency components are $\omega_{\mathrm{i},1}$ and $\omega_{\mathrm{i},2}$, and the resulting spectrum is the set of all the linear combinations of the input frequencies with integer coefficients:

$$\omega_{k_1,k_2} = k_1\omega_{\mathrm{i},1} + k_2\omega_{\mathrm{i},2} \qquad k_1, k_2 \text{ integers}. \tag{5.4}$$

  If the input frequencies are *incommensurate*, i.e. their ratio is not a rational number, the operation regime is usually called *quasi-periodic*.

In both cases, the system response can therefore be expressed as a superposition of harmonic components:

$$y(t) = \sum_{k=-\infty}^{+\infty} Y_k \mathrm{e}^{\mathrm{i}\omega_k t}, \tag{5.5}$$

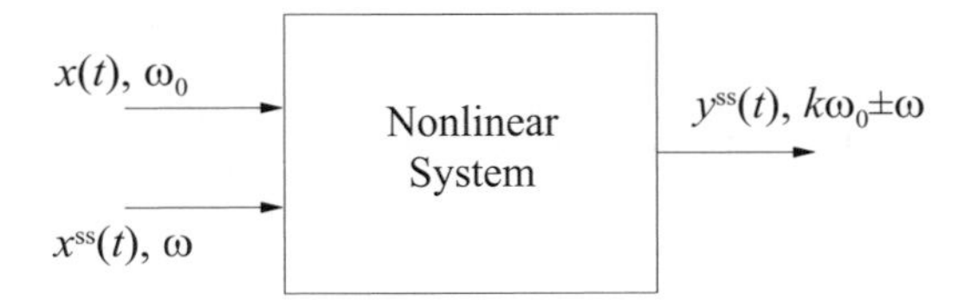

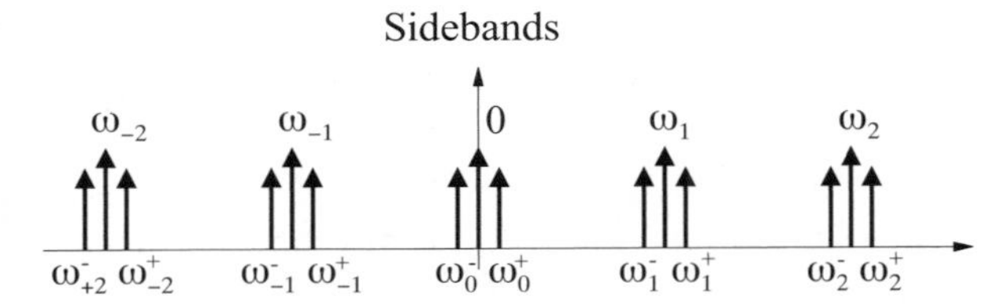

Fig. 5.2. Frequency conversion in a nonlinear system driven by a single tone LS excitation and a small-amplitude signal

where the summation index k must be interpreted as multidimensional, according to the number of input tones included in the excitation $x(t)$.

A practically important case of two-tone excitation occurs when one input signal is strong, the other weak. This can be interpretated as the superposition of a small-amplitude (weak) signal at frequency $\omega = \omega_{\mathrm{i},2}$ to the large-signal (strong) input at $\omega_0 = \omega_{\mathrm{i},1}$, i.e. the former acts as a linear perturbation of the large-signal steady state set up by the strong signal itself. In other words, system analysis is performed on a two-step basis: first, the LS operating point is calculated as a result of the strong signal only, leading to a spectral content ω_k according to (5.3). Then, the nonlinear system is linearized around the periodic steady state, and the response to the weak input signal is evaluated accordingly. This procedure, well-known to the field microwave circuit analysis, is termed small-signal large-signal (SSLS) analysis [74]. Concerning the spectral content, the output spectrum results as the set $\omega_k \pm \omega$, where $\omega_k = k\omega_0$ is the unperturbed LS spectrum, with the following interpretation: the input frequency ω is *converted* into an output set of spectral lines, usually referred to as *sidebands*, symmetrically placed with respect to each harmonic of the unperturbed LS regime (see Fig. 5.2). Due to linearity, the amplitude of the output sidebands is proportional to the amplitude of the (weak) input. This concept can be generalized if the weak input signal spectrum is already made of sidebands $\omega_k \pm \omega$; the same spectrum is found at the output, and the linear relationship between the input and output sideband amplitudes can be represented by means of the so-called *conversion matrix*. Notice that the conversion matrix is, in general, not diagonal, thus implying the occurrence of *frequency conversion* between each input and output sideband.

To formally derive the concept of frequency conversion within a system framework, let us consider a linear perturbation of the nonlinear system (5.1) driven in large-signal operation by a strong, periodic input signal. The sys-

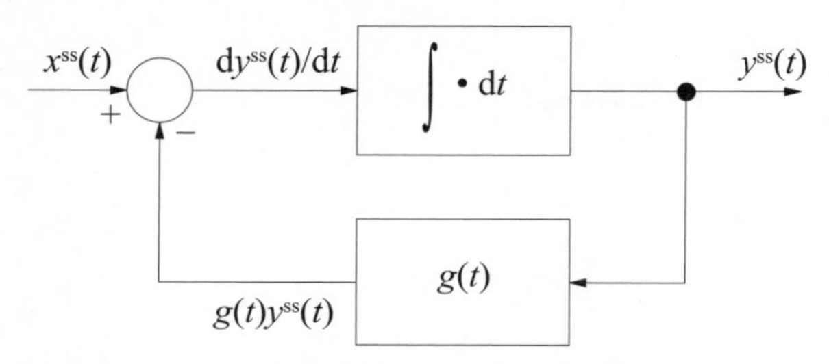

Fig. 5.3. Representation the nonlinear system in Fig. 5.1 linearized around the LS operating point

tem response can be obtained by SSLS analysis [74], i.e. through linearization around the LS solution (5.5), yielding a *linear periodically time-varying* (LPTV) system. A general theory of LPTV systems, based on the transfer function formalism, can be found, for example, in [128]. For the sake of our discussion, we confine the treatment to the simple case of a LPTV system defined by linearizing (5.1) (see Fig. 5.3):

$$\frac{dy^{ss}}{dt} + g(t)y^{ss}(t) = x^{ss}(t), \tag{5.6}$$

where $x^{ss}(t)$ is a forcing term (the input), and $g(t)$ is a periodic function corresponding to the memoryless part of the system:

$$g(t) = \left.\frac{df(\alpha)}{d\alpha}\right|_{\alpha=y(t)} = \sum_{k=-\infty}^{+\infty} G_k e^{i\omega_k t}, \tag{5.7}$$

where again $G_{-k} = G_k^*$ for any k, since $g(t)$ is a real function of time. According to SSLS theory, the input and output signal spectra are made of the set of *sideband* frequencies:

$$\omega_k^+ = \omega_k + \omega \qquad \omega_k^- = \omega_k - \omega, \tag{5.8}$$

where symbol + refers to the *upper* sideband, and symbol − to the *lower* sideband, displaced by an amount ω from the central frequency ω_k (see Fig. 5.2). For instance, the Fourier expansion of the output signal is:

$$y^{ss}(t) = \sum_{k=-\infty}^{+\infty} \left[Y_k^{ss,+} e^{i\omega_k^+ t} + Y_k^{ss,-} e^{i\omega_k^- t}\right], \tag{5.9}$$

where $Y_{-k}^{ss,-} = \left(Y_k^{ss,+}\right)^*$ and $Y_k^{ss,-} = \left(Y_{-k}^{ss,+}\right)^*$ for any integer k. These symmetry conditions allow us to express $y^{ss}(t)$ as the real part of the complex time-varying function:

$$y_c^{ss}(t) = \sum_{k=-\infty}^{+\infty} Y_k^{ss,+} e^{i\omega_k^+ t}, \tag{5.10}$$

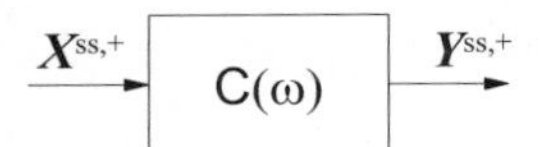

Fig. 5.4. System representation of the conversion matrix

thus making the set of coefficients $Y_k^{ss,+}$ sufficient to characterize the waveform in (5.9). Similar expressions hold for the input $x^{ss}(t)$.

Introducing into (5.6) the complex signals and the expansion (5.7), the sideband amplitudes $Y_k^{ss,+}$ are readily shown to satisfy the infinite algebraic linear system:

$$\mathrm{i}\omega_q^+ Y_q^{ss,+} + \sum_{n=-\infty}^{+\infty} G_{q-n} Y_n^{ss,+} = X_q^{ss,+}. \tag{5.11}$$

The inverse of the coefficient matrix of such an infinite set of equations is the *conversion matrix* [74] C of the LPTV linear system. If the LPTV system is not memoryless, the conversion matrix depends on the displacement frequency ω. By collecting the sideband amplitudes into infinite vectors, one has (see Fig. 5.4):

$$\boldsymbol{Y}^{ss,+} = \mathsf{C}(\omega) \cdot \boldsymbol{X}^{ss,+}, \tag{5.12}$$

where:

$$(\mathsf{C})_{i,j} = G_{i-j} + \mathrm{i}\omega_i^+ \delta_{i,j}. \tag{5.13}$$

In (5.13) δ is the Kronecker symbol. The conversion matrix approach (5.12) can be shown to apply to a generic LPTV system, e.g. see [74].

Conventionally, the conversion matrix relates the upper sideband complex amplitudes; however, a similar relationship holds for the lower sidebands. It can be readily shown that:

$$\boldsymbol{Y}^{ss,-} = \mathsf{C}^-(\omega) \cdot \boldsymbol{X}^{ss,-}, \tag{5.14}$$

where:

$$(\mathsf{C}^-)_{i,j} = G_{i-j} + \mathrm{i}\omega_i^- \delta_{i,j}. \tag{5.15}$$

In general, one has:

$$\mathsf{C}^-(\omega) = \mathsf{C}(-\omega). \tag{5.16}$$

The conversion matrix formalism is also the basis of LS noise analysis. From a physical standpoint, such an analysis implies two steps (see Fig. 5.5):

- As a first step, noise sources are amplitude-modulated by the periodic LS steady state [129]; this process is equivalent to the passage of the stationary

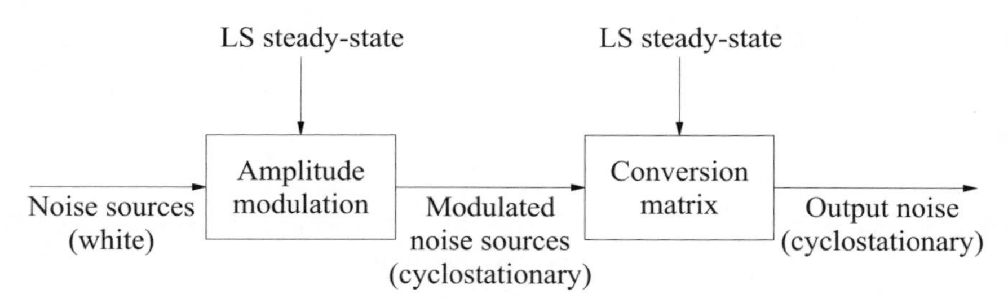

Fig. 5.5. Representation of the two-step procedure for noise analysis in forced large-signal operation

random process describing the noise sources in DC conditions described in Sect. 1.2 through a memoryless LPTV system of periodicity ω_0, thus implying noise frequency conversion[1] into the sideband set $\omega_k^{\pm}$. The frequency conversion process makes the corresponding stochastic processes *cyclostationary*, and therefore their correlation spectrum is completely defined by the *sideband correlation matrix* (SCM); for a discussion on this topic, see Sect. A.5.

- As a second step, the amplitude-modulated and frequency-converted (sideband) fluctuations can be interpreted as small-amplitude perturbations of the LS steady state; therefore, the resulting output fluctuations can be analyzed through the SSLS approach. This second frequency conversion through a LPTV system of the same periodicity ω_0 is described by a proper conversion matrix, and results again in a cyclostationary process (see [128] and Sect. A.5).

In both steps we need to evaluate the statistical properties of a cyclostationary random process filtered through a periodic time-varying linear system. By associating to a cyclostationary random process δx (δy) its spectral sideband representation $\delta \boldsymbol{X}$ ($\delta \boldsymbol{Y}$), see Sect. A.5, the transformation through an LPTV system can be formally expressed by means of the system conversion matrix C as:

$$\delta \boldsymbol{Y}(\omega) = \mathsf{C}(\omega) \cdot \delta \boldsymbol{X}(\omega). \tag{5.17}$$

By formally deriving the output SCM (see (A.84)) one directly obtains:

$$\mathsf{S}_{\delta y, \delta y} = \mathsf{C}(\omega) \cdot \mathsf{S}_{\delta x, \delta x} \cdot \mathsf{C}^{\dagger}(\omega). \tag{5.18}$$

For physics-based noise analysis, the conversion matrix is a generalization of the Green's function approach of Sect. 2.2, and therefore will be termed *conversion Green's function* (CGF).

[1] This topic will be addressed in Sect. 5.4.

5.2 Circuit-Oriented Large-Signal Noise Analysis

From a circuit standpoint, noise in large-signal conditions can be evaluated as the result of two analyses. First, the periodic, forced steady state of the circuit under large-signal excitation must be computed through a suitable technique. Then, small-amplitude random noise generators for each circuit element are introduced, linearly perturbing the periodic LS steady state, and the resulting induced noise voltages and currents are evaluated by means of a conversion matrix approach.

In order to efficiently perform the noiseless LS analysis, a frequency-domain technique, such as the harmonic balance (HB) method [74, 104, 130–132], has to be exploited. The basis of the HB technique is the dual frequency- and time-domain representation of periodic signals through a sampling approach [104].

According to the assumption of periodic LS steady state, each unknown (voltage or current) is made of a superposition of harmonics, and therefore can be expressed according to:

$$x(t) = \sum_{k=-N_L}^{N_L} X_k e^{i\omega_k t}. \tag{5.19}$$

The harmonic decomposition has been truncated to a finite number of harmonics N_L; this approximation is justified since, in practical circuits characterized by a global low-pass response, the spectral amplitude is a decreasing function of the harmonic order k for large k values.

Let us now define a set of $2N_L + 1$ time samples t_k distributed in the fundamental period $[0, T[$, where $T = 1/(2\pi\omega_0)$, and let us consider the collection $\boldsymbol{x}$ of the sampled variable values $x(t_k)$, so that (5.19) yields:

$$\boldsymbol{x} = \mathsf{T}^{-1} \cdot \boldsymbol{X}, \tag{5.20a}$$

$$\boldsymbol{X} = \mathsf{T} \cdot \boldsymbol{x}, \tag{5.20b}$$

where matrix T^{-1} is given by:

$$\mathsf{T}^{-1} = \begin{bmatrix} e^{i\omega_{-N_L} t_0} & \dots & e^{i\omega_0 t_0} & \dots & e^{i\omega_{N_L} t_0} \\ \vdots & & \vdots & & \vdots \\ e^{i\omega_{-N_L} t_{N_L}} & \dots & e^{i\omega_0 t_{N_L}} & \dots & e^{i\omega_{N_L} t_{N_L}} \\ \vdots & & \vdots & & \vdots \\ e^{i\omega_{-N_L} t_{2N_L}} & \dots & e^{i\omega_0 t_{2N_L}} & \dots & e^{i\omega_{N_L} t_{2N_L}} \end{bmatrix}. \tag{5.21}$$

In the strictly periodic case, the optimum choice of the time samples is uniform, since this makes the condition number of matrix T as high as possible [104]. Moreover, in this case (5.20a) and (5.20b) can be efficiently implemented according to a discrete Fourier transform algorithm. In the quasi-periodic case, several techniques are available to give an optimum time- and frequency-domain signal representation through sampling; the so-called frequency remapping technique has been proven to be superior to other techniques (such as the random sampling approach), since it allows efficient FFT algorithms to be exploited [127]. Another efficient method, not requiring frequency remapping, is the multidimensional DFT [133].

The time derivative of $x(t)$ defined as in (5.19) immediately yields the spectral components of $\dot{x}(t)$, denoted as $\dot{\boldsymbol{X}}$:

$$\dot{\boldsymbol{X}} = \mathsf{D} \cdot \boldsymbol{X}, \tag{5.22}$$

where:

$$\mathsf{D} = \mathrm{diag}\{\mathrm{i}\omega_k\} \tag{5.23}$$

is a diagonal matrix of dimension $2N_{\mathrm{L}} + 1$. Higher-order time derivatives can be treated similarly, through repeated multiplication of the diagonal matrix D.

The above formulation also enables us to evaluate in a straightforward way the frequency-domain representation for the memoryless nonlinear function $w[x(t)]$, in terms of $\boldsymbol{X}$. In fact, we define as $\boldsymbol{w}$ the collection of time-domain samples of $w[x(t)]$, i.e.

$$(\boldsymbol{w})_i = w[x(t_i)], \qquad i = 1, \ldots, 2N_L + 1, \tag{5.24}$$

for the sake of simplicity we will synthetically express (5.24) as $\boldsymbol{w} = \boldsymbol{w}[\boldsymbol{x}]$; it immediately follows:

$$\boldsymbol{w} = \boldsymbol{w}[\mathsf{T}^{-1} \cdot \boldsymbol{X}] \tag{5.25}$$

and therefore:

$$\boldsymbol{W} = \mathsf{T} \cdot \boldsymbol{w}[\mathsf{T}^{-1} \cdot \boldsymbol{X}]. \tag{5.26}$$

5.2.1 Frequency-Domain Large-Signal Analysis of Circuits

On the basis of the above discussion, the LS analysis of circuits including both linear and nonlinear (memoryless or reactive) elements under periodic excitation can be formulated directly by expressing in the frequency domain both the constitutive relationships of the elements and the topological Kirchhoff current and voltage laws [63]. For the sake of simplicity, we only consider a network including one-ports and controlled sources; the general case of multiports can be derived at the cost of an increased complexity in the notation.

Consider a network with N_{n} nodes and N_{s} sides. The voltages and currents for side k will be denoted as $v_k(t)$, $i_k(t)$ respectively; the sampled time-domain and frequency-domain representations will be denoted, respectively, as $\boldsymbol{v}_k$, $\boldsymbol{V}_k$ and $\boldsymbol{i}_k$, $\boldsymbol{I}_k$; one has:

$$\boldsymbol{V}_k = \mathsf{T} \cdot \boldsymbol{v}_k \tag{5.27a}$$
$$\boldsymbol{I}_k = \mathsf{T} \cdot \boldsymbol{i}_k. \tag{5.27b}$$

Let us discuss first the constitutive relationships for linear elements:

- **Linear resistor.** The constitutive relationship in the time domain is $v_k(t) = R_k i_k(t)$ where R_k is the resistor resistance. It follows immediately that $\boldsymbol{v}_k = R_k \boldsymbol{i}_k = \mathsf{R}_k \cdot \boldsymbol{i}_k$, where $\mathsf{R}_k = R_k \mathsf{I}$. Transforming into the frequency domain yields:

$$\boldsymbol{V}_k = \mathsf{R}_k \cdot \boldsymbol{I}_k. \tag{5.28}$$

- **Linear capacitor.** The constitutive relationship in the time domain can be expressed in terms of the capacitor charge $q_k(t)$ as $q_k(t) = C_k v_k(t)$ where C_k is the capacitance; thus $\boldsymbol{q}_k = C_k \boldsymbol{v}_k = \mathsf{C}_k \cdot \boldsymbol{v}_k$, where $\mathsf{C}_k = C_k \mathsf{I}$. Transforming into the frequency domain and taking into account (5.22) applied to $i_k(t) = \dot{q}_k(t)$ one obtains:

$$\boldsymbol{I}_k = \mathsf{D} \cdot \mathsf{C}_k \cdot \boldsymbol{V}_k. \tag{5.29}$$

- **Linear inductor.** The constitutive relationship in the time domain is expressed in terms of the inductor flux $\phi_k(t)$ as $\phi_k(t) = L_k i_k(t)$ where L_k is the inductance; thus $\boldsymbol{\phi}_k = L_k \boldsymbol{i}_k = \mathsf{L}_k \cdot \boldsymbol{i}_k$, where $\mathsf{L}_k = L_k \mathsf{I}$. Transforming into the frequency domain and taking into account (5.22) applied to $v_k(t) = \dot{\phi}_k(t)$ yields:

$$\boldsymbol{V}_k = \mathsf{D} \cdot \mathsf{L}_k \cdot \boldsymbol{I}_k. \tag{5.30}$$

Nonlinear elements can be described by combining the representations for a memoryless nonlinear function (5.25) and (5.26) and for the time derivative (5.22); one has:

- **Nonlinear resistor.** The constitutive relationship in the time domain is $v_k(t) = f_{R_k}[i_k(t)]$, where f_{R_k} is the nonlinear resistor characteristic curve. Applying (5.25) and (5.26) one immediately has $\boldsymbol{v}_k = \boldsymbol{f}_{R_k}(\boldsymbol{i}_k)$ and, in the frequency domain:

$$\boldsymbol{V}_k = \mathsf{T} \cdot \boldsymbol{f}_{R_k}(\mathsf{T}^{-1} \cdot \boldsymbol{I}_k). \tag{5.31}$$

- **Nonlinear capacitor.** In a similar way one has, in the time domain: $q_k(t) = f_{C_k}[v_k(t)]$, where f_{C_k} is the nonlinear capacitor characteristic curve. Applying (5.25), (5.26) and (5.22) one has:

$$\boldsymbol{I}_k = \mathsf{D} \cdot \mathsf{T} \cdot \boldsymbol{f}_{C_k}(\mathsf{T}^{-1} \cdot \boldsymbol{V}_k). \tag{5.32}$$

- **Nonlinear inductor.** Starting from the time-domain inductor characteristic $\phi_k(t) = f_{L_k}[i_k(t)]$ the frequency domain consitutitive relationship can be derived as:

$$\boldsymbol{V}_k = \mathsf{D} \cdot \mathsf{T} \cdot \boldsymbol{f}_{L_k}(\mathsf{T}^{-1} \cdot \boldsymbol{I}_k). \tag{5.33}$$

Controlled sources can be treated in a similar way, exploiting (5.26); for instance, a voltage-controlled current source with time-domain relationship $i_k(t) = f_{g_{\mathrm{m}k}}[v_l(t)]$, where v_l is the controlling voltage, can be expresssed in the frequency domain as:

$$\boldsymbol{I}_k = \mathsf{T} \cdot \boldsymbol{f}_{g_{\mathrm{m}k}}(\mathsf{T}^{-1} \cdot \boldsymbol{V}_l). \tag{5.34}$$

Finally, independent (voltage or current) sources are characterized by assigned values of their frequency components as:

$$\boldsymbol{I}_k = \boldsymbol{S}_{i,k}, \tag{5.35}$$

$$\boldsymbol{V}_k = \boldsymbol{S}_{v,k}, \tag{5.36}$$

where $\boldsymbol{S}_{i,k}$, $\boldsymbol{S}_{v,k}$ are known vectors.

The treatment of the topological relationships (Kirchhoff current and voltage laws) is trivial, since in the time domain they read:

$$\sum_k i_k(t) = 0, \tag{5.37}$$

where the summation is extended to all currents entering node n, and:

$$\sum_l v_l(t) = 0, \tag{5.38}$$

where the summation is extended to all voltages belonging to a closed path within the network; direct transformation yields:

$$\sum_k \boldsymbol{I}_k = \mathbf{0}, \tag{5.39a}$$

$$\sum_l \boldsymbol{V}_l = \mathbf{0}. \tag{5.39b}$$

Taking into account that $N_\mathrm{n} - 1$ linearly independent Kirchhoff current laws and $N_\mathrm{s} - N_\mathrm{n} + 1$ linearly independent Kirchhoff voltage laws can be written, one obtains N_s independent topological vector equations equivalent to $N_\mathrm{s}(2N_\mathrm{L} + 1)$ scalar equations. From the constitutive equations, one obtains in total N_s vector equations globally equivalent to $N_\mathrm{s}(2N_\mathrm{L} + 1)$ scalar equations. The total number of independent scalar relationships is therefore $2N_\mathrm{s}(2N_\mathrm{L} + 1)$, equal to the number of scalar unknowns associated to side voltages and currents; therefore the problem is well posed and its solution requires the numerical solution of a nonlinear algebraic system. More efficient (e.g. nodal) formulations, allowing the number of unknowns to be reduced, are discussed in [63]; for further details on the application to the HB analysis see, for example, [104].

5.2.2 Small-Signal Large-Signal Analysis of Circuits

Once the periodic steady state has been evaluated, small-signal large-signal (SSLS) analysis can be exploited in order to estimate the network response to a set of periodic, small-amplitude sources that linearly perturb the periodic steady state. Since the SSLS theory is linear, the superposition principle holds, and the analysis can be confined to the case of a periodic perturbing source of arbitrary angular frequency ω. The network solution in terms of side voltages and currents can now be generally expressed as:

$$v_k(t) + v_k^{\mathrm{ss}}(t), \tag{5.40a}$$
$$i_k(t) + i_k^{\mathrm{ss}}(t), \tag{5.40b}$$

where $v_k^{\mathrm{ss}}(t)$ and $i_k^{\mathrm{ss}}(t)$ are the SSLS response. A suitable frequency-domain representation for the SSLS signals, accounting for their sideband spectrum, has been already introduced in (5.10). In the present case, the infinite spectral representation of the complex signals associated to voltages and currents has to be truncated as:

$$v_{\mathrm{c}k}^{\mathrm{ss}}(t) = \sum_{l=-N_{\mathrm{S}}}^{+N_{\mathrm{S}}} V_{l,k}^{\mathrm{ss},+} \mathrm{e}^{\mathrm{i}\omega_l^+ t}, \tag{5.41}$$

$$i_{\mathrm{c}k}^{\mathrm{ss}}(t) = \sum_{l=-N_{\mathrm{S}}}^{+N_{\mathrm{S}}} I_{l,k}^{\mathrm{ss},+} \mathrm{e}^{\mathrm{i}\omega_l^+ t}. \tag{5.42}$$

The truncation order N_{S} is related to the number of harmonics, N_{L}, exploited for the steady state problem. To take into account all possible frequency conversions, one must have $2N_{\mathrm{S}} = N_{\mathrm{L}}$ [74]. The complex sideband amplitudes can be collected into complex column vectors, ordered from $-N_{\mathrm{S}}$ to N_{S}, which will be denoted as $\boldsymbol{V}_k^{\mathrm{ss},+}$ and $\boldsymbol{I}_k^{\mathrm{ss},+}$, respectively, for the voltage and current of the k-th side.

The constitutive and topological relationships of a newtork in SSLS operation can be completely expressed in terms of the upper sideband amplitudes defined above and of proper conversion matrices. Since SSLS operation is linear, we can immediately extend a linear or linearized relationship holding between real SSLS signals to their complex representation. Following the discussion of the HB LS network analysis, let us deal first with the SSLS constitutive relationships for linear elements:

- **Linear resistor.** Since the element is linear, the time-domain constitutive relationship for the small-signal perturbation is simply $v_k^{\mathrm{ss}}(t) = R_k i_k^{\mathrm{ss}}(t)$, where R_k is the resistor resistance, or, in terms of the complex signals, $v_{\mathrm{c}k}^{\mathrm{ss}}(t) = R_k i_{\mathrm{c}k}^{\mathrm{ss}}(t)$. It follows immediately that $\boldsymbol{V}_k^{\mathrm{ss},+} = R_k \boldsymbol{I}_k^{\mathrm{ss},+} = \mathsf{R}_k^{\mathrm{ss}} \cdot \boldsymbol{I}_k^{\mathrm{ss},+}$, where $\mathsf{R}_k^{\mathrm{ss}} = R_k \mathsf{I}$ is a diagonal matrix of dimension $2N_{\mathrm{S}} + 1$.

- **Linear capacitor.** The constitutive relationship in the time domain can be expressed in terms of the SSLS capacitor (complex) charge $q_{ck}^{ss}(t)$ as $q_{ck}^{ss}(t) = C_k v_{ck}^{ss}(t)$, where C_k is the capacitance; thus $\boldsymbol{Q}_k^{ss,+} = C_k \boldsymbol{V}_k^{ss,+} = \mathsf{C}_k^{ss} \cdot \boldsymbol{V}_k^{ss,+}$, where $\mathsf{C}_k^{ss} = C_k \mathsf{I}$ is a diagonal matrix of dimension $2N_S + 1$. Taking into account that $i_{ck}^{ss}(t) = \dot{q}_{ck}^{ss}(t)$ one immediately has $\boldsymbol{I}_k^{ss,+} = \mathsf{D}^{ss} \cdot \mathsf{C}_k^{ss} \cdot \boldsymbol{V}_k^{ss,+}$, where $\mathsf{D}^{ss} = \text{diag}\{i\omega_k^+\}$.
- **Linear inductor.** By duality, the linear constitutive relationship of the linear inductor results as $\boldsymbol{V}_k^{ss,+} = \mathsf{D}^{ss} \cdot \mathsf{L}_k^{ss} \cdot \boldsymbol{I}_k^{ss,+}$, where $\mathsf{L}_k^{ss} = L_k \mathsf{I}$ is a diagonal matrix of dimension $2N_S + 1$.

Nonlinear elements can be characterized by linearization around the instantaneous LS working point; one has:

- **Nonlinear resistor.** The constitutive relationship in the time domain can be expressed as:

$$v_k(t) + v_k^{ss}(t) = f_{R_k}[i_k(t) + i_k^{ss}(t)] \approx f_{R_k}[i_k(t)] + \left.\frac{\mathrm{d}f_{R_k}}{\mathrm{d}i_k}\right|_{i_k(t)} i_k^{ss}(t), \quad (5.43)$$

which immediately yields the SSLS relationship between complex signals:

$$v_{ck}^{ss}(t) = \left.\frac{\mathrm{d}f_{R_k}}{\mathrm{d}i_k}\right|_{i_k(t)} i_{ck}^{ss}(t) = r_k(t) i_{ck}^{ss}(t). \quad (5.44)$$

By exploiting the formalism introduced in Sect. 5.1 the SSLS constitutive relationship can be written as:

$$\boldsymbol{V}_k^{ss,+} = \mathsf{R}_k^{ss} \cdot \boldsymbol{I}_k^{ss,+}. \quad (5.45)$$

The resistance conversion matrix R_k^{ss} in (5.45) is defined as:

$$\left(\mathsf{R}_k^{ss}\right)_{i,j} = R_{i-j,k}, \quad (5.46)$$

where $R_{l,k}$ is the l-th complex amplitude of the Fourier expansion of the periodic function $r_k(t)$:

$$r_k(t) = \sum_{l=-\infty}^{\infty} R_{l,k} e^{i\omega_l t}. \quad (5.47)$$

In practice, the resistance conversion matrix has to be truncated to dimension $(2N_S + 1) \times (2N_S + 1)$.

- **Nonlinear capacitor.** In a similar way one has, in the time domain:

$$q_k(t) + q_k^{ss}(t) = f_{C_k}[v_k(t) + v_k^{ss}(t)] \approx f_{C_k}[v_k(t)] + \left.\frac{\mathrm{d}f_{C_k}}{\mathrm{d}v_k}\right|_{v_k(t)} v_k^{ss}(t), \quad (5.48)$$

leading to the SSLS relationship between complex signals:

$$q_{ck}^{\rm ss}(t) = \left.\frac{{\rm d}f_{C_k}}{{\rm d}v_k}\right|_{v_k(t)} v_{ck}^{\rm ss}(t) = c_k(t) v_{ck}^{\rm ss}(t). \tag{5.49}$$

The charge-voltage SSLS constitutive relationship can be written as:

$$\boldsymbol{Q}_k^{\rm ss,+} = \mathsf{C}_k^{\rm ss} \cdot \boldsymbol{V}_k^{\rm ss,+}, \tag{5.50}$$

where the capacitance conversion matrix $\mathsf{C}_k^{\rm ss}$ in (5.50) is defined as:

$$(\mathsf{C}_k^{\rm ss})_{i,j} = C_{i-j,k}, \tag{5.51}$$

$C_{l,k}$ being the l-th complex amplitude of the Fourier expansion of the periodic function $c_k(t)$:

$$c_k(t) = \sum_{l=-\infty}^{\infty} C_{l,k} {\rm e}^{{\rm i}\omega_l t}. \tag{5.52}$$

From (5.50) the constitutive relationship in terms of voltage and current upper sideband amplitudes immediately results as:

$$\boldsymbol{I}_k^{\rm ss,+} = \mathsf{D}^{\rm ss} \cdot \mathsf{C}_k^{\rm ss} \cdot \boldsymbol{V}_k^{\rm ss,+}, \tag{5.53}$$

where the capacitance conversion matrix is truncated to dimension $(2N_{\rm S}+1) \times (2N_{\rm S}+1)$.

- **Nonlinear inductor.** By duality, one has:

$$\boldsymbol{V}_k^{\rm ss,+} = \mathsf{D}^{\rm ss} \cdot \mathsf{L}_k^{\rm ss} \cdot \boldsymbol{I}_k^{\rm ss,+}, \tag{5.54}$$

where the inductance conversion matrix $\mathsf{L}_k^{\rm ss}$ in (5.54) is defined as:

$$(\mathsf{L}_k^{\rm ss})_{i,j} = L_{i-j,k}, \tag{5.55}$$

$L_{l,k}$ being the l-th complex amplitude of the Fourier expansion of the periodic function:

$$l_k(t) = \left.\frac{{\rm d}f_{L_k}}{{\rm d}v_k}\right|_{v_k(t)} = \sum_{l=-\infty}^{\infty} L_{l,k} {\rm e}^{{\rm i}\omega_l t}. \tag{5.56}$$

Controlled sources can be treated similarly; for instance, a voltage-controlled current source with time-domain relationship $i_k(t) = f_{g_{{\rm m}k}}[v_l(t)]$, where v_l is the controlling voltage, can be expresssed in the SSLS regime as:

$$i_k(t) + i_k^{\rm ss}(t) = f_{g_{{\rm m}k}}[v_l(t) + v_l^{\rm ss}(t)] \approx f_{g_{{\rm m}k}}[v_l(t)] + \left.\frac{{\rm d}f_{g_{{\rm m}k}}}{{\rm d}v_l}\right|_{v_l(t)} v_l^{\rm ss}(t), \tag{5.57}$$

which immediately yields the SSLS relationship between complex signals:

$$i_{ck}^{ss}(t) = \left.\frac{\mathrm{d}f_{g_{mk}}}{\mathrm{d}v_l}\right|_{v_l(t)} v_{cl}^{ss}(t) = g_{mk}(t)v_{cl}^{ss}(t). \tag{5.58}$$

The SSLS constitutive relationship results as:

$$\boldsymbol{I}_k^{ss,+} = \mathsf{G}_{mk}^{ss} \cdot \boldsymbol{V}_l^{ss,+}. \tag{5.59}$$

The transconductance conversion matrix G_{mk}^{ss} in (5.59) is defined as:

$$(\mathsf{G}_{mk}^{ss})_{i,j} = G_{m,i-j,k}, \tag{5.60}$$

where $G_{m,l,k}$ is the l-th complex amplitude of the Fourier expansion of the periodic function $g_{m,k}(t)$:

$$g_{m,k}(t) = \sum_{l=-\infty}^{\infty} G_{m,l,k} e^{i\omega_l t}. \tag{5.61}$$

Finally, independent (voltage or current) SSLS sources are characterized by assigned values of their sideband components as:

$$\boldsymbol{I}_k^{ss,+} = \boldsymbol{S}_{i,k}^{ss,+}, \tag{5.62}$$

$$\boldsymbol{V}_k^{ss,+} = \boldsymbol{S}_{v,k}^{ss,+}, \tag{5.63}$$

where $\boldsymbol{S}_{i,k}^{ss,+}$, $\boldsymbol{S}_{v,k}^{ss,+}$ are known vectors.

Topological relationships (Kirchhoff current and voltage laws), being linear, can be directly cast into the form:

$$\sum_k i_{ck}^{ss}(t) = 0, \tag{5.64}$$

where the summation is extended to all currents entering node n, and:

$$\sum_l v_{cl}^{ss}(t) = 0, \tag{5.65}$$

where the summation is extended to all voltages belonging to a closed path within the network; therefore one has:

$$\sum_k \boldsymbol{I}_k^{ss,+} = 0, \tag{5.66a}$$

$$\sum_l \boldsymbol{V}_l^{ss,+} = 0. \tag{5.66b}$$

In total, N_s independent topological vector equations can be derived; these are equivalent to $N_s(2N_S + 1)$ scalar equations. From the constitutive relationships, one obtains in total N_s vector equations globally equivalent to $N_s(2N_S + 1)$ scalar equations; the total number of independent scalar relationships is therefore $2N_s(2N_S + 1)$, equal to the number of scalar unknowns associated to the upper sideband complex amplitudes of side voltages and currents; this enables us to assess the well-posedness of the SSLS problem.

Since the SSLS problem is linear, the constitutive relationship of a subnetwork, together with the relevant topological equations, can be combined to provide a unified model at the subnetwork ports. Consider an N-port and define the arrays collecting all the complex sideband amplitudes of the port currents and voltages:

$$\boldsymbol{I}^{\text{ss},+} = \begin{bmatrix} \boldsymbol{I}_1^{\text{ss},+} \\ \vdots \\ \boldsymbol{I}_k^{\text{ss},+} \\ \vdots \\ \boldsymbol{I}_n^{\text{ss},+} \end{bmatrix}, \qquad \boldsymbol{V}^{\text{ss},+} = \begin{bmatrix} \boldsymbol{V}_1^{\text{ss},+} \\ \vdots \\ \boldsymbol{V}_k^{\text{ss},+} \\ \vdots \\ \boldsymbol{V}_n^{\text{ss},+} \end{bmatrix}. \tag{5.67}$$

The sideband amplitudes of port voltages $\boldsymbol{V}^{\text{ss},+}$ and currents $\boldsymbol{I}^{\text{ss},+}$ will be related by the Norton representation:

$$\boldsymbol{I}^{\text{ss},+} = \mathsf{Y}(\omega) \cdot \boldsymbol{V}^{\text{ss},+} + \boldsymbol{I}_0^{\text{ss},+}, \tag{5.68}$$

where Y is the admittance conversion matrix, $\boldsymbol{I}_0^{\text{ss},+}$ the sideband short-circuit currents. Similarly, one has the Thévenin representation:

$$\boldsymbol{V}^{\text{ss},+} = \mathsf{Z}(\omega) \cdot \boldsymbol{I}^{\text{ss},+} + \boldsymbol{V}_0^{\text{ss},+}, \tag{5.69}$$

where Z is the impedance conversion matrix, $\boldsymbol{V}_0^{\text{ss},+}$ the sideband open-circuit voltages. Notice that the impedance and admittance CMs depend on the sideband angular frequency ω. The above representation is often associated with a $N \times (2N_S + 1)$ port, referred to as a multi-frequency network, in which a port is introduced for each physical port and sideband, see Fig. 5.6 for the case $N = 2$.

5.2.3 Large-Signal Noise Analysis of Circuits

In small-signal operation, the current or voltage fluctuations of a generic N-port can exhibit, in practice, one of the following features:

1. fluctuations are described by white, stationary processes whose power spectrum is independent of the DC working point; this happens, for example, for a linear resistor (thermal noise);
2. fluctuations are described by white, stationary processes whose power spectrum depends on the DC working point; this happens, for example, for a diode (shot noise);

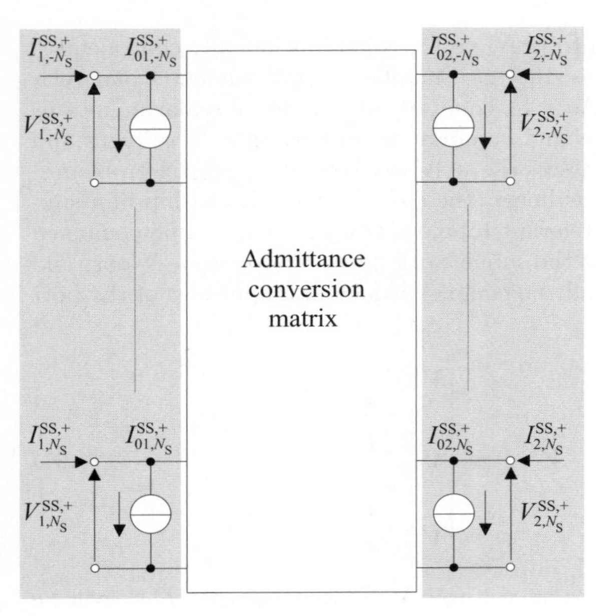

Fig. 5.6. Multi-frequency admittance representation of a 2-port

3. fluctuations are described by colored, stationary processes whose power spectrum depends on the DC working point; this happens, for example, for GR and $1/f$ noise in a resistor.

The above cases lead to a different treatment whenever the corresponding element is operated in LS conditions. In the first case, the statistical properties of fluctuations are not affected by the instantaneous variations of the working point. In the second case, fluctuations are amplitude-modulated by the instantaneous working point; if the LS regime is periodic, the resulting process is cyclostationary (see Sect. A.5) and can be described by a proper SCM, thus implying that only sideband fluctuations having the same angular frequency deviation ω with respect to the sideband central frequency are correlated, see (A.83). The principle of instantaneous modulation of white noise sources is thoroughly discussed, in a formal way, in [128]; such a modulation can be, as already noticed, interpreted as the effect of a linear, time-varying system whose input is white noise, the output modulated (sideband) noise.

The third case is still controversial, see the discussion in [128], although some generally accepted solutions are available. Basically, GR and $1/f$ noise processes possess intrinsic low-pass features, besides being dependent, in small-signal operation, from the DC working point. For what concerns GR noise, as discussed in Sect. 1.2.2, the fundamental physics-based approach clearly suggests that the microscopic noise source is white, while the colored (Lorentzian) behavior is generated by the model response; therefore, in large-signal operation the white source is first modulated (giving rise to modulated or sideband noise) and then low-pass filtered. Since the cutoff frequency of GR processes is low, such a filtering practically cancels out all sidebands, apart from the ones around DC. However, the propagation of such local fluctuations to the device terminals may again involve frequency conversion through the CGF, as described in Sect. 5.4. In this case, terminal GR and $1/f$ fluctuations exhibit a full sideband noise spectrum, and, therefore, are modulated by the instantaneous working point of the device. This can be expected to happen, for example, in a device including junctions, like a diode. Conversely, simpler structures (e.g. a linear resistor) are not expected to provide conversion at the CGF level; in this case the GR and $1/f$ fluctuations are actually modulated by the DC component of the LS steady state only.

Since $1/f$ processes do not possess a unique physical explanation, extension of the above remarks cannot be rigorously made in the general case (although it is certainly possible if $1/f$ noise is the result of a continuous superposition of GR sources). Nevertheless, it is commonly accepted [128, 134] that also for $1/f$ microscopic noise, modulation is only associated to the DC component of the LS solution, at least in radio-frequency (RF) and microwave operation, where the first harmonic typically is beyond 1 GHz.

Taking into account that noise is always a small-amplitude perturbation of a steady state, which is periodic in LS operation, we conclude that noise in the LS regime can be evaluated through a SSLS analysis, which can also include deterministic signals. In such conditions, noisy multi-ports are characterized by their impedance or admittance correlation matrix, as already discussed; noise generators can be readily introduced in a multi-frequency representation and characterized, in the cyclostationary case of interest, by the sideband correlation matrices. Namely, one has the Norton representation:

$$\boldsymbol{I}^{\mathrm{ss},+} = \mathsf{Y}(\omega) \cdot \boldsymbol{V}^{\mathrm{ss},+} + \boldsymbol{I}_0^{\mathrm{ss},+}, \tag{5.70}$$

where Y is the admittance CM and $\boldsymbol{I}_0^{\mathrm{ss},+} = \delta \boldsymbol{I}_{\mathrm{n}}$ is the sideband short-circuit current fluctuation vector described by the SCM $\mathsf{S}_{\delta i_{\mathrm{n}},\delta i_{\mathrm{n}}}$. Similarly, the Thévenin representation reads:

$$\boldsymbol{V}^{\mathrm{ss},+} = \mathsf{Z}(\omega) \cdot \boldsymbol{I}^{\mathrm{ss},+} + \boldsymbol{V}_0^{\mathrm{ss},+}, \tag{5.71}$$

where Z is the impedance CM and $\boldsymbol{V}_0^{\mathrm{ss},+} = \delta \boldsymbol{E}_{\mathrm{n}}$ is a sideband open-circuit voltage fluctuation vector described by the SCM $\mathsf{S}_{\delta e_{\mathrm{n}},\delta e_{\mathrm{n}}}$; the open- and short-circuit fluctuation SCMs are related as:

$$\mathsf{S}_{\delta e_{\mathrm{n}},\delta e_{\mathrm{n}}} = \mathsf{Z}(\omega) \cdot \mathsf{S}_{\delta i_{\mathrm{n}},\delta i_{\mathrm{n}}} \cdot \mathsf{Z}^{\dagger}(\omega). \tag{5.72}$$

Formally, each port of the multi-frequency, multi-port representation corresponds to a different (upper) sideband; as an example, Fig. 5.7 shows the

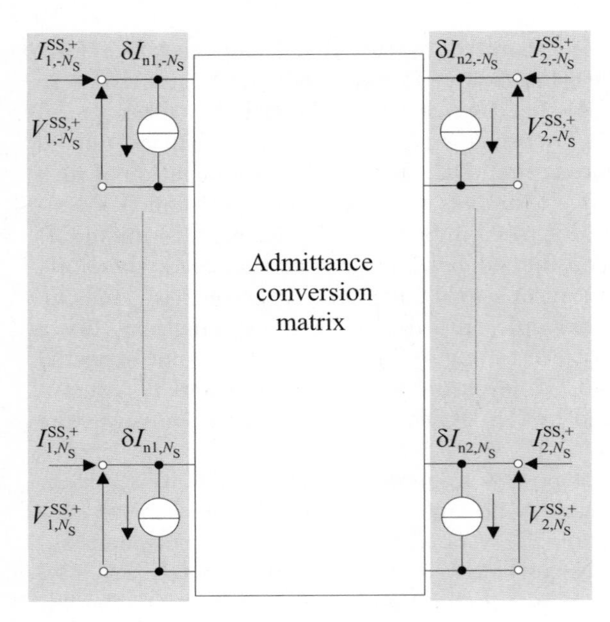

Fig. 5.7. Multi-frequency admittance representation of a noisy 2-port

relevant parallel representation for a two-port, where the input and output ports are expanded into a set of sideband ports and noise generators are introduced.

We now discuss a few important examples of SCM characterization of noisy devices:

1. For a linear resistor affected by thermal noise, the SCM is scalar, i.e. according to the Nyquist theorem:
$$\mathsf{S}_{\delta i_n,\delta i_n}(\omega) = 4k_B T G \mathsf{I}, \tag{5.73a}$$
$$\mathsf{S}_{\delta e_n,\delta e_n}(\omega) = 4k_B T R \mathsf{I}, \tag{5.73b}$$
where G is the resistor conductance, R the resistance. Notice that the SCM is frequency independent and diagonal, thus implying that sideband fluctuations are white and uncorrelated.
2. An arbitrary N-port including linear resistors, linear/nonlinear inductor and capacitors, linear/nonlinear dependent sources satisfies a generalized form of the Nyquist theorem:
$$\mathsf{S}_{\delta i_n,\delta i_n}(\omega) = 2k_B T \left[\mathsf{Y}(\omega) + \mathsf{Y}^\dagger(\omega)\right], \tag{5.74a}$$
$$\mathsf{S}_{\delta e_n,\delta e_n}(\omega) = 2k_B T \left[\mathsf{Z}(\omega) + \mathsf{Z}^\dagger(\omega)\right]. \tag{5.74b}$$
In this case sideband fluctuations are possibly colored (owing to the effect of reactive elements) and generally correlated for the same angular

frequency deviation ω. Notice that no rigorous extension of the Nyquist theorem is possible if the network includes nonlinear devices, albeit characterized by their SSLS correlation matrix [128].

3. A nonlinear one-port that exhibits white shot noise in SS operation (e.g. a semiconductor diode) can be characterized by a proper amplitude modulation of the small-signal short-circuit fluctuations. Let us define as I_k the k-th frequency component of the steady state LS current $i(t)$; the SCM of the short-circuit noise current can be put in the form [128, 129]:

$$\mathsf{S}_{\delta i_{\mathrm{n}},\delta i_{\mathrm{n}}}(\omega) = 2q\mathsf{I}, \tag{5.75}$$

where:

$$(\mathsf{I})_{i,j} = I_{i-j}. \tag{5.76}$$

The result can be extended to N-ports charaterized by shot noise behaviobehaviorr, like bipolar transistors.

4. Finally, consider a one-port characterized, in SS operation, by a colored noise source with spectrum:

$$S_{\delta i_{\mathrm{n}},\delta i_{\mathrm{n}}}(\omega) = |\tilde{h}(\omega)|^2 f(I_0) \tag{5.77}$$

where I_0 is the DC current. This case corresponds, for example, to GR and $1/f$ noise. In such case, several models have been proposed [128]. We only consider a theoretical model wherein amplitude modulation is followed by low-pass filtering; in this case, exploiting the analyisis in [128] and expressing the result in terms of SCM according to (A.83), it can be shown that:

$$\left(\mathsf{S}_{\delta i_{\mathrm{n}},\delta i_{\mathrm{n}}}(\omega)\right)_{i,j} = \tilde{h}(\omega + i\omega_0)\tilde{h}^*(\omega + j\omega_0)F_{i-j}, \tag{5.78}$$

where F_k is the k-th frequency component of $f(i(t))$. If $\tilde{h}(\omega)$ is a low-pass function with cutoff frequency lower than ω_0, one clearly has $\left(\mathsf{S}_{\delta i_{\mathrm{n}},\delta i_{\mathrm{n}}}(\omega)\right)_{i,j} \approx 0$ unless $i = j = 0$ and $\left(\mathsf{S}_{\delta i_{\mathrm{n}},\delta i_{\mathrm{n}}}(\omega)\right)_{0,0} = |\tilde{h}(\omega)|^2 F_0$. In other words, the noise source is only modulated by the DC component of the steady state. Other models (in which filtering precedes modulation) lead to a full sideband noise spectrum; clearly, the choice of either model is device dependent and should be made in accordance with experimental results.

Once the noise elements have been characterized as discussed above, the SSLS solution of the network in terms of the side voltage and current fluctuations can be expressed as a linear superposition of the internal noise generators through proper conversion matrices; the resulting SCM can then be evaluated by application of (5.18).

5.3 Physics-Based Large-Signal and SSLS Device Analysis

The periodic LS regime is particularly important in the design of devices for analog applications, above all in the RF and microwave fields. Examples are bipolar and field-effect transistors for quasi-linear applications, such as power amplifiers, or nonlinear applications, such as mixers and frequency multipliers. As in LS circuit analysis, direct time-domain techniques do not perform efficiently in this case, besides being unsuited to the development of noise characterization. During the last few years, LS analysis techniques for analog circuit simulation such as the HB method have been successfully applied to the LS physics-based numerical device modeling, see [135–139]. Further techniques, such as the SSLS analysis of frequency conversion, have been extended to device simulation both within the framework of quasi-2D approaches [11, 134] and for the full drift-diffusion (DD) model [10]. Finally, SSLS analysis is the basis for physics-based noise modeling as well.

In the following treatment, we will focus on a simple bipolar drift-diffusion model, consisting of Poisson's equation and of the electron and hole continuity equations. The extension to higher-order transport models is straightforward as a matter of principle, although in practice extremely intensive from a computational standpoint. Furthermore, as discussed in Sect. 2.2.5, even in small-signal operation the modeling of the microscopic noise sources within a full hydrodynamic model is not completely established yet, let alone in LS conditions.

We assume that the DD model has been spatially discretized according to the approach described in Sect. 3.1.1, leading to the nodal equations reported in (3.20). In the following discussion, we exploit the same definitions for the total numbers of internal and external nodes, and of contact electrical variables, as discussed in Sect. 3.1.2.

5.3.1 LS Analysis

Let us discuss first the time-domain equations of the spatially discretized DD system under LS excitation. As is well-known, the LS regime requires the equations of the embedding circuit to be solved together with the physics-based model [10, 127]. The discretized DD model leads to the following sets of nonlinear, dynamic nodal equations. The first corresponds to the $3N_\mathrm{i}$ discretized equations for the internal nodes:

$$\boldsymbol{f}_\varphi\left(\boldsymbol{\varphi}_\mathrm{i}, \boldsymbol{\varphi}_\mathrm{x}, \boldsymbol{n}_\mathrm{i}, \boldsymbol{p}_\mathrm{i}\right) = \boldsymbol{0}, \tag{5.79a}$$

$$\boldsymbol{f}_n\left(\boldsymbol{\varphi}_\mathrm{i}, \boldsymbol{\varphi}_\mathrm{x}, \boldsymbol{n}_\mathrm{i}, \boldsymbol{n}_\mathrm{x}, \boldsymbol{p}_\mathrm{i}; \dot{\boldsymbol{n}}_\mathrm{i}\right) = \boldsymbol{0}, \tag{5.79b}$$

$$\boldsymbol{f}_p\left(\boldsymbol{\varphi}_\mathrm{i}, \boldsymbol{\varphi}_\mathrm{x}, \boldsymbol{n}_\mathrm{i}, \boldsymbol{p}_\mathrm{i}, \boldsymbol{p}_\mathrm{x}; \dot{\boldsymbol{p}}_\mathrm{i}\right) = \boldsymbol{0}, \tag{5.79c}$$

where $\boldsymbol{f}_\varphi$, $\boldsymbol{f}_n$, $\boldsymbol{f}_p$ denote the discretized Poisson, electron continuity and hole continuity equations, respectively. Discretization of the boundary conditions on the external, contact nodes (see (3.27), (3.28), (3.32) and (3.33)) leads to the $3N_\mathrm{x}$ equations:

$$\boldsymbol{b}_\varphi\left(\boldsymbol{\varphi}_\mathrm{x}, \boldsymbol{v}_\mathrm{c}\right) = \boldsymbol{0}, \tag{5.79d}$$

$$\boldsymbol{b}_n\left(\boldsymbol{\varphi}_\mathrm{i}, \boldsymbol{\varphi}_\mathrm{x}, \boldsymbol{n}_\mathrm{i}, \boldsymbol{n}_\mathrm{x}, \boldsymbol{p}_\mathrm{x}; \dot{\boldsymbol{n}}_\mathrm{x}\right) = \boldsymbol{0}, \tag{5.79e}$$

$$\boldsymbol{b}_p\left(\boldsymbol{\varphi}_\mathrm{i}, \boldsymbol{\varphi}_\mathrm{x}, \boldsymbol{n}_\mathrm{x}, \boldsymbol{p}_\mathrm{i}, \boldsymbol{p}_\mathrm{x}; \dot{\boldsymbol{p}}_\mathrm{x}\right) = \boldsymbol{0}, \tag{5.79f}$$

which, in order, denote the boundary conditions relative to the Poisson and continuity equations. Boundary conditions on the non-contact device boundaries are already included in the equations for the internal nodes.

Finally, $2N_\mathrm{c}$ equations are further needed to close the system in the $3(N_\mathrm{i}+N_\mathrm{x}) + 2N_\mathrm{c}$ unknowns. Such equations express the contact currents $\boldsymbol{i}_\mathrm{c}$ as a function of the nodal unknowns (see (3.26) and (3.23)):

$$\boldsymbol{c}\left(\boldsymbol{\varphi}_\mathrm{i}, \boldsymbol{\varphi}_\mathrm{x}, \boldsymbol{n}_\mathrm{i}, \boldsymbol{n}_\mathrm{x}, \boldsymbol{p}_\mathrm{i}, \boldsymbol{p}_\mathrm{x}, \boldsymbol{i}_\mathrm{c}; \dot{\boldsymbol{n}}_\mathrm{x}, \dot{\boldsymbol{p}}_\mathrm{x}\right) = \boldsymbol{0}, \tag{5.79g}$$

and the constitutive relationships of the external circuit connected to the N_c ungrounded device terminals:

$$\boldsymbol{e}\left(\boldsymbol{v}_\mathrm{c}, \boldsymbol{i}_\mathrm{c}; \frac{\mathrm{d}}{\mathrm{d}t}\right) = \boldsymbol{s}, \tag{5.79h}$$

where $\boldsymbol{s}(t)$ is the set of periodic LS applied (voltage or current) sources.

According to the assumption of periodic LS steady state, each unknown (nodal and circuit) is made up of a superposition of harmonics, and therefore can be expressed according to (5.19) as a truncated complex Fourier series. Such a complex representation is used in the formulation because it is well suited to introducing the SSLS analysis; from a practical standpoint, a direct (real) Fourier representation is more convenient [127], and the present implementation, discussed in Sect. 5.5, was made accordingly. For the quasi-periodic case, the same comments apply as in Sect. 5.2.

The HB method can be applied to solve the nonlinear algebraic system (5.79) [104], thus directly extending the network treatment discussed in Sect. 5.2. Let us introduce an expanded frequency-domain unknown vector collecting the frequency components of all node and circuit unknowns (i.e. including $2N_\mathrm{L} + 1$ scalar terms for each unknown), see Fig. 5.8. Moreover, scalar functions are expanded into their $(2N_\mathrm{L} + 1)$ frequency components as well. According to this notation, system (5.79) is translated, in the frequency domain, into the following set of $(2N_\mathrm{L}+1)[3N_\mathrm{i}+N_\mathrm{x})+2N_\mathrm{c}]$ algebraic equations:

$$\boldsymbol{F}_\varphi\left(\boldsymbol{\Phi}_\mathrm{i}, \boldsymbol{\Phi}_\mathrm{x}, \boldsymbol{N}_\mathrm{i}, \boldsymbol{P}_\mathrm{i}\right) = \boldsymbol{0}, \tag{5.80a}$$

$$\boldsymbol{F}_n\left(\boldsymbol{\Phi}_\mathrm{i}, \boldsymbol{\Phi}_\mathrm{x}, \boldsymbol{N}_\mathrm{i}, \boldsymbol{N}_\mathrm{x}, \boldsymbol{P}_\mathrm{i}; \mathsf{D}_\mathrm{i} \cdot \boldsymbol{N}_\mathrm{i}\right) = \boldsymbol{0}, \tag{5.80b}$$

$$\boldsymbol{F}_p\left(\boldsymbol{\Phi}_\mathrm{i}, \boldsymbol{\Phi}_\mathrm{x}, \boldsymbol{N}_\mathrm{i}, \boldsymbol{P}_\mathrm{i}, \boldsymbol{P}_\mathrm{x}; \mathsf{D}_\mathrm{i} \cdot \boldsymbol{P}_\mathrm{i}\right) = \boldsymbol{0}, \tag{5.80c}$$

$$\boldsymbol{B}_\varphi\left(\boldsymbol{\Phi}_\mathrm{x}, \boldsymbol{V}_\mathrm{c}\right) = \boldsymbol{0}, \tag{5.80d}$$

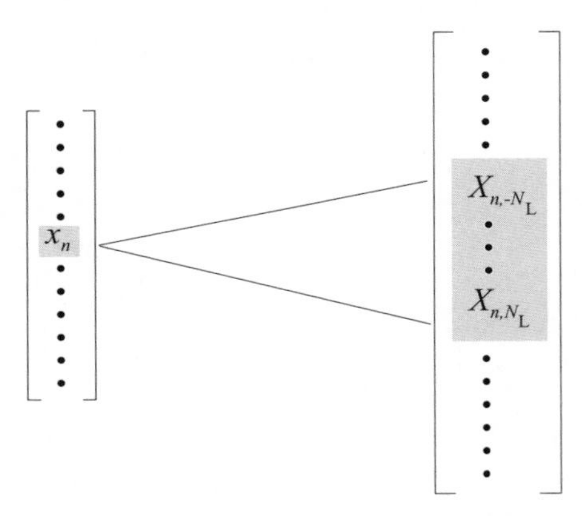

Fig. 5.8. Expanded vector for the frequency components of vector dynamical equations

$$\boldsymbol{B}_n\left(\boldsymbol{\Phi}_{\mathrm{i}}, \boldsymbol{\Phi}_{\mathrm{x}}, \boldsymbol{N}_{\mathrm{i}}, \boldsymbol{N}_{\mathrm{x}}, \boldsymbol{P}_{\mathrm{x}}; \mathsf{D}_{\mathrm{x}} \cdot \boldsymbol{N}_{\mathrm{x}}\right) = \mathbf{0}, \tag{5.80e}$$

$$\boldsymbol{B}_p\left(\boldsymbol{\Phi}_{\mathrm{i}}, \boldsymbol{\Phi}_{\mathrm{x}}, \boldsymbol{N}_{\mathrm{x}}, \boldsymbol{P}_{\mathrm{i}}, \boldsymbol{P}_{\mathrm{x}}; \mathsf{D}_{\mathrm{x}} \cdot \boldsymbol{P}_{\mathrm{x}}\right) = \mathbf{0}, \tag{5.80f}$$

$$\boldsymbol{C}\left(\boldsymbol{\Phi}_{\mathrm{i}}, \boldsymbol{\Phi}_{\mathrm{x}}, \boldsymbol{N}_{\mathrm{i}}, \boldsymbol{N}_{\mathrm{x}}, \boldsymbol{P}_{\mathrm{i}}, \boldsymbol{P}_{\mathrm{x}}, \boldsymbol{I}_{\mathrm{c}}; \mathsf{D}_{\mathrm{x}} \cdot \boldsymbol{N}_{\mathrm{x}}, \mathsf{D}_{\mathrm{x}} \cdot \boldsymbol{P}_{\mathrm{x}}\right) = \mathbf{0}, \tag{5.80g}$$

$$\boldsymbol{E}\left(\boldsymbol{V}_{\mathrm{c}}, \boldsymbol{I}_{\mathrm{c}}; \mathsf{D}\right) = \boldsymbol{S}, \tag{5.80h}$$

where D_{i} and D_{x} are block diagonal matrices made, respectively, of N_{i} and N_{x} replicas of D, defined in (5.23). The solution of the nonlinear HB system can be performed by means of Newton or quasi-Newton iteration, see [127] for a discussion of the specific problems arising in device modeling. Owing to the quite impressive size of the HB system arising in 2D (let alone 3D) simulation, direct solution methods are often impracticable and iterative techniques must be exploited [127].

5.3.2 SSLS Analysis

In order to introduce the SSLS analysis, let us apply the equivalence theorem [63] to the external device terminals, so as to replace the circuit equation (5.79h) with a linear relationship with equivalent current (or voltage) sources $\boldsymbol{s}_{\mathrm{eq}}$ corresponding to the terminal currents (or voltages) evaluated in the LS steady state:

$$\boldsymbol{e}'(\boldsymbol{v}_{\mathrm{c}}, \boldsymbol{i}_{\mathrm{c}}) = \boldsymbol{s}_{\mathrm{eq}}. \tag{5.81}$$

If we now superimpose a small-amplitude harmonic perturbation $\boldsymbol{s}_{\mathrm{eq}}^{\mathrm{ss}}$ onto the LS steady state, each model unknown (nodal and circuit) can be decomposed as the sum of the LS steady state solution and a small perturbation x^{ss}: $x(t) = x_0(t) + x^{\mathrm{ss}}(t)$. After linearization, (5.79) can be cast into the following form. The linearized Poisson, electron and hole continuity equations can be expressed as:

$$\mathsf{f}_{\varphi,\boldsymbol{\varphi}_\mathrm{i}} \cdot \boldsymbol{\varphi}_\mathrm{i}^\mathrm{ss} + \mathsf{f}_{\varphi,\boldsymbol{\varphi}_\mathrm{x}} \cdot \boldsymbol{\varphi}_\mathrm{x}^\mathrm{ss} + \mathsf{f}_{\varphi,\boldsymbol{n}_\mathrm{i}} \cdot \boldsymbol{n}_\mathrm{i}^\mathrm{ss} + \mathsf{f}_{\varphi,\boldsymbol{p}_\mathrm{i}} \cdot \boldsymbol{p}_\mathrm{i}^\mathrm{ss} = \mathbf{0}, \tag{5.82a}$$

$$\mathsf{f}_{n,\boldsymbol{\varphi}_\mathrm{i}} \cdot \boldsymbol{\varphi}_\mathrm{i}^\mathrm{ss} + \mathsf{f}_{n,\boldsymbol{\varphi}_\mathrm{x}} \cdot \boldsymbol{\varphi}_\mathrm{x}^\mathrm{ss} + \mathsf{f}_{n,\boldsymbol{n}_\mathrm{i}} \cdot \boldsymbol{n}_\mathrm{i}^\mathrm{ss} + \mathsf{f}_{n,\boldsymbol{n}_\mathrm{x}} \cdot \boldsymbol{n}_\mathrm{x}^\mathrm{ss} + \mathsf{f}_{n,\boldsymbol{p}_\mathrm{i}} \cdot \boldsymbol{p}_\mathrm{i}^\mathrm{ss} + \mathsf{f}_{n,\dot{\boldsymbol{n}}_\mathrm{i}} \cdot \dot{\boldsymbol{n}}_\mathrm{i}^\mathrm{ss} = \mathbf{0}, \tag{5.82b}$$

$$\mathsf{f}_{p,\boldsymbol{\varphi}_\mathrm{i}} \cdot \boldsymbol{\varphi}_\mathrm{i}^\mathrm{ss} + \mathsf{f}_{p,\boldsymbol{\varphi}_\mathrm{x}} \cdot \boldsymbol{\varphi}_\mathrm{x}^\mathrm{ss} + \mathsf{f}_{p,\boldsymbol{n}_\mathrm{i}} \cdot \boldsymbol{n}_\mathrm{i}^\mathrm{ss} + \mathsf{f}_{p,\boldsymbol{p}_\mathrm{i}} \cdot \boldsymbol{p}_\mathrm{i}^\mathrm{ss} + \mathsf{f}_{p,\boldsymbol{p}_\mathrm{x}} \cdot \boldsymbol{p}_\mathrm{x}^\mathrm{ss} + \mathsf{f}_{p,\dot{\boldsymbol{p}}_\mathrm{i}} \cdot \dot{\boldsymbol{p}}_\mathrm{i}^\mathrm{ss} = \mathbf{0}. \tag{5.82c}$$

The time-varying matrices f are strictly related to the small-signal matrices A introduced in Sect. 3.1.2. The boundary conditions lead to the SSLS equations:

$$\mathsf{b}_{\varphi,\boldsymbol{\varphi}_\mathrm{x}} \cdot \boldsymbol{\varphi}_\mathrm{x}^\mathrm{ss} + \mathsf{b}_{\varphi,\boldsymbol{v}_\mathrm{c}} \cdot \boldsymbol{v}_\mathrm{c}^\mathrm{ss} = \mathbf{0}, \tag{5.82d}$$

$$\mathsf{b}_{n,\boldsymbol{\varphi}_\mathrm{i}} \cdot \boldsymbol{\varphi}_\mathrm{i}^\mathrm{ss} + \mathsf{b}_{n,\boldsymbol{\varphi}_\mathrm{x}} \cdot \boldsymbol{\varphi}_\mathrm{x}^\mathrm{ss} + \mathsf{b}_{n,\boldsymbol{n}_\mathrm{i}} \cdot \boldsymbol{n}_\mathrm{i}^\mathrm{ss} + \mathsf{b}_{n,\boldsymbol{n}_\mathrm{x}} \cdot \boldsymbol{n}_\mathrm{x}^\mathrm{ss} + \mathsf{b}_{n,\boldsymbol{p}_\mathrm{x}} \cdot \boldsymbol{p}_\mathrm{x}^\mathrm{ss} + \mathsf{b}_{n,\dot{\boldsymbol{n}}_\mathrm{x}} \cdot \dot{\boldsymbol{n}}_\mathrm{x}^\mathrm{ss} = \mathbf{0}, \tag{5.82e}$$

$$\mathsf{b}_{p,\boldsymbol{\varphi}_\mathrm{i}} \cdot \boldsymbol{\varphi}_\mathrm{i}^\mathrm{ss} + \mathsf{b}_{p,\boldsymbol{\varphi}_\mathrm{x}} \cdot \boldsymbol{\varphi}_\mathrm{x}^\mathrm{ss} + \mathsf{b}_{p,\boldsymbol{n}_\mathrm{x}} \cdot \boldsymbol{n}_\mathrm{x}^\mathrm{ss} + \mathsf{b}_{p,\boldsymbol{p}_\mathrm{i}} \cdot \boldsymbol{p}_\mathrm{i}^\mathrm{ss} + \mathsf{b}_{p,\boldsymbol{p}_\mathrm{x}} \cdot \boldsymbol{p}_\mathrm{x}^\mathrm{ss} + \mathsf{b}_{p,\dot{\boldsymbol{p}}_\mathrm{x}} \cdot \dot{\boldsymbol{p}}_\mathrm{x}^\mathrm{ss} = \mathbf{0}, \tag{5.82f}$$

while the contact current definitions and the external circuit constitutive equations result in:

$$\mathsf{c}_{\boldsymbol{\varphi}_\mathrm{i}} \cdot \boldsymbol{\varphi}_\mathrm{i}^\mathrm{ss} + \mathsf{c}_{\boldsymbol{\varphi}_\mathrm{x}} \cdot \boldsymbol{\varphi}_\mathrm{x}^\mathrm{ss} + \mathsf{c}_{\boldsymbol{n}_\mathrm{i}} \cdot \boldsymbol{n}_\mathrm{i}^\mathrm{ss} + \mathsf{c}_{\boldsymbol{n}_\mathrm{x}} \cdot \boldsymbol{n}_\mathrm{x}^\mathrm{ss} + \mathsf{c}_{\boldsymbol{p}_\mathrm{i}} \cdot \boldsymbol{p}_\mathrm{i}^\mathrm{ss} + \mathsf{c}_{\boldsymbol{p}_\mathrm{x}} \cdot \boldsymbol{p}_\mathrm{x}^\mathrm{ss} + \mathsf{c}_{\dot{\boldsymbol{n}}_\mathrm{x}} \cdot \dot{\boldsymbol{n}}_\mathrm{x}^\mathrm{ss} + \mathsf{c}_{\dot{\boldsymbol{p}}_\mathrm{x}} \cdot \dot{\boldsymbol{p}}_\mathrm{x}^\mathrm{ss} + \mathsf{c}_{\boldsymbol{i}_\mathrm{c}} \cdot \boldsymbol{n}_\mathrm{c}^\mathrm{ss} = \mathbf{0}, \tag{5.82g}$$

$$\mathsf{e}'_{\boldsymbol{v}_\mathrm{c}} \cdot \boldsymbol{v}_\mathrm{c}^\mathrm{ss} + \mathsf{e}'_{\boldsymbol{i}_\mathrm{c}} \cdot \boldsymbol{i}_\mathrm{c}^\mathrm{ss} = \boldsymbol{s}_\mathrm{eq}^\mathrm{ss}. \tag{5.82h}$$

In (5.82) the first factor of each term is the (matrix) gradient of the corresponding model equation evaluated in the LS steady state. The vector system (5.82) is analogous to the scalar example of LPTV (5.6). Notice, however, that Poisson's equation (and its boundary condition) is memoryless, and that in (5.82h) the matrices e' are diagonal, with diagonal elements equal to either 0 or 1. If the equivalent source for terminal n is a current (voltage) generator, the (n,n) element of $\mathsf{e}'_{\boldsymbol{i}_\mathrm{c}}$ ($\mathsf{e}'_{\boldsymbol{v}_\mathrm{c}}$) is 1, and the (n,n) element of $\mathsf{e}'_{\boldsymbol{v}_\mathrm{c}}$ ($\mathsf{e}'_{\boldsymbol{i}_\mathrm{c}}$) is 0.

According to the procedure outlined in Sect. 5.2.2, each unknown is expanded into the sideband formulation:

$$x_\mathrm{c}^\mathrm{ss}(t) = \sum_{k=-N_\mathrm{S}}^{N_\mathrm{S}} X_k^{\mathrm{ss},+} \mathrm{e}^{\mathrm{i}\omega_k^+ t}. \tag{5.83}$$

Then, system (5.82) is expressed in the spectral domain leading to a conversion matrix formulation akin to (5.12):

$$\mathsf{C}_{\mathsf{f}_{\varphi,\boldsymbol{\varphi}_\mathrm{i}}} \cdot \boldsymbol{\Phi}_\mathrm{i}^{\mathrm{ss},+} + \mathsf{C}_{\mathsf{f}_{\varphi,\boldsymbol{\varphi}_\mathrm{x}}} \cdot \boldsymbol{\Phi}_\mathrm{x}^{\mathrm{ss},+} + \mathsf{C}_{\mathsf{f}_{\varphi,\boldsymbol{n}_\mathrm{i}}} \cdot \boldsymbol{N}_\mathrm{i}^{\mathrm{ss},+} + \mathsf{C}_{\mathsf{f}_{\varphi,\boldsymbol{p}_\mathrm{i}}} \cdot \boldsymbol{P}_\mathrm{i}^{\mathrm{ss},+} = \mathbf{0}, \tag{5.84a}$$

$$\mathsf{C}_{\mathsf{f}_{n,\varphi_\mathrm{i}}} \cdot \boldsymbol{\Phi}_\mathrm{i}^{\mathrm{ss},+} + \mathsf{C}_{\mathsf{f}_{n,\varphi_\mathrm{x}}} \cdot \boldsymbol{\Phi}_\mathrm{x}^{\mathrm{ss},+} + \mathsf{C}_{\mathsf{f}_{n,n_\mathrm{i}}} \cdot \boldsymbol{N}_\mathrm{i}^{\mathrm{ss},+} + \mathsf{C}_{\mathsf{f}_{n,n_\mathrm{x}}} \cdot \boldsymbol{N}_\mathrm{x}^{\mathrm{ss},+} + \mathsf{C}_{\mathsf{f}_{n,p_\mathrm{i}}} \cdot \boldsymbol{P}_\mathrm{i}^{\mathrm{ss},+} + \mathsf{C}_{\mathsf{f}_{n,\dot{n}_\mathrm{i}}} \cdot \mathsf{D}_\mathrm{i}^{\mathrm{ss}} \cdot \boldsymbol{N}_\mathrm{i}^{\mathrm{ss},+} = \mathbf{0}, \tag{5.84b}$$

$$\mathsf{C}_{\mathsf{f}_{p,\varphi_\mathrm{i}}} \cdot \boldsymbol{\Phi}_\mathrm{i}^{\mathrm{ss},+} + \mathsf{C}_{\mathsf{f}_{p,\varphi_\mathrm{x}}} \cdot \boldsymbol{\Phi}_\mathrm{x}^{\mathrm{ss},+} + \mathsf{C}_{\mathsf{f}_{p,n_\mathrm{i}}} \cdot \boldsymbol{N}_\mathrm{i}^{\mathrm{ss},+} + \mathsf{C}_{\mathsf{f}_{p,p_\mathrm{i}}} \cdot \boldsymbol{P}_\mathrm{i}^{\mathrm{ss},+} + \mathsf{C}_{\mathsf{f}_{p,p_\mathrm{x}}} \cdot \boldsymbol{P}_\mathrm{x}^{\mathrm{ss},+} + \mathsf{C}_{\mathsf{f}_{p,\dot{p}_\mathrm{i}}} \cdot \mathsf{D}_\mathrm{i}^{\mathrm{ss}} \cdot \boldsymbol{P}_\mathrm{i}^{\mathrm{ss},+} = \mathbf{0}, \tag{5.84c}$$

$$\mathsf{C}_{\mathsf{b}_{\varphi,\varphi_\mathrm{x}}} \cdot \boldsymbol{\Phi}_\mathrm{x}^{\mathrm{ss},+} + \mathsf{C}_{\mathsf{b}_{\varphi,v_\mathrm{c}}} \cdot \boldsymbol{V}_\mathrm{c}^{\mathrm{ss},+} = \mathbf{0}, \tag{5.84d}$$

$$\mathsf{C}_{\mathsf{b}_{n,\varphi_\mathrm{i}}} \cdot \boldsymbol{\Phi}_\mathrm{i}^{\mathrm{ss},+} + \mathsf{C}_{\mathsf{b}_{n,\varphi_\mathrm{x}}} \cdot \boldsymbol{\Phi}_\mathrm{x}^{\mathrm{ss},+} + \mathsf{C}_{\mathsf{b}_{n,n_\mathrm{i}}} \cdot \boldsymbol{N}_\mathrm{i}^{\mathrm{ss},+} + \mathsf{C}_{\mathsf{b}_{n,n_\mathrm{x}}} \cdot \boldsymbol{N}_\mathrm{x}^{\mathrm{ss},+} + \mathsf{C}_{\mathsf{b}_{n,p_\mathrm{x}}} \cdot \boldsymbol{P}_\mathrm{x}^{\mathrm{ss},+} + \mathsf{C}_{\mathsf{b}_{n,\dot{n}_\mathrm{x}}} \cdot \mathsf{D}_\mathrm{x}^{\mathrm{ss}} \cdot \boldsymbol{N}_\mathrm{x}^{\mathrm{ss},+} = \mathbf{0}, \tag{5.84e}$$

$$\mathsf{C}_{\mathsf{b}_{p,\varphi_\mathrm{i}}} \cdot \boldsymbol{\Phi}_\mathrm{i}^{\mathrm{ss},+} + \mathsf{C}_{\mathsf{b}_{p,\varphi_\mathrm{x}}} \cdot \boldsymbol{\Phi}_\mathrm{x}^{\mathrm{ss},+} + \mathsf{C}_{\mathsf{b}_{p,n_\mathrm{x}}} \cdot \boldsymbol{N}_\mathrm{x}^{\mathrm{ss},+} + \mathsf{C}_{\mathsf{b}_{p,p_\mathrm{i}}} \cdot \boldsymbol{P}_\mathrm{i}^{\mathrm{ss},+} + \mathsf{C}_{\mathsf{b}_{p,p_\mathrm{x}}} \cdot \boldsymbol{P}_\mathrm{x}^{\mathrm{ss},+} + \mathsf{C}_{\mathsf{b}_{p,\dot{p}_\mathrm{x}}} \cdot \mathsf{D}_\mathrm{x}^{\mathrm{ss}} \cdot \boldsymbol{P}_\mathrm{x}^{\mathrm{ss},+} = \mathbf{0}, \tag{5.84f}$$

$$\mathsf{C}_{\mathsf{c}_{\varphi_\mathrm{i}}} \cdot \boldsymbol{\Phi}_\mathrm{i}^{\mathrm{ss},+} + \mathsf{C}_{\mathsf{c}_{\varphi_\mathrm{x}}} \cdot \boldsymbol{\Phi}_\mathrm{x}^{\mathrm{ss},+} + \mathsf{C}_{\mathsf{c}_{n_\mathrm{i}}} \cdot \boldsymbol{N}_\mathrm{i}^{\mathrm{ss},+} + \mathsf{C}_{\mathsf{c}_{n_\mathrm{x}}} \cdot \boldsymbol{N}_\mathrm{x}^{\mathrm{ss},+} + \mathsf{C}_{\mathsf{c}_{p_\mathrm{i}}} \cdot \boldsymbol{P}_\mathrm{i}^{\mathrm{ss},+} + \mathsf{C}_{\mathsf{c}_{p_\mathrm{x}}} \cdot \boldsymbol{P}_\mathrm{x}^{\mathrm{ss},+} + \mathsf{C}_{\mathsf{c}_{\dot{n}_\mathrm{x}}} \cdot \mathsf{D}_\mathrm{x}^{\mathrm{ss}} \cdot \boldsymbol{N}_\mathrm{x}^{\mathrm{ss},+} + \mathsf{C}_{\mathsf{c}_{\dot{p}_\mathrm{x}}} \cdot \mathsf{D}_\mathrm{x}^{\mathrm{ss}} \cdot \boldsymbol{P}_\mathrm{x}^{\mathrm{ss},+} + \mathsf{C}_{\mathsf{c}_{i_\mathrm{c}}} \cdot \boldsymbol{I}_\mathrm{c}^{\mathrm{ss},+} = \mathbf{0}, \tag{5.84g}$$

$$\mathsf{C}_{\mathsf{e}'_{v_\mathrm{c}}} \cdot \boldsymbol{V}_\mathrm{c}^{\mathrm{ss},+} + \mathsf{C}_{\mathsf{e}'_{i_\mathrm{c}}} \cdot \boldsymbol{I}_\mathrm{c}^{\mathrm{ss},+} = \boldsymbol{S}_\mathrm{eq}^{\mathrm{ss},+}. \tag{5.84h}$$

The conversion matrices C in (5.84) are assembled according to the following rule, here presented, as an example, for $\mathsf{C}_{\mathsf{f}_{\varphi,\varphi_\mathrm{i}}}$. For each node and for each nodal equation, each element of the matrix $\mathsf{f}_{\varphi,\varphi_\mathrm{i}}$ is converted into the frequency domain, resulting in a CM according to the memoryless part of (5.11), thus making $\mathsf{C}_{\mathsf{f}_{\varphi,\varphi_\mathrm{i}}}$ a matrix of matrices, see Fig. 5.9. For instance, for node j and nodal equation i, the elements of the conversion submatrix are:

$$\left(\mathsf{C}_{\mathsf{f}_{\varphi_i,\varphi_{\mathrm{i},j}}}\right)_{q,n} = \left(\boldsymbol{F}_{\varphi_i,\varphi_{\mathrm{i},j}}\right)_{q-n}, \tag{5.85}$$

where q, n are sideband indices, and $\boldsymbol{F}_{\varphi_i,\varphi_{\mathrm{i},j}}$ is the vector of frequency components of $f_{\varphi_i,\varphi_{\mathrm{i},j}}(t)$. The memory part of (5.82) has been treated in accordance with (5.11) introducing into (5.84) matrices $\mathsf{D}_\mathrm{i}^{\mathrm{ss}}$ and $\mathsf{D}_\mathrm{x}^{\mathrm{ss}}$. These are block diagonal matrices, replicating N_i and N_x times, respectively, the fundamental matrix $\mathsf{D}^{\mathrm{ss}} = \mathrm{diag}\{\mathrm{i}\omega_k^+\}$.

Since, as already noted, the LS solution must be carried out to the order $N_\mathrm{L} = 2N_\mathrm{S}$ to fully account for all of the possible conversions among the sidebands up to the order N_S, system (5.84) is a set of $(2N_\mathrm{S}+1)[3(N_\mathrm{i}+N_\mathrm{x})+2N_\mathrm{c}]$ complex equations, to be compared with the $(2N_\mathrm{L}+1)[3(N_\mathrm{i}+N_\mathrm{x})+2N_\mathrm{c}]$ real equations to be solved in the LS case. From a computational standpoint, we emphasize that if Newton's method is used to solve the LS equations, the harmonic components of each element of the matrix gradients in (5.82) are readily evaluated from the Jacobian matrix in the last iteration of the Newton loop [140].

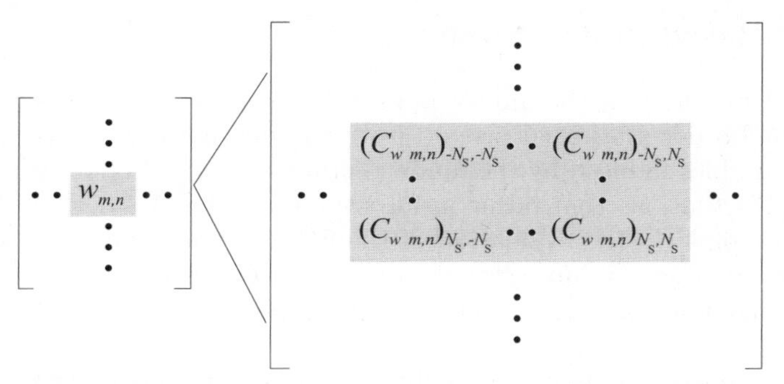

Fig. 5.9. Expanded matrix for the frequency components of function (matrix) gradients

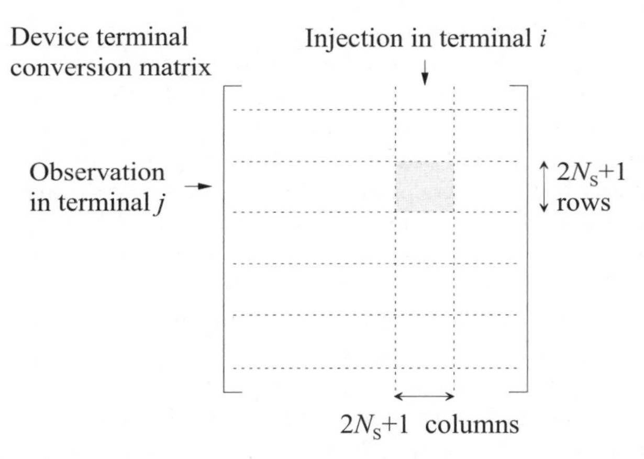

Fig. 5.10. Evaluation of the device terminal conversion matrix

In order to enable evaluation of the device terminal CM, the source term $\boldsymbol{S}_{\mathrm{eq}}^{\mathrm{ss},+}$ in (5.84) must include an independent excitation for each sideband and each terminal; this implies that $\boldsymbol{S}_{\mathrm{eq}}^{\mathrm{ss},+}$ is a unit diagonal matrix of dimension $(2N_{\mathrm{S}}+1) \times N_{\mathrm{c}}$. In the source matrix, each block of $2N_{\mathrm{S}}+1$ columns corresponds to the excitation of one device terminal (e.g. terminal i) for every sideband, and the solution of (5.84) enables us to evaluate all the blocks of the device CM corresponding to injection in terminal i, i.e. the device conversion submatrices between terminal i and every device terminal (see Fig. 5.10). In particular, the device Y CM is evaluated when the forcing term is a voltage source, and the Z CM when the forcing term is a current source. The evaluation of hybrid CMs is similar.

5.4 LS Physics-Based Noise Analysis

As already discussed in Chap. 2, the aim of physics-based noise analysis is to evaluate the second-order statistical properties of the fluctuations of the terminal electrical variables induced by the fundamental microscopic velocity and carrier number fluctuations that occur inside the device. For LS device operation, this corresponds to the evaluation of the SCM of the equivalent short-circuit current (or open-circuit voltage, or any set corresponding to hybrid representations) noise generators at the device ports, as shown in Fig. 5.7.

Also in (forced) LS operation, microscopic fluctuations can be introduced into the model as impressed Langevin stochastic forcing terms appearing in the right-hand side of the electron and hole continuity equations and (if present) in the trap rate equations. The approach is a natural extension of the conventional noise analysis in SS operation, with some significant differences. First, microscopic noise sources are modulated by the LS instantaneous steady state, thereby becoming cyclostationary processes (rather than stationary as in the SS case). Second, the small-amplitude assumption for the noise source still allows a linear analysis of the system response to the stochastic excitation, but linearization takes place around the noiseless steady state LS working point; in other words, the linearized system is periodically time-varying, and must therefore be treated according to the SSLS (conversion matrix) formalism. Due to linearity, the induced fluctuations at the device terminals can be evaluated by means of a convolution integral of the microscopic fluctuations with suitable Green's functions of the linearized system, thus extending to the LS case the Green's function approach described in Chap. 2. For the sake of simplicity, we shall confine the discussion to a simple bipolar DD model, without any trap-level rate equation. The extension to models including traps is straightforward. The relevant Green's functions relate a unit injection in the electron or hole continuity equation to the open-circuit voltage or short-circuit current variations at the device terminals, depending on the boundary conditions enforced on the linearized system. Both microscopic and terminal fluctuations are expressed as sidebands, so that the Green's functions turn out to be conversion matrices, thus justifying the name of conversion Green's functions. In this way, the device response, expressed according to the sideband formalism, is evaluated as a function of the relevant Green's functions, depending on the variable chosen as output. Since microscopic noise sources are distributed within the device volume, the output is given by a spatial convolution integral whose kernel is the (space-dependent) CGF, thus extending the result (5.12). For instance, the short-circuit current induced at terminal i is evaluated with the superposition integral:

$$\delta \boldsymbol{I}_i^{\mathrm{ss},+} = \sum_{\alpha=n,p} \left[\int_{\Omega} \mathsf{G}_{\alpha,i}(\boldsymbol{r};\omega) \cdot \boldsymbol{\Gamma}_{\alpha}^{+}(\boldsymbol{r}) \, \mathrm{d}\boldsymbol{r} \right.$$

$$+ \sum_{\kappa=x,y,z} \int_{\Omega} \frac{\partial}{\partial \kappa} \mathsf{G}_{\alpha,i}(\boldsymbol{r};\omega) \cdot \delta \boldsymbol{J}^{+}_{\alpha,\kappa}(\boldsymbol{r}) \, \mathrm{d}\boldsymbol{r} \Bigg] , \tag{5.86}$$

where Ω is the device volume, the Green's function $\mathsf{G}_{\alpha,i}$ is the conversion matrix (CGF) for continuity equation α and terminal i, and ω is the sideband displacement frequency, see Sect. A.5. In (5.86), the first integral is the response to population fluctuations, the second to current density noise sources; $\boldsymbol{\Gamma}^{+}_{\alpha}(\boldsymbol{r})$ and $\delta \boldsymbol{J}^{+}_{\alpha,\kappa}(\boldsymbol{r})$ are the vectors of sideband amplitudes for the (modulated) microscopic noise sources representing population and current density[2] fluctuations, respectively (see Sect. 1.2).

Finally, supposing that microscopic fluctuations are spatially uncorrelated, the SCM of the short-circuit noise currents for terminals i and j is derived, by properly extending (5.18), as:

$$\mathsf{S}_{\delta i_i,\delta i_j} = \sum_{\alpha,\beta=n,p} \Bigg\{ \int_{\Omega} \mathsf{G}_{\alpha,i}(\boldsymbol{r};\omega) \cdot \mathsf{K}_{\gamma_\alpha,\gamma_\beta}(\boldsymbol{r}) \cdot \mathsf{G}^{\dagger}_{\beta,j}(\boldsymbol{r};\omega) \, \mathrm{d}\boldsymbol{r}$$

$$+ \sum_{\kappa_1,\kappa_2=x,y,z} \int_{\Omega} \frac{\partial}{\partial \kappa_1} \mathsf{G}_{\alpha,i}(\boldsymbol{r};\omega) \cdot \mathsf{K}_{\delta J_{\alpha,\kappa_1},\delta J_{\alpha,\kappa_2}}(\boldsymbol{r}) \cdot \frac{\partial}{\partial \kappa_2} \mathsf{G}^{\dagger}_{\alpha,j}(\boldsymbol{r};\omega) \, \mathrm{d}\boldsymbol{r} \Bigg\} . \tag{5.87}$$

The physics based noise analysis is therefore carried out in two steps, depicted in Fig. 5.5. The first consists in assessing the statistical properties of the microscopic noise sources as a result of the time-varying, LS noiseless steady state. The second step is the propagation of the microscopic noise sources to the device terminals by means of the relevant CGFs, thus expressing the terminal noise SCM according to a generalization of the Green's function technique for noise analysis in the DC regime [1,6]. The numerical evaluation of the CGF can be carried out by applying SSLS analysis to the discretized Langevin system.

5.4.1 The LS-Modulated Microscopic Noise Sources

In the first place the statistical properties of the spatially uncorrelated microscopic fluctuations must be evaluated in the LS regime. According to the discussion in Sect. 5.2.3, fluctuations described in SS operation through white processes related to the fast microscopic dynamics of particles, with characteristic times typically of the order of less than 1 ps, can be considered as slowly modulated by the LS instantaneous working point [10,129]. Such an amplitude modulation can be modeled as the passage through a time-varying, memoryless linear system, characterized by a CM independent of the input frequency. Following this, the local noise source becomes a cyclostationary noise process which can be characterized by a SCM, as outlined in Sect. 5.1,

[2] Actually, $\delta \boldsymbol{J}^{+}_{\alpha,\kappa}$ is the κ-spatial component of the vector microscopic noise source.

implying that frequency components belonging to different sidebands are correlated only if their frequency deviation from the respective LS frequency is the same.

Let us first consider diffusion noise. Electron and hole velocity fluctuations are physically uncorrelated, and must be expressed in terms of equivalent current-density fluctuations in order to be included as forcing terms for the continuity equations (see Sect. 1.2). Let us define as $\delta J_{\alpha,\kappa}$ ($\alpha = n, p$ and $\kappa = x, y, z$) the κ (spatial) component of the vector current density fluctuation for carrier α. The elements of the SCM of the velocity fluctuations for spatial components κ_1 and κ_2 are given by:

$$\left(\mathsf{K}_{\delta J_{n,\kappa_1},\delta J_{n,\kappa_2}}(\boldsymbol{r})\right)_{q,n} = 2q^2\left[D_{n,\kappa_1\kappa_2} + D^*_{n,\kappa_2\kappa_1}\right](\boldsymbol{N}(\boldsymbol{r}))_{q-n}, \tag{5.88a}$$

$$\left(\mathsf{K}_{\delta J_{p,\kappa_1},\delta J_{p,\kappa_2}}(\boldsymbol{r})\right)_{q,n} = 2q^2\left[D_{p,\kappa_1\kappa_2} + D^*_{p,\kappa_2\kappa_1}\right](\boldsymbol{P}(\boldsymbol{r}))_{q-n}, \tag{5.88b}$$

for electrons and holes, respectively. In (5.88), $D_{\kappa_1\kappa_2}$ is the (κ_1, κ_2) component of the diffusivity tensor for the corresponding carrier, and $\boldsymbol{N}(\boldsymbol{r}), \boldsymbol{P}(\boldsymbol{r})$ are the frequency components of the LS electron and hole-density distributions.

Concerning population fluctuations, in accordance with the simple bipolar DD model we are considering, we treat explicitly only the case of band-to-band transitions. The extension to trap-assisted processes is straightforward. The microscopic noise source appearing in the continuity equation for carrier α ($\alpha = n, p$) is denoted as γ_α. The SCM of population fluctuations can be obtained by amplitude modulation of the DC expression (1.94), yielding:

$$\left(\mathsf{K}_{\gamma_n,\gamma_n}\right)_{q,n} = 2\left[\left(\boldsymbol{R}_n^{\mathrm{dir}}(\boldsymbol{r})\right)_{q-n} + \left(\boldsymbol{G}_n^{\mathrm{dir}}(\boldsymbol{r})\right)_{q-n}\right], \tag{5.89a}$$

$$\left(\mathsf{K}_{\gamma_p,\gamma_p}\right)_{q,n} = 2\left[\left(\boldsymbol{R}_p^{\mathrm{dir}}(\boldsymbol{r})\right)_{q-n} + \left(\boldsymbol{G}_p^{\mathrm{dir}}(\boldsymbol{r})\right)_{q-n}\right], \tag{5.89b}$$

$$\left(\mathsf{K}_{\gamma_n,\gamma_p}\right)_{q,n} = 2\left[\left(\boldsymbol{R}_n^{\mathrm{dir}}(\boldsymbol{r})\right)_{q-n} + \left(\boldsymbol{G}_n^{\mathrm{dir}}(\boldsymbol{r})\right)_{q-n}\right], \tag{5.89c}$$

where $\boldsymbol{G}_\alpha^{\mathrm{dir}}(\boldsymbol{r}), \boldsymbol{R}_\alpha^{\mathrm{dir}}(\boldsymbol{r})$ are the frequency components of the LS generation and recombination rates distributions, respectively.

Colored microscopic fluctuations, primarily the $1/f$ noise source (1.121) and the approximate equivalent source for GR noise (1.104), are treated according to (5.78):

1. **1/f noise.** The phenomenological source yields ($\beta = n, p$):

$$\left(\mathsf{K}_{\delta J_{\beta,\kappa_1},\delta J_{\beta,\kappa_2}}(\boldsymbol{r};\omega)\right)_{q,n} = \frac{2\pi\alpha_{\mathrm{H}\beta}}{\sqrt{(\omega+\omega_q)(\omega+\omega_n)}}(\boldsymbol{F}_{1,\beta}(\boldsymbol{r}))_{q-n}\,\delta_{\kappa_1,\kappa_2}, \tag{5.90}$$

where δ is Kronecker's symbol and $\boldsymbol{F}_{1,\beta}(\boldsymbol{r})$ is defined as the vector of frequency components of $\left|\boldsymbol{J}_\beta(\boldsymbol{r})\right|^2/\beta(\boldsymbol{r})$;

2. **Approximate equivalent GR noise sources for band-to-band transitions.** Application of (5.78) to (1.104) and (1.111) readily gives $(\alpha, \beta = n, p)$:

$$\left(\mathsf{K}_{\delta J_{\alpha,\kappa_1},\delta J_{\beta,\kappa_2}}(\boldsymbol{r};\omega)\right)_{q,n} = \frac{2q^2\tau_{\mathrm{eq}}^2}{\sqrt{[1+(\omega+\omega_q)^2\tau_{\mathrm{eq}}^2][1+(\omega+\omega_n)^2\tau_{\mathrm{eq}}^2]}} \times \left(\boldsymbol{F}_{2,\alpha\beta}(\boldsymbol{r})\right)_{q-n}, \tag{5.91}$$

where $\boldsymbol{F}_{2,\alpha\beta}(\boldsymbol{r})$ is defined as the vector of frequency components of $v_{\alpha,\kappa_1}(\boldsymbol{r})v_{\beta,\kappa_2}(\boldsymbol{r})\left(R_{\alpha}^{\mathrm{dir}}(\boldsymbol{r}) + G_{\alpha}^{\mathrm{dir}}(\boldsymbol{r})\right)$.

In both cases, $\kappa_1, \kappa_2 = x, y, z$ denote the spatial components of the current density fluctuations considered. As already remarked in Sect. 5.2.3, the low-pass nature of the colored microscopic SS noise sources makes the LS microscopic fluctuations practically modulated only by the DC component of the LS steady state.

5.4.2 Evaluation of the CGF

The CGFs are evaluated from the discretized linear system (5.84) already introduced for the SSLS analysis, though for noise analysis the linearized system must be slightly modified. As no external forcing term is applied to the device terminals, (5.84h) is no longer needed, while an auxiliary equation must be added to enforce short-circuit or open-circuit boundary conditions, according to whether the short-circuit or the open-circuit CGF is sought. To fix the ideas, we will refer in the following discussion to short-circuit boundary conditions. A unit diagonal forcing term, i.e. an independent sideband-by-sideband excitation, is then added, for each node, to the right-hand side of each continuity equation, enabling the evaluation of the CGFs. The resulting system of equations is:

$$\mathsf{C}_{\mathsf{f}_{\varphi,\boldsymbol{\varphi}_\mathrm{i}}} \cdot \mathsf{\Phi}_\mathrm{i}^{\mathrm{ss},+} + \mathsf{C}_{\mathsf{f}_{\varphi,\boldsymbol{\varphi}_\mathrm{x}}} \cdot \mathsf{\Phi}_\mathrm{x}^{\mathrm{ss},+} + \mathsf{C}_{\mathsf{f}_{\varphi,\boldsymbol{n}_\mathrm{i}}} \cdot \mathsf{N}_\mathrm{i}^{\mathrm{ss},+} + \mathsf{C}_{\mathsf{f}_{\varphi,\boldsymbol{p}_\mathrm{i}}} \cdot \mathsf{P}_\mathrm{i}^{\mathrm{ss},+} = 0, \tag{5.92a}$$

$$\begin{aligned}\mathsf{C}_{\mathsf{f}_{n,\boldsymbol{\varphi}_\mathrm{i}}} \cdot \mathsf{\Phi}_\mathrm{i}^{\mathrm{ss},+} + \mathsf{C}_{\mathsf{f}_{n,\boldsymbol{\varphi}_\mathrm{x}}} \cdot \mathsf{\Phi}_\mathrm{x}^{\mathrm{ss},+} + \mathsf{C}_{\mathsf{f}_{n,\boldsymbol{n}_\mathrm{i}}} \cdot \mathsf{N}_\mathrm{i}^{\mathrm{ss},+} + \mathsf{C}_{\mathsf{f}_{n,\boldsymbol{n}_\mathrm{x}}} \cdot \mathsf{N}_\mathrm{x}^{\mathrm{ss},+} \\ + \mathsf{C}_{\mathsf{f}_{n,\boldsymbol{p}_\mathrm{i}}} \cdot \mathsf{P}_\mathrm{i}^{\mathrm{ss},+} + \mathsf{C}_{\mathsf{f}_{n,\dot{\boldsymbol{n}}_\mathrm{i}}} \cdot \mathsf{D}_\mathrm{i}^{\mathrm{ss}} \cdot \mathsf{N}_\mathrm{i}^{\mathrm{ss},+} = \mathsf{S}_n\delta_{\alpha,n},\end{aligned} \tag{5.92b}$$

$$\begin{aligned}\mathsf{C}_{\mathsf{f}_{p,\boldsymbol{\varphi}_\mathrm{i}}} \cdot \mathsf{\Phi}_\mathrm{i}^{\mathrm{ss},+} + \mathsf{C}_{\mathsf{f}_{p,\boldsymbol{\varphi}_\mathrm{x}}} \cdot \mathsf{\Phi}_\mathrm{x}^{\mathrm{ss},+} + \mathsf{C}_{\mathsf{f}_{p,\boldsymbol{n}_\mathrm{i}}} \cdot \mathsf{N}_\mathrm{i}^{\mathrm{ss},+} + \mathsf{C}_{\mathsf{f}_{p,\boldsymbol{p}_\mathrm{i}}} \cdot \mathsf{P}_\mathrm{i}^{\mathrm{ss},+} \\ + \mathsf{C}_{\mathsf{f}_{p,\boldsymbol{p}_\mathrm{x}}} \cdot \mathsf{P}_\mathrm{x}^{\mathrm{ss},+} + \mathsf{C}_{\mathsf{f}_{p,\dot{\boldsymbol{p}}_\mathrm{i}}} \cdot \mathsf{D}_\mathrm{i}^{\mathrm{ss}} \cdot \mathsf{P}_\mathrm{i}^{\mathrm{ss},+} = \mathsf{S}_p\delta_{\alpha,p},\end{aligned} \tag{5.92c}$$

$$\mathsf{C}_{\mathsf{b}_{\varphi,\boldsymbol{\varphi}_\mathrm{x}}} \cdot \mathsf{\Phi}_\mathrm{x}^{\mathrm{ss},+} + \mathsf{C}_{\mathsf{b}_{\varphi,\boldsymbol{v}_\mathrm{c}}} \cdot \mathsf{V}_\mathrm{c}^{\mathrm{ss},+} = 0, \tag{5.92d}$$

$$\begin{aligned}\mathsf{C}_{\mathsf{b}_{n,\boldsymbol{\varphi}_\mathrm{i}}} \cdot \mathsf{\Phi}_\mathrm{i}^{\mathrm{ss},+} + \mathsf{C}_{\mathsf{b}_{n,\boldsymbol{\varphi}_\mathrm{x}}} \cdot \mathsf{\Phi}_\mathrm{x}^{\mathrm{ss},+} + \mathsf{C}_{\mathsf{b}_{n,\boldsymbol{n}_\mathrm{i}}} \cdot \mathsf{N}_\mathrm{i}^{\mathrm{ss},+} + \mathsf{C}_{\mathsf{b}_{n,\boldsymbol{n}_\mathrm{x}}} \cdot \mathsf{N}_\mathrm{x}^{\mathrm{ss},+} \\ + \mathsf{C}_{\mathsf{b}_{n,\boldsymbol{p}_\mathrm{x}}} \cdot \mathsf{P}_\mathrm{x}^{\mathrm{ss},+} + \mathsf{C}_{\mathsf{b}_{n,\dot{\boldsymbol{n}}_\mathrm{x}}} \cdot \mathsf{D}_\mathrm{x}^{\mathrm{ss}} \cdot \mathsf{N}_\mathrm{x}^{\mathrm{ss},+} = 0,\end{aligned} \tag{5.92e}$$

$$\mathsf{C}_{\mathsf{b}_{p,\varphi_\mathrm{i}}} \cdot \Phi_\mathrm{i}^{\mathrm{ss},+} + \mathsf{C}_{\mathsf{b}_{p,\varphi_\mathrm{x}}} \cdot \Phi_\mathrm{x}^{\mathrm{ss},+} + \mathsf{C}_{\mathsf{b}_{p,n_\mathrm{x}}} \cdot \mathsf{N}_\mathrm{x}^{\mathrm{ss},+} + \mathsf{C}_{\mathsf{b}_{p,p_\mathrm{i}}} \cdot \mathsf{P}_\mathrm{i}^{\mathrm{ss},+}$$
$$+ \mathsf{C}_{\mathsf{b}_{p,p_\mathrm{x}}} \cdot \mathsf{P}_\mathrm{x}^{\mathrm{ss},+} + \mathsf{C}_{\mathsf{b}_{p,\dot{p}_\mathrm{x}}} \cdot \mathsf{D}_\mathrm{x}^{\mathrm{ss}} \cdot \mathsf{P}_\mathrm{x}^{\mathrm{ss},+} = 0, \tag{5.92f}$$

$$\mathsf{C}_{\mathsf{c}_{\varphi_\mathrm{i}}} \cdot \Phi_\mathrm{i}^{\mathrm{ss},+} + \mathsf{C}_{\mathsf{c}_{\varphi_\mathrm{x}}} \cdot \Phi_\mathrm{x}^{\mathrm{ss},+} + \mathsf{C}_{\mathsf{c}_{n_\mathrm{i}}} \cdot \mathsf{N}_\mathrm{i}^{\mathrm{ss},+} + \mathsf{C}_{\mathsf{c}_{n_\mathrm{x}}} \cdot \mathsf{N}_\mathrm{x}^{\mathrm{ss},+} + \mathsf{C}_{\mathsf{c}_{p_\mathrm{i}}} \cdot \mathsf{P}_\mathrm{i}^{\mathrm{ss},+}$$
$$+ \mathsf{C}_{\mathsf{c}_{p_\mathrm{x}}} \cdot \mathsf{P}_\mathrm{x}^{\mathrm{ss},+} + \mathsf{C}_{\mathsf{c}_{\dot{n}_\mathrm{x}}} \cdot \mathsf{D}_\mathrm{x}^{\mathrm{ss}} \cdot \mathsf{N}_\mathrm{x}^{\mathrm{ss},+} + \mathsf{C}_{\mathsf{c}_{\dot{p}_\mathrm{x}}} \cdot \mathsf{D}_\mathrm{x}^{\mathrm{ss}} \cdot \mathsf{P}_\mathrm{x}^{\mathrm{ss},+}$$
$$+ \mathsf{C}_{\mathsf{c}_{i_\mathrm{c}}} \cdot \mathsf{G}_\alpha = 0, \tag{5.92g}$$

$$\mathsf{V}_\mathrm{c}^{\mathrm{ss},+} = 0. \tag{5.92h}$$

In (5.92) we have replaced the unknowns $\boldsymbol{I}_\mathrm{c}^{\mathrm{ss},+}$ with G_α corresponding to a global CGF collecting the CGFs $\mathsf{G}_{\alpha,i}$ defined for each terminal; δ is the Kronecker symbol, expressing that the source is injected either in the electron or in the hole continuity equation, to solve for the electron or hole CGF, respectively. Source terms S_n^+ and S_p^+ correspond to the unit source terms, homogeneous to an injected current, and are unit diagonal matrices of size $(2N_\mathrm{S}+1) \times N_\mathrm{i}$. Notice that the source term is explicitly placed in the internal nodes of the discretized model only, since injection in a terminal node results in a unit diagonal CGF for that terminal current, and to zero for the other terminal currents. In order to directly evaluate the CGFs, one factorization and $2(2N_\mathrm{S}+1)N_\mathrm{i}$ back-substitutions of the linear system (5.92) are therefore needed, thus making LS noise analysis a formidable task, unless proper numerical techniques are implemented. These can be interpreted as a straightforward extension of the generalized adjoint approach described in Sect. 3.3.2. This reduces the computational intensity to a number of system back-solves equal to $N_\mathrm{c}(2N_\mathrm{S}+1)$, i.e. the same number of back-solves needed for the evaluation of the terminal device CM by means of SSLS analysis.

5.5 Results

From a practical standpoint, physics-based modeling of noise in large-signal conditions is, at least for non-autonomous circuits, significant in strongly nonlinear devices, i.e. those which exhibit frequency conversion and harmonic generation. Devices operating in quasi-linear conditions, such as in power amplifiers are, on the other hand, less interesting simply because power systems operate well above the device noise floor.

On the basis of the above remarks, two application areas relevant to the large-signal noise modeling are passive and active mixers and frequency multipliers. A FET mixer noise figure evaluation based on a quasi-2D large-signal physics-based noise model was reported in [11]; the results are in good agreement with measured data.

We report here a case-study on a RF varactor diode frequency doubler (see Fig. 5.11), discussed in more detail in [10]. The circuit includes a reverse-biased silicon symmetrical *pn* diode (doping 10^{17} cm^{-3}, length 25 μm, area

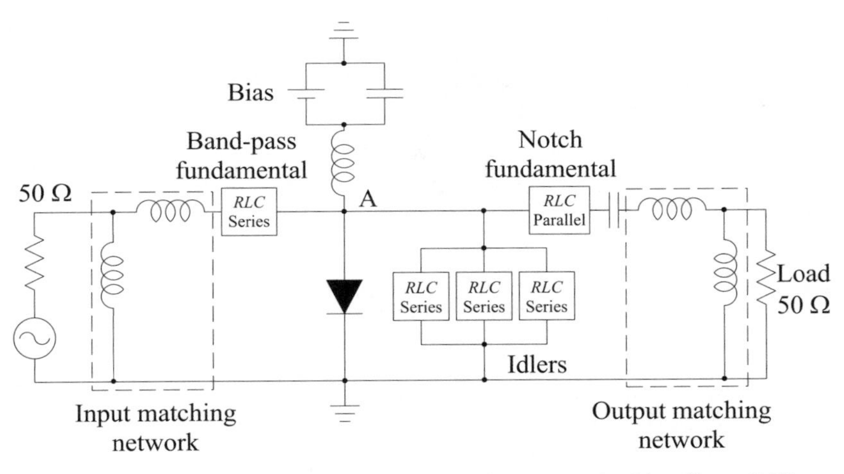

Fig. 5.11. Circuit scheme for the varactor frequency doubler (from [10])

10^4 μm^2), input and output matching networks, and idler resonators to suppress harmonics. The doubler circuit was optimized for frequency conversion from 1 to 2 GHz (fundamental to second harmonic) on the basis of a circuit-oriented diode model extracted from DC and small-signal simulations. The HB LS simulation of the diode (including the embedding circuit), was then performed for $N_{\mathrm{L}} = 4$ harmonics, plus DC; 350 grid points were used in the 1D device mesh.

The SCM of the short-circuit noise current sources of the diode was then evaluated and compared with circuit analysis based on a conventional noise diode model (derived according to [128,129]); the result is shown, for some of the SCM elements, in Fig. 5.12. The agreement is generally good, particularly for the diagonal elements.

The noise power spectrum on the load resistor is shown in Fig. 5.13, separating the total noise and the contribution from the source internal impedance; the agreement with the circuit model is again good. Notice that at the higher harmonics noise is found to be suppressed by the idler resonators, while, at the second harmonic, the contribution of the diode is slightly smaller than that of the generator resistance.

From physics-based simulation, the noise sideband correlation matrix is obtained as the spatial integral of a distributed, generalized noise density, (the integrand functions in (5.87)), whose inspection enables us to identify the device regions where noise (diagonal elements of the SCM) and noise frequency conversion (off-diagonal elements of the SCM) occur. In the varactor diode, the dominating contribution is due to thermal noise originating either in the p or n sides, seen in Fig. 5.14 as a constant plateau; the injection contribution is negligible. Notice that, in LS operation, the instantaneous working point of the diode goes from -9.4 V to -1.2 V; the noise density can thus be

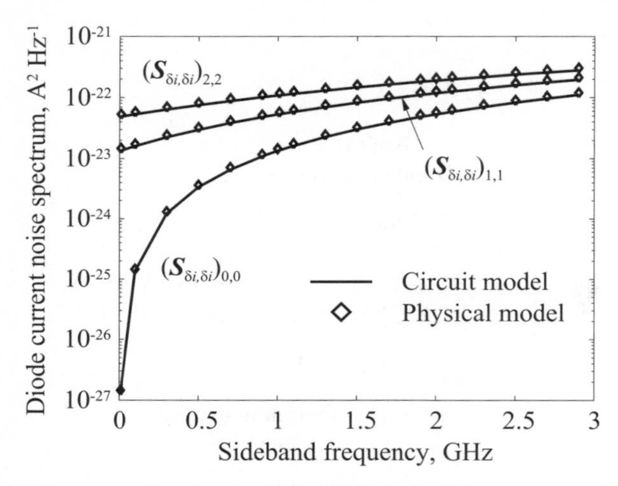

Fig. 5.12. Comparison between the frequency dependence of the diode noise current correlation spectra evaluated through circuit and physics-based analysis for the varactor frequency doubler. Only the power spectra of sidebands 0, 1 and 2 are shown (from [10])

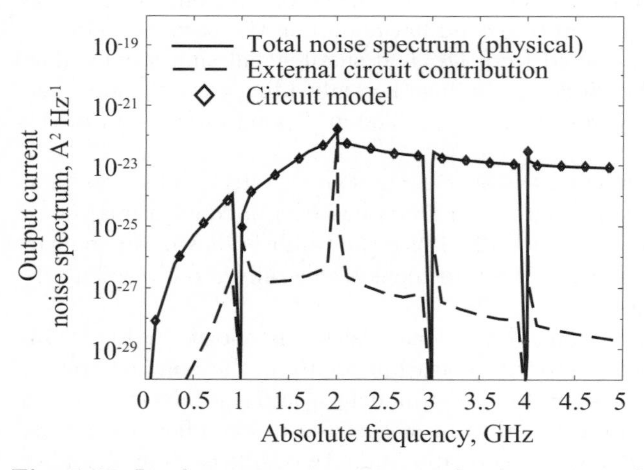

Fig. 5.13. Load current noise spectrum for the varactor doubler (from [10])

seen as the average between a set of reverse-biased operating points, at least in a quasi-static sense. This interpretation is consistent with a simple circuit interpretation, according to which the shot-like diode noise is negligible in reverse bias, thus making the noise of the parasitic resistance dominant. A further discussion concerning the behavior of a resistive doubler based on a forward-bias diode can be found in [10]; the analysis confirms, as expected, that noise conversion only takes place in the regions of the devices hosting reactive or resistive nonlinearities.

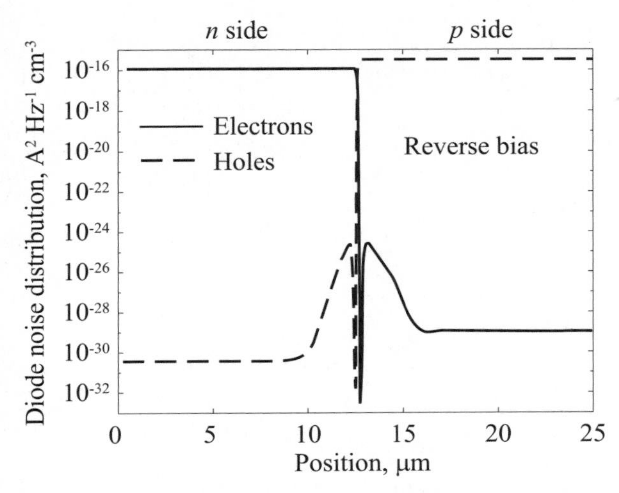

Fig. 5.14. Microscopic electron and hole noise distribution for the power spectrum of sideband 2 in the varactor frequency doubler (from [10])

A. Appendix: Review of Probability Theory and Random Processes

This Appendix is devoted to a brief introduction to the mathematics exploited in noise analysis, with particular attention to basic probability concepts, random processes and their linear transformations. The treatment is based mainly on [14, 27, 141]; a discussion on the statistical properties of random processes in quasi-periodic large-signal operation can also be found in [128].

A.1 Fundamentals of Probability Theory

Probability theory was born in the 16th century as an attempt to give a mathematical foundation to gambling. However, an axiomatic foundation, based on set theory, was given only in 1933 by Kolmogorov. The basic concept exploited to define probability is the *set of events*, i.e. a collection of all the possible *events* whose probability distribution is sought. Such events can represent almost everything: in our case they will be expressed mainly as vectors of integer, real or complex numbers. Let Ω indicate the set of all possible events, and $A \subseteq \Omega$ a subset corresponding to the events $\omega \in A$ under consideration. In order to properly define the probability of A, $P(A)$, the set of events must form a σ-algebra, i.e. it must be closed under set union and intersection. The *probability* $P(A)$ is then a real function of A satisfying the following three axioms:

1. $P(A) \geq 0$ for all A;
2. $P(\Omega) = 1$;
3. provided $\{A_i\}$ is a countable set of non-overlapping sets in the space of events (i.e. $A_i \cap A_j = \varnothing$ for $i \neq j$), the following property must hold:

$$P\left(\bigcup_i A_i\right) = \sum_i P(A_i).$$

From these, one easily derives that $P(\varnothing) = 0$ and $P(\overline{A}) = 1 - P(A)$, where $\overline{A}$ is the complement of A in Ω. Further, for two events A and B not necessarily disjoint one has

$$P(A \cup B) = P(A) + P(B) - P(A \cap B), \tag{A.1}$$

where $P(A \cap B)$ is the *joint probability* of events A and B, i.e. the probability associated to the joint occurrence of the two events. On the other hand, one can define the *conditional probability* of event A subordinated to the occurrence of event B as:

$$P(A|B) = \frac{P(A \cap B)}{P(B)}. \tag{A.2}$$

Event A is *independent* of B if the occurrence of event B has no effect on A; in this case one clearly has $P(A|B) = P(A)$, and therefore $P(A \cap B) = P(A)P(B)$. The property of statistical indepence is reflexive, since one has also $P(B|A) = P(B)$; in other words, if A is independent of B, B is also independent of A.

Random Variables and Probability Density. A *random variable* is defined as a real function $X(\cdot)$ of the elements of the set of events Ω. In order to provide a meaningful statistical definition of a random variable, such a function must be well behaved enough that the probability associated to the event $A_x = \{\omega \in \Omega : X(\omega) < x\}$ is defined for any real x. This guarantees the existence of the *distribution function* $F_X(x)$ associated to the random variable X:

$$F_X(x) = P(X < x) = P(A_x). \tag{A.3}$$

The distribution function is easily shown to be a positive, nondecreasing function limited by the values:

$$\lim_{x \to -\infty} F_X(x) = 0 \qquad \lim_{x \to +\infty} F_X(x) = 1. \tag{A.4}$$

The *probability density* $f_X(x)$ associated to a random variable X is often introduced as the first derivative of the distribution function, so that the following relationships hold

$$f_X(x) = \frac{\mathrm{d}}{\mathrm{d}x} F_X(x), \qquad F_X(x) = \int_{-\infty}^{x} f_X(x')\,\mathrm{d}x'. \tag{A.5}$$

The probability density satisfies properties:

$$f_X(x) \geq 0 \qquad \text{for all } x, \tag{A.6}$$

$$\int_{-\infty}^{+\infty} f_X(x)\,\mathrm{d}x = 1. \tag{A.7}$$

The probability density can be exploited to evaluate the probability associated to the random variable taking values in an interval $[x_1, x_2]$, since one has:

$$P\left(x_1 \leq X < x_2\right) = F_X(x_1) - F_X(x_2) = \int_{x_1}^{x_2} f_X(x)\,\mathrm{d}x. \tag{A.8}$$

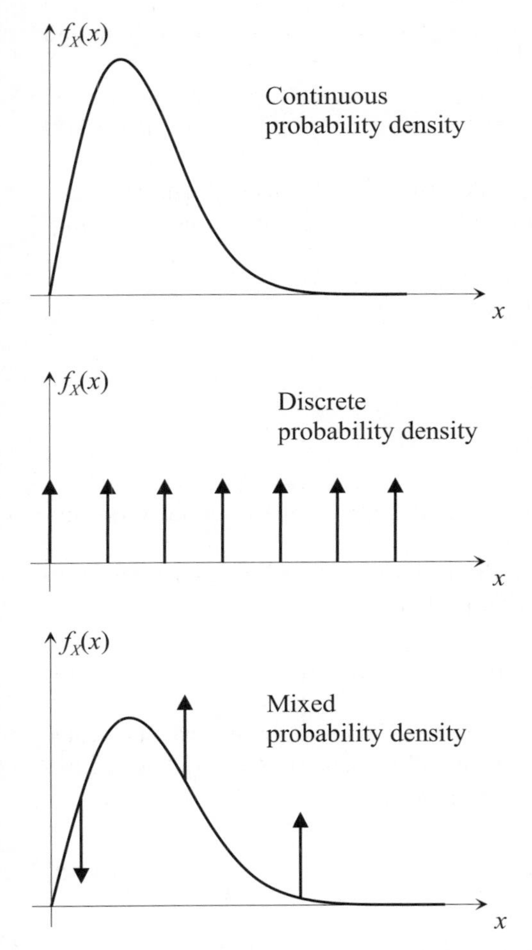

Fig. A.1. Representation of the three types of random variables in terms of their respective probability density: continuous (*top*), discrete (*middle*) or mixed (*bottom*)

The continuity properties of a random variable are strictly related to those of its probability density function: the latter can be *continuous*, *discrete* (i.e. made of a superposition of Dirac delta functions only) or *mixed*; the related random variable takes values, correspondingly, in a continuous, countable or mixed subset of real numbers (see Fig. A.1).

Two types of random variables, one continuous and one discrete, have paramount importance in noise theory, i.e. those characterized by the following distributions:

1. the *normal* or *Gaussian* random variable, characterized by a Gaussian probability density:

$$f_X(x) = \frac{1}{\sqrt{2\pi}\sigma} \mathrm{e}^{-[(x-\eta)^2/2\sigma^2]}, \tag{A.9}$$

   where the parameters η and σ, called the *mean* or *average* and the *variance*, respectively, will be discussed below; the corresponding probability distribution is given by (A.5):

$$F_X(x) = \frac{1}{2} + \frac{1}{2}\,\mathrm{erf}\left(\frac{x-\eta}{\sigma}\right), \tag{A.10}$$

   where erf is the error function:

$$\mathrm{erf}(\alpha) = \frac{2}{\sqrt{\pi}} \int_0^{\alpha} \mathrm{e}^{-x^2/2}\, \mathrm{d}x. \tag{A.11}$$

   Figure A.2 shows a representation of the Gaussian probability density and distribution for $\eta = 0$ and $\sigma = 1$.
2. the *Poisson* random variable, a discrete quantity which can take (non-negative) integer values only, with probability:

$$P(X = k) = \mathrm{e}^{-a} \frac{a^k}{k!} \qquad k = 0, 1, \ldots \qquad a > 0; \tag{A.12}$$

   the probability distribution is therefore constant between two successive integers, and the corresponding probability density is a sequence of impulses (Dirac delta functions, see Fig. A.3):

$$f_X(x) = \mathrm{e}^{-a} \sum_{k=0}^{+\infty} \frac{a^k}{k!} \delta(x-k). \tag{A.13}$$

A *complex* random variable is defined by requiring that its real and imaginary parts are real random variables according to the previous definition and conditions.

The concepts of distribution and density functions can also be easily extended to the case of several random variables. For instance, for two random variables X and Y the *joint* distribution function is defined as:

$$F_{X,Y}(x, y) = P\,(X < x \text{ and } Y < y)\,, \tag{A.14}$$

while the corresponding joint probability density is:

$$f_{X,Y}(x, y) = \frac{\partial^2 F_{X,Y}(x, y)}{\partial x \partial y}. \tag{A.15}$$

The *marginal* distribution and density are then defined from the joint functions as:

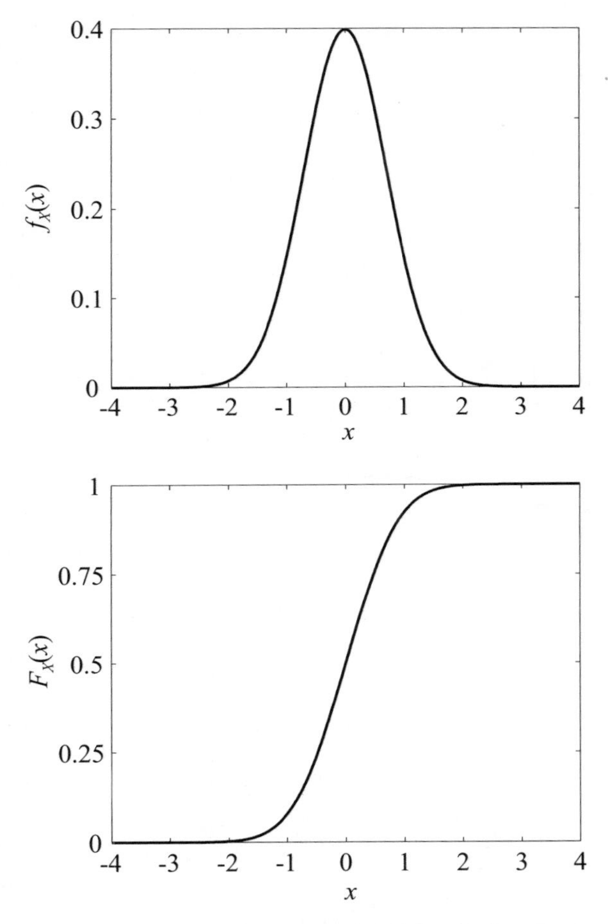

Fig. A.2. Probability density (*above*) and distribution (*below*) for a Gaussian random variable with $\eta = 0$ and $\sigma = 1$

$$F_X(x) = \lim_{y \to +\infty} F_{X,Y}(x, y), \tag{A.16}$$

$$f_X(x) = \int_{-\infty}^{+\infty} f_{X,Y}(x, y) \, \mathrm{d}y. \tag{A.17}$$

The *conditional* distribution and density are, by definition:

$$F_{X|Y}(x|Y < y) = \frac{F_{X,Y}(x, y)}{F_Y(y)}, \tag{A.18}$$

$$f_{X|Y}(x|Y = y) = \frac{f_{X,Y}(x, y)}{f_Y(y)}. \tag{A.19}$$

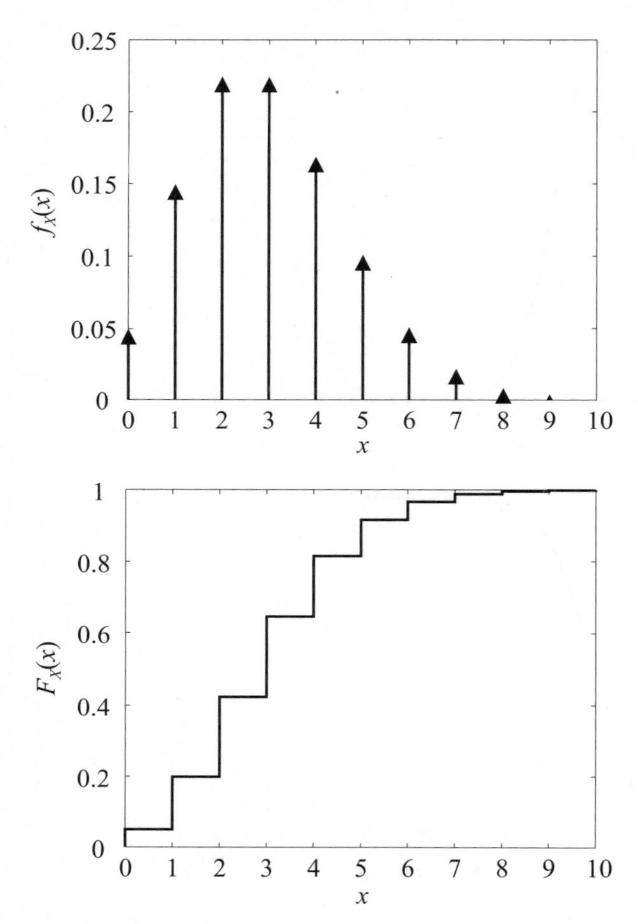

Fig. A.3. Probability density (*above*) and distribution (*below*) for a Poisson random variable with $a = 3$

Finally, it can be shown that the statistical independence of two random variables is equivalent to the following condition on the joint distribution or probability density:

$$F_{X,Y}(x, y) = F_X(x)F_Y(y), \tag{A.20}$$

$$f_{X,Y}(x, y) = f_X(x)f_Y(y). \tag{A.21}$$

Functions of Random Variables. A random variable Y obtained from a random variable X through the application of a deterministic function g has statistical properties given by

$$f_Y(y) = \frac{f_X\left[g^{-1}(y)\right]}{\left|g'\left[g^{-1}(y)\right]\right|}, \tag{A.22}$$

where g' is the first derivative of function g, provided that g and g^{-1} are differentiable. In the general case of vector random variables $\boldsymbol{Y}$, $\boldsymbol{X}$ and of a vector deterministic function $\boldsymbol{g}$, the joint probability densities are related through:

$$f_{\boldsymbol{Y}}(\boldsymbol{y}) = \frac{f_{\boldsymbol{X}}\left[\boldsymbol{g}^{-1}(\boldsymbol{y})\right]}{\det\left\{\mathsf{J}\left[\boldsymbol{g}^{-1}(\boldsymbol{y})\right]\right\}}, \tag{A.23}$$

where J is the Jacobian of $\boldsymbol{g}$, and $\det\{\mathsf{J}\}$ its determinant.

The application of a function has no effect on the conditional probability, meaning that the conditional probability density of $\boldsymbol{Z}$ conditioned to a value $\boldsymbol{Y} = \boldsymbol{g}(\boldsymbol{X})$ is given by:

$$f_{\boldsymbol{Z}|\boldsymbol{Y}}(\boldsymbol{z}|\boldsymbol{Y} = \boldsymbol{y}) = f_{\boldsymbol{Z}|\boldsymbol{X}}(\boldsymbol{z}|\boldsymbol{X} = \boldsymbol{g}^{-1}(\boldsymbol{y})). \tag{A.24}$$

Expectation, Moments and Correlation. The *expected value* $\langle X\rangle$ of a random variable X is a number representing the average of the variable values over the sample space, weighted by the probability associated to the occurrence of the variable values:

$$\langle X\rangle = \sum_{\omega\in\Omega} \omega P(\omega), \tag{A.25}$$

where the sum must be interpreted in an extended way, i.e. as an integral if the random variable is continuous. The expected value can be expressed in terms of the probability density as follows:

$$\langle X\rangle = \int_{-\infty}^{+\infty} x f_X(x)\,\mathrm{d}x. \tag{A.26}$$

The expectation operator is linear, meaning that given two random variables X and Y, one has $\langle\alpha X + \beta Y\rangle = \alpha\langle X\rangle + \beta\langle Y\rangle$ for any real or complex numbers α, β. Further, the expected value of a function of a random variable can be evaluated through the *fundamental theorem of expectation*:

$$\langle g(X)\rangle = \int_{-\infty}^{+\infty} g(x) f_X(x)\,\mathrm{d}x. \tag{A.27}$$

The concept of *conditional expectation* is introduced as the expected value of a random variable X assuming the occurrence of an event B; therefore, the conditional probability density is involved:

$$\langle X\rangle_B = \int_{-\infty}^{+\infty} x f_{X|B}(x)\,\mathrm{d}x. \tag{A.28}$$

The *characteristic function* associated to the random variable is the expected value of the complex random variable $\exp(\mathrm{i}\omega X)$:

$$\Phi_X(\omega) = \langle \mathrm{e}^{\mathrm{i}\omega X} \rangle = \int_{-\infty}^{+\infty} \mathrm{e}^{\mathrm{i}\omega x} f_X(x)\, \mathrm{d}x, \tag{A.29}$$

i.e., the conjugate of the Fourier transform of the probability density. The characteristic function is often exploited to evaluate the *moment of order n* of X, defined as $\langle X^n \rangle$, since one has:

$$\langle X^n \rangle = \frac{1}{\mathrm{i}^n} \left. \frac{\mathrm{d}^n \Phi_X}{\mathrm{d}\omega^n} \right|_{\omega=0} . \tag{A.30}$$

The first moment of variable X is its *mean* or *average*, $m_X = \langle X \rangle$. The *variance* σ_X^2 is the expected value of the square of the variable fluctuation with respect to its mean value, i.e. $\sigma_X^2 = \left\langle (X - m_X)^2 \right\rangle = \langle X^2 \rangle - m_X^2$. The moments of the fluctuation of a random variable with respect to its average are often called *central moments*, and therefore the variance is the *second central moment* of X. The mean represents the center of mass of the probability density, while the variance is a measure of the spreading of the probability density around its center m_X. For instance, for the Gaussian distribution (A.9) one has:

$$m_X = \eta, \qquad \sigma_X = \sigma. \tag{A.31}$$

Notice that for a Gaussian process, every moment of order higher than 2 is zero, therefore the first two moments completely define the probability density, and consequently the statistical properties of the random variable itself.

The second moment of a random variable X is called *(auto) correlation*

$$R_{X,X} = \langle X X^* \rangle = \int_{-\infty}^{+\infty} \int_{-\infty}^{+\infty} |x|^2 f_X(x)\, \mathrm{d}x, \tag{A.32}$$

where the complex conjugate applies for complex random variables only. By extension, the second joint moment of two random variables X, Y is their *(cross) correlation*

$$R_{X,Y} = \langle X Y^* \rangle = \int_{-\infty}^{+\infty} \int_{-\infty}^{+\infty} x y^* f_{X,Y}(x, y)\, \mathrm{d}x \mathrm{d}y. \tag{A.33}$$

On the other hand, the second central moment is the *(cross) covariance* $K_{X,Y} = \langle (X - m_X)(Y - m_Y)^* \rangle = R_{X,Y} - m_X m_Y^*$. Two random variables are *uncorrelated* if $K_{X,Y} = 0$, but are *orthogonal* if $R_{X,Y} = 0$. Notice that if two random variables are statistically independent, they are also uncorrelated, but not orthogonal unless at least one has zero average. Further, two uncorrelated random variables are not necessarily also statistically independent, unless they are *jointly Gaussian*, i.e. if any linear combination of them is a Gaussian random variable. Based on the cross-covariance, the *correlation coefficient* is defined as

$$C_{X,Y} = \frac{K_{X,Y}}{\sigma_X \sigma_Y}. \tag{A.34}$$

For the general case of a vector of n random variables $\boldsymbol{X} = [X_1, \ldots, X_n]^{\mathrm{T}}$, the *(auto) correlation matrix* is defined as $\mathsf{R}_{\boldsymbol{X},\boldsymbol{X}} = \left\langle \boldsymbol{X}\boldsymbol{X}^\dagger \right\rangle$, where $\dagger$ is the Hermitian conjugate, and the *(self) covariance matrix* is

$$\mathsf{K}_{\boldsymbol{X},\boldsymbol{X}} = \left\langle (\boldsymbol{X} - \boldsymbol{m}_{\boldsymbol{X}})(\boldsymbol{X} - \boldsymbol{m}_{\boldsymbol{X}})^\dagger \right\rangle = \mathsf{R}_{\boldsymbol{X},\boldsymbol{X}} - \boldsymbol{m}_{\boldsymbol{X}} \boldsymbol{m}_{\boldsymbol{X}}^\dagger. \tag{A.35}$$

A.2 Random Processes

A *random process* $X(\cdot,\cdot)$ is a random variable[1] depending on an independently varying parameter, and therefore is a function of two distinct entities: the set of events, and the parameter space. The latter can be continuous or discrete, although in our applications we shall always deal, at least as an approximation, with the continuous case; further, the dimension of the parameter space can be one, for time-dependent random processes, or higher than one, e.g. for space- and time-dependent random processes. In general, time dependence is always present in our random processes, therefore we shall confine this treatment to the simplest case; the additional complexity introduced by the space dependence can be treated in a straightforward manner. Notice that a random process depending on time is often called a *stochastic process*.

For any fixed element ω of the set of events Ω, the deterministic function of time $X(\cdot,\omega)$ represents a *realization* of the random process. On the other hand, each time value t defines a random variable $X(t,\cdot)$ whose statistical properties represent the characteristics of the random process itself. For the sake of simplicity, the dependence of a random process on the set of events will be dropped from the notation: $X(t,\omega) = X(t)$; further, the probability density associated to $X(t,\cdot)$ will be denoted as $f_{X(t)}(x) = f_X(x,t)$.

For any random process, two kinds of averaging procedure can be defined:

- the *ensemble average* or *mean* $m_X(t)$, introduced as the mean of the random variable $X(t)$: $m_X(t) = \langle X(t) \rangle$. In general, the ensemble average is a deterministic function of time;

[1] Complex random processes can also be defined according to the definition of complex random variables.

- the *time average* or *mean*, which is a random variable, obtained as the time integral of each realization of the random process:

$$\overline{X(t)} = \lim_{T \to +\infty} \frac{1}{T} \int_{-\infty}^{+\infty} X(t) \, \mathrm{d}t.$$

Among random processes, a very important class is that of *ergodic* processes. This concept is quite involved (e.g. see [27,141] for a discussion), but, broadly speaking, can be stated as follows: the statistical properties of an ergodic random process can be obtained from a single realization through time averages. From this definition, it is apparent that a random process can exhibit several degrees of ergodicity, depending on the degree of the statistical properties which can be derived from time-averaging. In the simplest case, a random process is *ergodic of the mean* if its time average is deterministic and equal to the ensemble average, which is, therefore, also time-independent. We shall assume our processes satisfy such an assumption.

The *autocorrelation function* of random process X is the correlation of the two random variables $X(t_1)$ and $X(t_2)$:

$$R_{X,X}(t_1, t_2) = \langle X(t_1) X^*(t_2) \rangle, \tag{A.36}$$

where again the complex conjugate makes sense for complex random processes only. Extending the definition given for random variables, the concept of *autocovariance function* can be exploited as well:

$$\begin{aligned} K_{X,X}(t_1, t_2) &= \langle (X(t_1) - m_X(t_1))(X(t_2) - m_X(t_2))^* \rangle \\ &= R_{X,X}(t_1, t_2) - m_X(t_1) m_X^*(t_2). \end{aligned} \tag{A.37}$$

Definitions (A.36) and (A.37) can be extended to yield the *cross-correlation* and *cross-covariance functions* between two random processes X and Y:

$$R_{X,Y}(t_1, t_2) = \langle X(t_1) Y^*(t_2) \rangle, \tag{A.38}$$

$$\begin{aligned} K_{X,Y}(t_1, t_2) &= \langle (X(t_1) - m_X(t_1))(Y(t_2) - m_Y(t_2))^* \rangle \\ &= R_{X,Y}(t_1, t_2) - m_X(t_1) m_Y^*(t_2). \end{aligned} \tag{A.39}$$

A random process X is *Gaussian* if any n-tuple of time samples $\boldsymbol{X} = [X(t_1) \ldots X(t_n)]^{\mathrm{T}}$ is a set of jointly Gaussian random variables. This is the simplest random process, since its statistical properties are completely defined by the mean $\boldsymbol{m_X}$ and the covariance matrix $\mathsf{K}_{\boldsymbol{X},\boldsymbol{X}}$. Because of this simplicity, and of the fundamental property that a linear transformation of a Gaussian random process yields as output again a Gaussian random process (see Sect. A.4), they are widely exploited to represent fluctuation phenomena.

Markov, Stationary and Cyclostationary Random Processes. A random process is a *Markov process* if the conditional probability density of the process conditioned to an initial value at time t_0 is independent of the previous history of the process itself. Analytically, provided $\boldsymbol{X}(t)$ is the (vector) stochastic process under consideration, and $t_1, \ldots, t_n$, $\tau_1, \ldots, \tau_n$ are nondecreasing time samples ($t_1 \geq t_2 \geq \cdots \geq t_n \geq \tau_1 \geq \cdots \geq \tau_n$), the conditional probability density must satisfy ($\boldsymbol{x}_i = \boldsymbol{X}(t_i)$ and $\boldsymbol{y}_i = \boldsymbol{X}(\tau_i)$):

$$f_{\boldsymbol{X}}(\boldsymbol{x}_1, t_1; \ldots; \boldsymbol{x}_n, t_n | \boldsymbol{y}_1, \tau_1; \ldots; \boldsymbol{y}_n, \tau_n) = f_{\boldsymbol{X}}(\boldsymbol{x}_1, t_1; \ldots; \boldsymbol{x}_n, t_n | \boldsymbol{y}_1, \tau_1). \tag{A.40}$$

The class of Markov processes is of paramount importance in noise analysis, thus its main features will be further discussed in Sect. A.6.

Another class of stochastic processes with great practical importance is characterized by the following two properties:

1. the mean is independent of time:

$$m_X(t) = m_X; \tag{A.41}$$

2. the autocorrelation function, and therefore the autocovariance function as well, depends only on the time difference $\tau = t_1 - t_2$:

$$R_{X,X}(t_1, t_2) = R_{X,X}(\tau) = \langle X(t+\tau) X^*(t) \rangle \qquad \text{for any } t. \tag{A.42}$$

Such processes are called *wide sense stationary* (WSS). The autocorrelation function of a real WSS process has the following properties:

- $R_{X,X}$ is an even function of τ: $R_{X,X}(\tau) = R_{X,X}(-\tau)$;
- $R_{X,X}$ has its maximum at the origin: $R_{X,X}(\tau) \leq R_{X,X}(0)$;
- $R_{X,X}(\tau)$ is a continuous function of τ provided it is continuous for $\tau = 0$;

and admits of a simple physical interpretation: it is a measure of the correlation between the two random functions obtained by sampling the random process at two times separated by τ. Therefore, the second property is easily interpreted as a consequence of the reduced correlation between the two samples as the time difference increases. Further, intuition suggests that $R_{X,X}(\tau)$ goes to zero, as $|\tau|$ increases, faster as the realizations of $X(t)$ exhibit rapid variations. A WSS random process whose autocorrelation function is proportional to $\delta(\tau)$ is called *white noise*. A graphic interpretation of such considerations can be found in Fig. A.4.

Two random processes $X(t)$ and $Y(t)$ are *jointly* WSS if any linear combination of them is WSS. This definition can be proved to be equivalent to three conditions:

1. the mean of both processes is independent of time:

$$m_X(t) = m_X, \qquad m_Y(t) = m_Y; \tag{A.43}$$

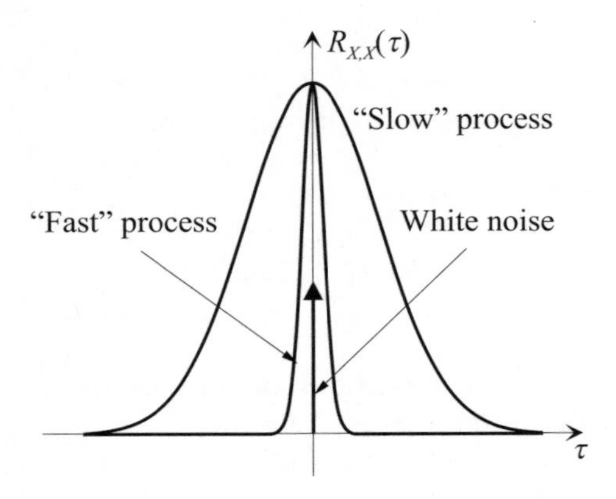

Fig. A.4. τ-dependence of the autocorrelation function for a WSS random process, and representation of the autocorrelation function of a white noise

2. the autocorrelation functions of both processes depend only on the time difference $\tau = t_1 - t_2$:

$$R_{X,X}(t_1, t_2) = R_{X,X}(\tau), \qquad R_{Y,Y}(t_1, t_2) = R_{Y,Y}(\tau); \tag{A.44}$$

3. the cross-correlation function depends only on the time difference $\tau = t_1 - t_2$:

$$R_{X,Y}(t_1, t_2) = R_{X,Y}(\tau) = \langle X(t + \tau)Y^*(t)\rangle \qquad \text{for any } t. \tag{A.45}$$

As an example, let us consider a physical system amenable to be modeled as a source of particles emitted at random times and a collector which absorbs them. The number $N(t)$ of particles collected at time t is a discrete random process, termed a *Poisson process,* whose realizations are staircase functions with unit steps occurring at random times, the instants when the particle hits the collector (see Fig. A.5).

Since the mean of $N(t)$ is clearly an increasing function of time, the Poisson process is not stationary. On the other hand, the time derivative $X(t) = \mathrm{d}N/\mathrm{d}t$ is a stationary random process called *Poisson increments,* whose realizations are collections of Dirac's delta functions (see Fig. A.6). The mean and autocorrelation of $X(t)$ are [27]:

$$m_X = \lambda, \qquad R_{X,X}(\tau) = \lambda^2 + \lambda\delta(\tau) \tag{A.46}$$

where λ is the mean number of particles collected per unit time; the (zero average) fluctuations $\delta X(t) = X(t) - \lambda$ of the Poisson increments are a white noise process whose correlation function is proportional to the increments mean:

$$m_{\delta X} = 0, \qquad R_{\delta X,\delta X}(\tau) = \lambda\delta(\tau). \tag{A.47}$$

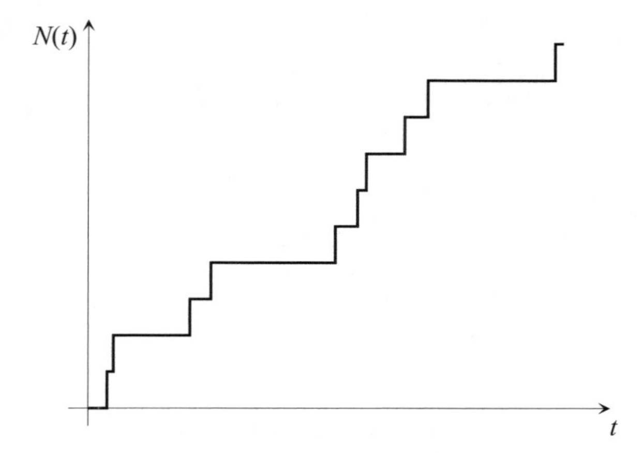

Fig. A.5. Realization of a Poisson process

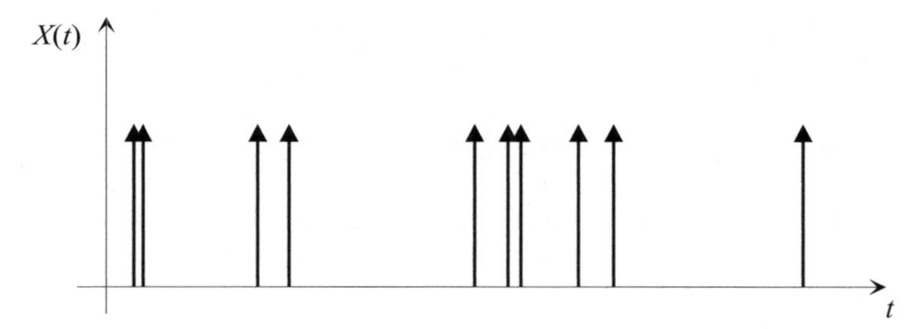

Fig. A.6. Realization of a Poisson increments process

This result is also known as *Campbell's theorem*, see [27]. Random processes satisfying the aforementioned conditions are called *shot noise*. In circuit and device noise applications, Campbell's theorem is often exploited to model short-circuit current noise. Let us consider identical charged particles, with charge q. The corresponding collected current is therefore $i(t) = qX(t)$, and the current fluctuations $\delta i(t) = i(t) - \langle i \rangle = q\delta X(t)$ are characterized by

$$R_{\delta i,\delta i}(\tau) = q^2 R_{\delta X,\delta X}(\tau) = q^2 \lambda \delta(\tau) = q\langle i \rangle \delta(\tau). \tag{A.48}$$

Among non-stationary processes, a particular class has great importance for noise modeling in (quasi-) periodic large signal operation of semiconductor devices. Such processes exhibit a *periodic* dependence on time of their statistical properties. A random process is *wide sense cyclostationary* if, for any times t, t_1 and t_2, the following properties hold for a fixed period T:

$$m_X(t) = m_X(t+T), \qquad R_{X,X}(t_1,t_2) = R_{X,X}(t_1+T, t_2+T). \tag{A.49}$$

The condition on the periodicity of the autocorrelation function can be expressed also in the following form:

$$R_{X,X}\left(t+\frac{\tau}{2},t-\frac{\tau}{2}\right)=R_{X,X}\left(t+\frac{\tau}{2}+T,t-\frac{\tau}{2}+T\right); \tag{A.50}$$

this enables us to represent the autocorrelation function as a Fourier series with respect to variable t:

$$R_{X,X}\left(t+\frac{\tau}{2},t-\frac{\tau}{2}\right)=\sum_{n=-\infty}^{+\infty}R_{X,X}^{(n)}(\tau)\mathrm{e}^{\mathrm{i}n\omega_0 t}, \tag{A.51}$$

$$R_{X,X}^{(n)}(\tau)=\frac{1}{T}\int_{-T/2}^{T/2}R_{X,X}\left(t+\frac{\tau}{2},t-\frac{\tau}{2}\right)\mathrm{e}^{-\mathrm{i}n\omega_0 t}\,\mathrm{d}t, \tag{A.52}$$

where $\omega_0=2\pi/T$. The Fourier coefficient $R_{X,X}^{(n)}(\tau)$ is called the *cyclic autocorrelation function* of the cyclostationary process.

These definitions can be extended to *quasi-cyclostationary* processes, i.e. processes whose periodicity is not exact but characterized by the occurrence of a superposition of several periodic components T_i $(i=1,2,3,\ldots)$. Equation (A.51) is still valid, but the summation index is now a vector $\boldsymbol{n}=[n_1,n_2,n_3,\ldots]^{\mathrm{T}}$ and (A.52) is replaced by:

$$R_{X,X}^{(\boldsymbol{n})}(\tau)=\lim_{T\to+\infty}\frac{1}{T}\int_{-T/2}^{T/2}R_{X,X}\left(t+\frac{\tau}{2},t-\frac{\tau}{2}\right)\mathrm{e}^{-\mathrm{i}\sum_i n_i\omega_i t}\,\mathrm{d}t, \tag{A.53}$$

where $\omega_i=2\pi/T_i$. Further details on cyclostationary stochastic processes and their transformation through linear time-invariant and linear periodic time-varying systems are reported in Sect. A.5.

A.3 Correlation Spectra and Generalized Harmonic Analysis of Stochastic Processes

Let us consider first the case of (jointly) WSS processes $X(t)$ and $Y(t)$. The *bilateral autocorrelation (power) spectrum* of $X(t)$ is defined as the Fourier transform of the autocorrelation function:

$$\hat{S}_{X,X}(\omega)=\int_{-\infty}^{+\infty}R_{X,X}(\tau)\mathrm{e}^{-\mathrm{i}\omega\tau}\,\mathrm{d}\tau, \tag{A.54}$$

$$R_{X,X}(\tau)=\frac{1}{2\pi}\int_{-\infty}^{+\infty}\hat{S}_{X,X}(\omega)\mathrm{e}^{\mathrm{i}\omega\tau}\,\mathrm{d}\omega, \tag{A.55}$$

where the integrals must be interpreted in an extended way (distribution topology). The concept can be generalized to the *cross-correlation (bilateral) spectrum* of two WSS processes as:

$$\hat{S}_{X,Y}(\omega)=\int_{-\infty}^{+\infty}R_{X,Y}(\tau)\mathrm{e}^{-\mathrm{i}\omega\tau}\,\mathrm{d}\tau, \tag{A.56}$$

$$R_{X,Y}(\tau)=\frac{1}{2\pi}\int_{-\infty}^{+\infty}\hat{S}_{X,Y}(\omega)\mathrm{e}^{\mathrm{i}\omega\tau}\,\mathrm{d}\omega. \tag{A.57}$$

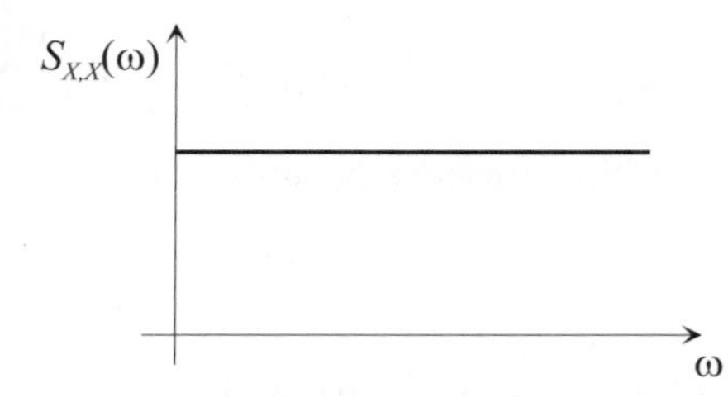

Fig. A.7. Power spectrum of a white noise WSS process

Since, for real processes, the correlation function is a real and even function of τ, the corresponding spectra are even functions of ω, and therefore all the relevant information is included in the spectrum for $\omega \geq 0$ only. As a consequence, the *unilateral* spectra are often exploited in noise analysis:

$$S_{X,X}(\omega) = 2\hat{S}_{X,X}(\omega), \qquad S_{X,Y}(\omega) = 2\hat{S}_{X,Y}(\omega) \qquad \text{for } \omega \geq 0. \tag{A.58}$$

For a white noise WSS process, characterized by $R_{X,X}(\tau) \propto \delta(\tau)$, the autocorrelation spectrum is constant and independent of frequency (see Fig. A.7). For instance, Campbell's theorem (A.48) reads

$$S_{\delta i,\delta i}(\omega) = 2q\langle i \rangle. \tag{A.59}$$

Let us consider now a WSS stochastic process characterized by a periodic autocorrelation function, i.e. such that $R_{X,X}(\tau + T) = R_{X,X}(\tau)$ for some period T. The deterministic function $R_{X,X}(\tau)$ can be expanded in Fourier series:

$$R_{X,X}(\tau) = \sum_{n=-\infty}^{+\infty} R^{(n)} \mathrm{e}^{\mathrm{i}n\omega_0\tau}, \tag{A.60}$$

where $\omega_0 = 2\pi/T$ and:

$$R^{(n)} = \frac{1}{T} \int_0^T R_{X,X}(\tau) \mathrm{e}^{-\mathrm{i}n\omega_0\tau} \, \mathrm{d}\tau. \tag{A.61}$$

The unilateral power spectrum of $X(t)$ is therefore:

$$S_{X,X}(\omega) = 4\pi \sum_{n=-\infty}^{+\infty} R^{(n)} \delta(\omega - n\omega_0). \tag{A.62}$$

For non-stationary random processes $X_1(t)$ and $X_2(t)$, the (bilateral) correlation spectrum is defined as the double Fourier transform of the correlation function:

$$\hat{G}_{X_1,X_2}(\omega_1, \omega_2) = \int_{-\infty}^{+\infty} \int_{-\infty}^{+\infty} R_{X_1,X_2}(t_1, t_2) \mathrm{e}^{-\mathrm{i}(\omega_1 t_1 - \omega_2 t_2)} \, \mathrm{d}t_1 \mathrm{d}t_2, \tag{A.63}$$

$$R_{X_1,X_2}(t_1,t_2) = \frac{1}{4\pi^2} \int_{-\infty}^{+\infty} \int_{-\infty}^{+\infty} \hat{G}_{X_1,X_2}(\omega_1,\omega_2) e^{i(\omega_1 t_1 - \omega_2 t_2)} \, d\omega_1 d\omega_2. \quad \text{(A.64)}$$

According to this generalization, for jointly WSS processes it can be shown that $R_{X_1,X_2}(t_1,t_2) = R_{X_1,X_2}(t_1 - t_2)$ implies:

$$\hat{G}_{X_1,X_2}(\omega_1,\omega_2) = 2\pi \hat{S}_{X_1,X_2}(\omega_1) \delta(\omega_1 - \omega_2). \quad \text{(A.65)}$$

The physical interpretation of the correlation spectrum depends on the definition, whose deep mathematical meaning is quite complex [27], of the Fourier transform of a random process, i.e. on the *generalized harmonic analysis* of random processes. Formally one has:

$$\tilde{X}(\omega) = \int_{-\infty}^{+\infty} X(t) e^{-i\omega t} \, dt; \quad \text{(A.66)}$$

starting from this definition, the following relations can be proved:

$$m_{\tilde{X}}(\omega) = \int_{-\infty}^{+\infty} m_X(t) e^{-i\omega t} \, dt, \quad \text{(A.67)}$$

$$\hat{G}_{X_1,X_2}(\omega_1,\omega_2) = \left\langle \tilde{X}_1(\omega_1) \tilde{X}_2^*(\omega_2) \right\rangle, \quad \text{(A.68)}$$

i.e. the correlation spectrum is the correlation between the spectral components of the stochastic processes evaluated at the two frequencies ω_1 and ω_2. According to this interpretation, WSS processes are characterized by spectral components uncorrelated unless $\omega_1 = \omega_2$ (see (A.65)); non-stationary processes, on the other hand, are characterized by correlated frequency components (see Fig. A.8).

From (A.68) and (A.65) one immediately has:

$$\hat{S}_{X_1,X_2}(\omega) = \hat{S}^*_{X_2,X_1}(\omega). \quad \text{(A.69)}$$

A.4 Linear Transformations of Stochastic Processes

Let us consider a vector stochastic process $\boldsymbol{Y}(t)$, obtained from a vector stochastic process $\boldsymbol{X}(t)$ through a linear transformation characterized by the (matrix) impulse response $\mathsf{h}(t,u)$ (see Fig. A.9):

$$\boldsymbol{Y}(t) = \int_{-\infty}^{+\infty} \mathsf{h}(t,u) \cdot \boldsymbol{X}(u) \, du. \quad \text{(A.70)}$$

Since the averaging procedure is a linear operator, the following properties are easily derived:

$$\langle \boldsymbol{Y} \rangle = \int_{-\infty}^{+\infty} \mathsf{h}(t,u) \cdot \langle \boldsymbol{X}(u) \rangle \, du, \quad \text{(A.71)}$$

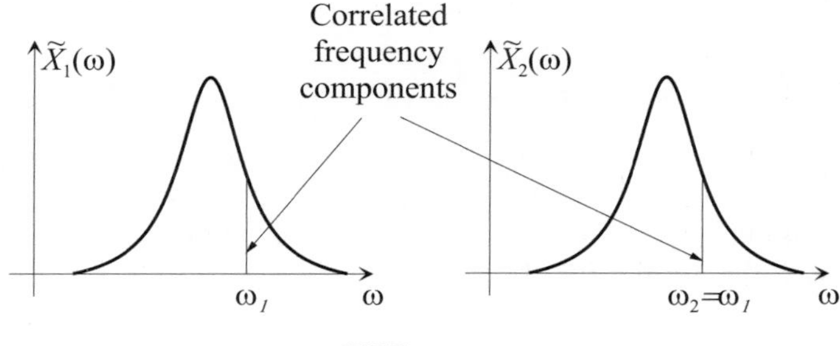

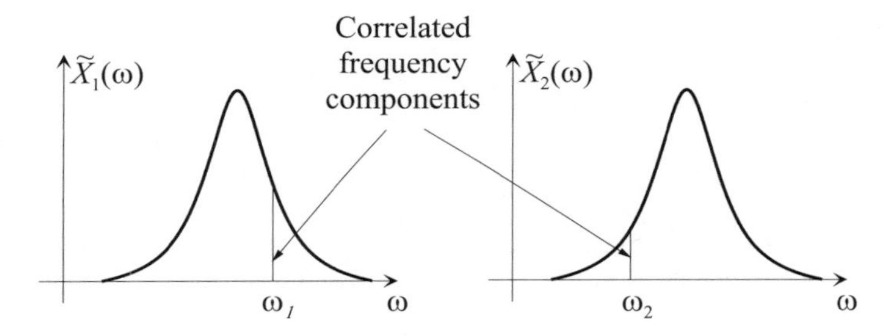

Fig. A.8. Spectral domain interpretation of (A.68) for WSS (*above*) and non-stationary (*below*) stochastic processes

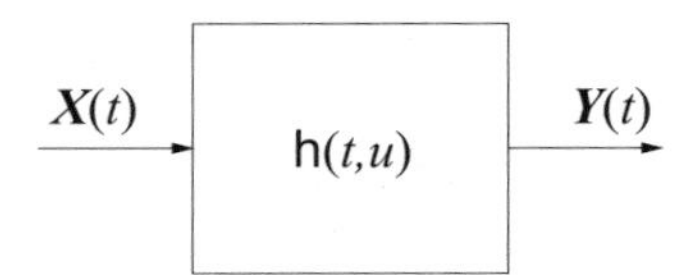

Fig. A.9. Representation of a linear transformation of a stochastic process $\boldsymbol{X}$

$$\mathsf{R}_{\boldsymbol{Y}_1,\boldsymbol{Y}_2}(t_1,t_2) = \int_{-\infty}^{+\infty}\int_{-\infty}^{+\infty} \mathsf{h}_1(t_1,u_1)\cdot\mathsf{R}_{\boldsymbol{X}_1,\boldsymbol{X}_2}(u_1,u_2)\cdot\mathsf{h}_2^{\dagger}(t_2,u_2)\; du_1 du_2, \tag{A.72}$$

where $\boldsymbol{Y}_1$ ($\boldsymbol{Y}_2$) is obtained by filtering $\boldsymbol{X}_1$ ($\boldsymbol{X}_2$) through a linear system with impulse response h_1 (h_2).

The effect of filtering by means of a linear system is conveniently expressed in the frequency domain provided the linear system itself is *time-invariant*, i.e. if $\mathsf{h}(t,u) = \mathsf{h}(t-u)$. Exploiting the generalized harmonic analysis described in Sect. A.3, condition

$$\tilde{\boldsymbol{Y}}_i(\omega) = \tilde{\mathsf{h}}_i(\omega)\cdot\tilde{\boldsymbol{X}}_i(\omega) \qquad i = 1,2 \tag{A.73}$$

and (A.68) yield

$$\hat{\mathsf{G}}_{\boldsymbol{Y}_1,\boldsymbol{Y}_2}(\omega_1,\omega_2) = \tilde{\mathsf{h}}_1(\omega_1) \cdot \hat{\mathsf{G}}_{\boldsymbol{X}_1,\boldsymbol{X}_2}(\omega_1,\omega_2) \cdot \tilde{\mathsf{h}}_2^{\dagger}(\omega_2). \tag{A.74}$$

If the input processes are jointly WSS, the same property holds for $\boldsymbol{Y}_1(t)$ and $\boldsymbol{Y}_2(t)$ (this can be seen from (A.72), for time-invariant linear transformations, through a simple change of variables). Therefore, exploiting (A.65) the following relationship is established:

$$\mathsf{S}_{\boldsymbol{Y}_1,\boldsymbol{Y}_2}(\omega) = \tilde{\mathsf{h}}_1(\omega) \cdot \mathsf{S}_{\boldsymbol{X}_1,\boldsymbol{X}_2}(\omega) \cdot \tilde{\mathsf{h}}_2^{\dagger}(\omega). \tag{A.75}$$

A.5 Cyclostationary Stochastic Processes

Cyclostationary stochastic processes have paramount importance in the noise analysis of devices and circuits operated in the (quasi-) periodic harmonic regime, as discussed in [128] and Chap. 5. In noise analysis, cyclostationarity usually arises from amplitude modulation of a WSS stochastic process $x(t)$ by means of a real, (quasi-) periodic function of time $h(t)$. Due to (quasi-) periodicity, $h(t)$ can be represented, exactly or approximately, as a superposition of harmonics corresponding to a discrete set of angular frequencies ω_k:

$$h(t) = \sum_k H_k \mathrm{e}^{\mathrm{i}\omega_k t}. \tag{A.76}$$

Since $h(t)$ is real, for any index k, af index k' must exist such that

$$\omega_{k'} = -\omega_k, \qquad H_{k'} = H_k^*. \tag{A.77}$$

The modulated process $y(t) = h(t)x(t)$ is characterized by a mean value:

$$m_y(t) = h(t)m_x, \tag{A.78}$$

and by the autocorrelation function:

$$\begin{aligned} R_{y,y}(t_1,t_2) &= \langle y(t_1)y^*(t_2)\rangle \\ &= \sum_{k,l} H_k H_l^* \mathrm{e}^{\mathrm{i}\omega_k t_1 - \mathrm{i}\omega_l t_2} \langle x(t_1)x^*(t_2)\rangle \\ &= \sum_{k,l} H_k H_l^* \mathrm{e}^{\mathrm{i}\omega_k t_1 - \mathrm{i}\omega_l t_2} R_{x,x}(t_1 - t_2). \end{aligned} \tag{A.79}$$

Therefore, condition (A.49) is met and $y(t)$ results to be a (quasi-) cyclostationary process with periodicity defined by the same property of $h(t)$.

The corresponding bilateral autocorrelation spectrum is evaluated according to definition (A.63):

$$\begin{aligned}
\hat{G}_{y,y}(\omega_1, \omega_2) & \\
&= \sum_{k,l} H_k H_l^* \int_{-\infty}^{+\infty} \int_{-\infty}^{+\infty} R_{x,x}(t_1 - t_2) \mathrm{e}^{-\mathrm{i}[(\omega_1 - \omega_k)t_1 - (\omega_2 - \omega_l)t_2]} \, \mathrm{d}t_1 \mathrm{d}t_2 \\
&= \sum_{k,l} H_k H_l^* \int_{-\infty}^{+\infty} \int_{-\infty}^{+\infty} R_{x,x}(\tau) \mathrm{e}^{-\mathrm{i}[(\omega_1 - \omega_k)(\tau + t_2) - (\omega_2 - \omega_l)t_2]} \, \mathrm{d}\tau \mathrm{d}t_2 \\
&= \sum_{k,l} H_k H_l^* \int_{-\infty}^{+\infty} R_{x,x}(\tau) \mathrm{e}^{-\mathrm{i}(\omega_1 - \omega_k)\tau} \, \mathrm{d}\tau \\
&\qquad \times \int_{-\infty}^{+\infty} \mathrm{e}^{-\mathrm{i}[(\omega_1 - \omega_k) - (\omega_2 - \omega_l)]t_2} \, \mathrm{d}t_2 \\
&= 2\pi \sum_{k,l} H_k H_l^* \hat{S}_{x,x}(\omega_1 - \omega_k) \delta[(\omega_1 - \omega_k) - (\omega_2 - \omega_l)],
\end{aligned} \tag{A.80}$$

i.e. the spectral components $\tilde{y}(\omega_1)$ and $\tilde{y}(\omega_2)$ are correlated if and only if they have the same distance $\omega = \omega_1 - \omega_k = \omega_2 - \omega_l$ from one of the fundamental frequencies of the modulating function (see Fig. A.10 for a graphical interpretation). In other words, only the frequency components of $y(t)$ satisfying this property are correlated.

The neighborhood of any fundamental frequency ω_i, characterized by a local distance ω (see Fig. A.11), is called the *upper (lower) sideband* of ω_i if $\omega > 0$ ($\omega < 0$). Usually, sidebands are defined as non-overlapping intervals, i.e. $\omega \leq \omega_0/2$. According to this description, the autocorrelation spectrum of $y(t)$ is completely determined by a correlation matrix whose elements, dependent on ω only, express the correlation spectra between the amplitude of the various sidebands. Such a matrix is called a *sideband correlation matrix* (SCM). In the above example (see (A.80)) the (k, l) SCM element is defined as:

$$(\mathsf{S}_{y,y}(\omega))_{k,l} = H_k H_l^* \hat{S}_{x,x}(\omega). \tag{A.81}$$

The SCM has, in principle, infinite dimension, but can be suitably truncated. Finally, notice that for $\omega > 0$ ($\omega < 0$) the SCM describes the correlation of upper (lower) sidebands; in this particular case $\mathsf{S}_{x,x}(\omega) = \mathsf{S}_{x,x}(-\omega)$ and therefore the SCM element describing the correlation of upper and lower sidebands is the same.

Equation (A.80) is a particular case of the general expression for the autocorrelation spectrum of a cyclostationary stochastic process $x(t)$, since from (A.51):

$$\hat{G}_{x,x}(\omega_1, \omega_2) = 2\pi \sum_{n=-\infty}^{+\infty} \hat{S}_{x,x}^{(n)} \left(\omega_1 - n\frac{\omega_0}{2} \right) \delta(\omega_1 - \omega_2 - n\omega_0), \tag{A.82}$$

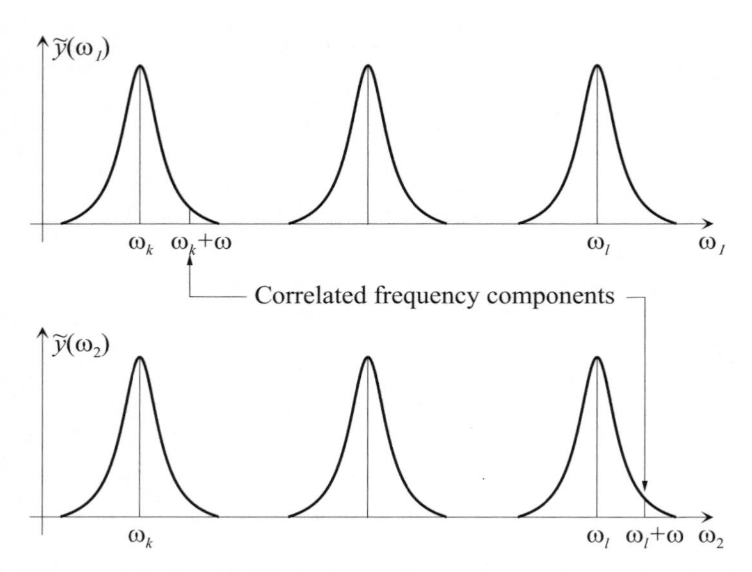

Fig. A.10. Correlation between the spectral components of a stochastic process obtained through amplitude modulation of a stationary random process by means of a (quasi-) periodic deterministic function

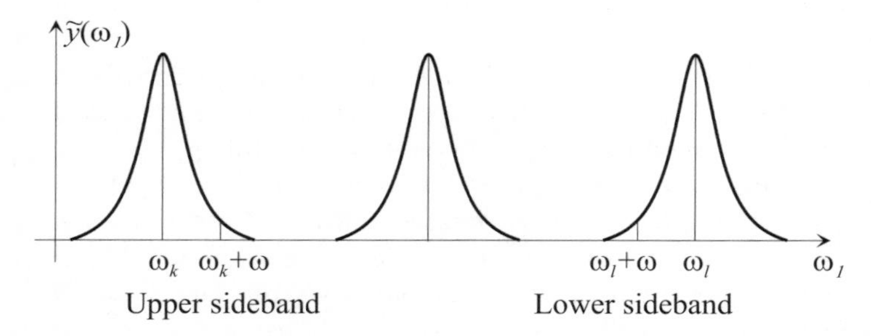

Fig. A.11. Definition of upper and lower sideband of a fundamental frequency ω_i

where $\hat{S}_{x,x}^{(n)}(\omega)$, the *cyclic autocorrelation spectrum*, is the Fourier transform of the cyclic autocorrelation function $R_{x,x}^{(n)}(\tau)$. From (A.82) we notice that the autocorrelation spectrum is zero unless $\omega_1 - \omega_2 - n\omega_0 = 0$. Assume we select two upper sidebands with absolute angular frequencies $\omega_1 = \omega + k\omega_0$ and $\omega_2 = \omega + l\omega_0$, respectively; the above requirement implies that $n = k - l$. In this case the (k, l) SCM element will be defined as:

$$(\mathsf{S}_{x,x}(\omega))_{k,l} = \hat{S}_{x,x}^{(k-l)}\left(\omega + \frac{k+l}{2}\omega_0\right). \tag{A.83}$$

The SCM of a cyclostationary process can be given a spectral interpretation as follows (see, for a rigorous treatment, [142, 143]). Given a cyclostationary process $x(t)$, its Fourier trasform $\tilde{x}$ can be interpreted as an array of band-limited functions $X_n(\omega)$ where $-\omega_0/2 < \omega < \omega_0/2$ is the sideband angular frequency and n denotes the n-th sideband. Then, the SCM element (k, l) simply is:

$$(\mathsf{S}_{x,x}(\omega))_{k,l} = \langle X_k(\omega) X_l^*(\omega) \rangle. \tag{A.84}$$

Finally, the property of (quasi-) cyclostationarity is conserved by time-invariant linear transformations. In fact, provided $x_1(t)$ and $x_2(t)$ are two jointly cyclostationary stochastic processes, the cross-correlation function of the stochastic processes $y_1(t)$ and $y_2(t)$ generated through two linear, time-invariant transformations with impulse response h_1 and h_2:

$$y_1(t) = \int_{-\infty}^{+\infty} h_1(t-\tau) x_1(\tau)\, \mathrm{d}\tau, \qquad y_2(t) = \int_{-\infty}^{+\infty} h_2(t-\tau) x_2(\tau)\, \mathrm{d}\tau, \tag{A.85}$$

is given by

$$\begin{aligned} R_{y_1,y_2}(t_1, t_2) &= \int_{-\infty}^{+\infty} \int_{-\infty}^{+\infty} h_1(t_1 - \tau_1) R_{x_1,x_2}(\tau_1, \tau_2) h_2^*(\tau_2)\, \mathrm{d}\tau_1 \mathrm{d}\tau_2 \\ &= \int_{-\infty}^{+\infty} \int_{-\infty}^{+\infty} h_1(u_1) R_{x_1,x_2}(t_1 - u_1, t_2 - u_2) h_2^*(u_2)\, \mathrm{d}u_1 \mathrm{d}u_2 \end{aligned} \tag{A.86}$$

and is therefore periodic. In general, it can be shown that cyclostationarity is also conserved by a time-periodic linear transformation having the same period as the input process, see [142, 143].

A.6 A Glimpse of Markov Stochastic Processes

This section provides a very introductory description to the mathematical properties and analysis techniques of Markov processes. A more rigorous treatment can be found in the literature, e.g. in [14, 15, 144].

The time evolution of a Markov process $\boldsymbol{X}(t)$ can be studied according to two different, although equivalent, approaches. The first exploits classical analysis, deriving equations for the time-varying probability density $f_{\boldsymbol{X}}(x, t)$. Details on this technique can be found in [14, 144]. The second, although somewhat more appealing to an intuitive view, is much more complex from a mathematical standpoint: it amounts to adding a zero-average stochastic forcing term to the (partial) differential equation describing the system dynamics, therefore converting such an equation into a *stochastic (partial) differential equation* [15], which require the introduction of proper calculus rules[2] to be correctly treated. We shall describe both techniques, highlighting the conditions to be satisified to prove their equivalence.

[2] Two schemes have been proposed: Ito's calculus and Stratonovich's calculus, which basically differ for the definition of integration exploited. Details on this can be found in [15].

A general property of Markov processes is that they satisfy the Chapman–Kolmogorov equation for the conditional probability density:

$$f_{\boldsymbol{X}}(\boldsymbol{x}_1, t_1|\boldsymbol{x}_3, t_3) = \int f_{\boldsymbol{X}}(\boldsymbol{x}_1, t_1|\boldsymbol{x}_2, t_2) f_{\boldsymbol{X}}(\boldsymbol{x}_2, t_2|\boldsymbol{x}_3, t_3)\ \mathrm{d}\boldsymbol{x}_2, \tag{A.87}$$

which, under some mild conditions, can be expressed in differential form (forward *differential* Chapman–Kolmogorov equation, according to the terminology in [14]):

$$\begin{aligned} \frac{\partial f_{\boldsymbol{X}}(\boldsymbol{z}, t|\boldsymbol{y}, t_0)}{\partial t} =& -\sum_i \frac{\partial}{\partial z_i}\left[A_i(\boldsymbol{z}, t) f_{\boldsymbol{X}}(\boldsymbol{z}, t|\boldsymbol{y}, t_0)\right] \\ &+ \frac{1}{2}\sum_{i,j} \frac{\partial^2}{\partial z_i \partial z_j}\left[B_{ij}(\boldsymbol{z}, t) f_{\boldsymbol{X}}(\boldsymbol{z}, t|\boldsymbol{y}, t_0)\right] \\ &+ \int \left[W_{\boldsymbol{x},\boldsymbol{z}} f_{\boldsymbol{X}}(\boldsymbol{x}, t|\boldsymbol{y}, t_0) - W_{\boldsymbol{z},\boldsymbol{x}} f_{\boldsymbol{X}}(\boldsymbol{z}, t|\boldsymbol{y}, t_0)\right]\ \mathrm{d}\boldsymbol{x}, \end{aligned} \tag{A.88}$$

where

- $W_{\boldsymbol{z},\boldsymbol{x}}$ is the probability per unit time of an instantaneous transition at time t from $\boldsymbol{z}$ to $\boldsymbol{x}$ (*transition rate*):

$$W_{\boldsymbol{z},\boldsymbol{x}} = \lim_{\Delta t \to 0} \frac{f_{\boldsymbol{X}}(\boldsymbol{x}, t + \Delta t|\boldsymbol{z}, t)}{\Delta t};$$

- functions A_i and B_{ij} are given by:

$$\begin{aligned} A_i(\boldsymbol{z}, t) &= \lim_{\Delta t \to 0} \int (x_i - z_i) \frac{f_{\boldsymbol{X}}(\boldsymbol{x}, t + \Delta t|\boldsymbol{z}, t)}{\Delta t}\ \mathrm{d}\boldsymbol{x} \\ B_{ij}(\boldsymbol{z}, t) &= \lim_{\Delta t \to 0} \int (x_i - z_i)(x_j - z_j) \frac{f_{\boldsymbol{X}}(\boldsymbol{x}, t + \Delta t|\boldsymbol{z}, t)}{\Delta t}\ \mathrm{d}\boldsymbol{x}; \end{aligned}$$

- the solution must satisfy the initial condition

$$f_{\boldsymbol{X}}(\boldsymbol{z}, t_0|\boldsymbol{y}, t_0) = \delta(\boldsymbol{y} - \boldsymbol{z}). \tag{A.89}$$

Two limiting cases of (A.88) have a great importance in physical applications:

- the *master equation* (ME), for $A_i = B_{ij} = 0$:

$$\frac{\partial f_{\boldsymbol{X}}(\boldsymbol{z}, t|\boldsymbol{y}, t_0)}{\partial t} = \int \left[W_{\boldsymbol{x},\boldsymbol{z}} f_{\boldsymbol{X}}(\boldsymbol{x}, t|\boldsymbol{y}, t_0) - W_{\boldsymbol{z},\boldsymbol{x}} f_{\boldsymbol{X}}(\boldsymbol{z}, t|\boldsymbol{y}, t_0)\right]\ \mathrm{d}\boldsymbol{x} \tag{A.90}$$

  which describes a *jump process*, i.e. a process corresponding to discontinuous paths in the state space. Therefore, this approach is specifically useful to describe discrete processes;

- the *Fokker–Planck equation*, for $W_{\boldsymbol{x},\boldsymbol{z}} = 0$:

$$\frac{\partial f_{\boldsymbol{X}}(\boldsymbol{z}, t|\boldsymbol{y}, t_0)}{\partial t} = -\sum_i \frac{\partial}{\partial z_i} \left[A_i(\boldsymbol{z}, t) f_{\boldsymbol{X}}(\boldsymbol{z}, t|\boldsymbol{y}, t_0)\right] + \frac{1}{2}\sum_{i,j} \frac{\partial^2}{\partial z_i \partial z_j} \left[B_{ij}(\boldsymbol{z}, t) f_{\boldsymbol{X}}(\boldsymbol{z}, t|\boldsymbol{y}, t_0)\right] \tag{A.91}$$

  which describes a *diffusion process*, i.e. a process corresponding to continuous paths in the state space. Therefore, this equation is exploited in classical or semi-classical descriptions, as those applied to model carrier transport in a semiconductor.

Provided one can derive from statistical mechanics physical principles the ME governing the time evolution of the conditional probability density, this can be approximated by a diffusion process, i.e. a Fokker–Planck equation, by means of the Kramers–Moyal expansion:

$$\frac{\partial f_{\boldsymbol{X}}(\boldsymbol{z}, t|\boldsymbol{y}, t_0)}{\partial t} = \sum_{n=1}^{+\infty} \frac{(-1)^n}{n!} \left(\frac{\partial}{\partial \boldsymbol{z}}\right)^n : \left[\mathsf{F}_n(\boldsymbol{z}) f_{\boldsymbol{X}}(\boldsymbol{z}, t|\boldsymbol{y}, t_0)\right], \tag{A.92}$$

where $\mathsf{F}_n(\boldsymbol{z})$, the n-th order Fokker–Planck moment, is a rank-n tensor defined by

$$\mathsf{F}_n(\boldsymbol{z}) = \int (\boldsymbol{x} - \boldsymbol{z})^n \, W_{\boldsymbol{z},\boldsymbol{x}} \, \mathrm{d}\boldsymbol{x}$$

and ":" means a sum over all corresponding indices so that a scalar results, e.g. for vectors and rank-2 tensors:

$$\boldsymbol{a} : \boldsymbol{b} = \sum_i a_i b_i,$$
$$\mathsf{A} : \mathsf{B} = \sum_{i,j} A_{ij} B_{ij}.$$

The approximation sought is obtained provided one can assume, at least approximately, $\mathsf{F}_n = 0$ for $n \geq 3$, since this reduces (A.92) to (A.91) wherein

$$\boldsymbol{A}(\boldsymbol{z}) = \boldsymbol{F}_1(\boldsymbol{z}),$$
$$\mathsf{B}(\boldsymbol{z}) = \mathsf{F}_2(\boldsymbol{z}).$$

The mean value of the Markov stochastic process $\boldsymbol{z}$ satisfies the dynamic equation (see [16] for a proof)

$$\frac{\partial \langle \boldsymbol{z} \rangle_{\boldsymbol{y}}}{\partial t} = \langle \boldsymbol{F}_1(\boldsymbol{z}) \rangle_{\boldsymbol{y}}, \tag{A.93}$$

where subscript $\boldsymbol{y}$ stands for a mean performed on the conditional probability density $f_{\boldsymbol{X}}(\boldsymbol{z}, t|\boldsymbol{y}, t_0)$.

This Fokker–Planck approximation is useful since it is equivalent [16, 17] to the stochastic (partial) differential equation (a Langevin equation) for the Markov stochastic process $\boldsymbol{z}$:

$$\frac{\partial \boldsymbol{z}}{\partial t} = \boldsymbol{F}_1[\boldsymbol{z}(t)] + \boldsymbol{\xi}(t), \tag{A.94}$$

where the Langevin force $\boldsymbol{\xi}(t)$ is a white stochastic process whose correlation spectrum is defined by the second Fokker–Planck moment evaluated in the average steady state $\langle \boldsymbol{z} \rangle_{\boldsymbol{y}}$:

$$\mathsf{S}_{\boldsymbol{\xi},\boldsymbol{\xi}} = 2\mathsf{F}_2\left(\langle \boldsymbol{z} \rangle_{\boldsymbol{y}}\right). \tag{A.95}$$

It should be noted that in some cases this procedure must be carried out with some further care due to complications arising from the presence of a spatial dependence in the stochastic processes. Details on this can be found, for example, in Chap. 8 of [14].

References

1. W. Shockley, J. A. Copeland, R. P. James: The impedance field method of noise calculation in active semiconductor devices, In *Quantum theory of atoms, molecules, and the solid-state*, ed. by Per-Olov Lowdin (Academic Press, New York, 1966) pp. 537–563
2. P. A. Layman: *The analysis of thermal and $1/f$ noise in MOS devices, circuits and systems*, PhD thesis (University of Waterloo, 1989)
3. G. Ghione, E. Bellotti, F. Filicori: Physical noise modelling of majority-carrier devices: an adjoint network approach, In *Proc. International Electron Device Meeting*, Washington, USA, 1989, pp. 351–355
4. G. Ghione, F. Filicori: A computationally efficient unified approach to the numerical analysis of the sensitivity and noise of semiconductor devices, IEEE Trans. Computer-Aided Design **CAD–12** (1993) pp. 425–438
5. F. Bonani, M. R. Pinto, R. K. Smith, G. Ghione: An efficient approach to multi-dimensional impedance field noise simulation of bipolar devices, In *Proc. 13th Int. Conf. on Noise in Physical Systems and 1/f Noise*, ed. by V. Bareikis, R. Katilius (World Scientific, Singapore, 1995) pp. 379–382
6. F. Bonani, G. Ghione, M. R. Pinto, R. K. Smith: An efficient approach to noise analysis through multidimensional physics-based models, IEEE Trans. Electron Devices **ED-45** (1998) pp. 261–269
7. H. Happy, A. Cappy: *Helena: HEMT electrical properties and noise analysis software and user's manual* (Artech House, Norwood, 1993)
8. A. Cappy, W. Heinrich: High-frequency FET noise performances: a new approach, IEEE Trans. Electron Devices **ED-36** (1989) pp. 403–409
9. F. Bonani, G. Ghione, S. Donati, L. Varani, L. Reggiani: A general framework for the noise analysis of semiconductor devices operating in nonlinear (large-signal quasi-periodic) conditions, In *Proc. 14th Int. Conf. on Noise in Physical Systems and 1/f Noise*, ed. by C. Claeys, E. Simoen (World Scientific, Singapore, 1997) pp. 144–147
10. F. Bonani, S. Donati Guerrieri, G. Ghione, M. Pirola: A TCAD approach to the physics-based modelling of frequency conversion and noise in semiconductor devices under large-signal forced operation, accepted for publication in IEEE Trans. Electron Devices **ED-48** (2001) pp. 966–977
11. F. Danneville, G. Dambrine, A. Cappy: Noise modelling in MESFET and HEMT mixers using a uniform noisy line model, IEEE Trans. Electron Devices **ED-45** (1998) pp. 2207–2212
12. H. C. de Graaf, F. M. Klaassen: *Compact transistor modelling for circuit design* (Springer-Verlag, Wien, 1990)
13. M. S. Gupta: Conductance fluctuations in mesoscopic conductors at low temperatures, IEEE Trans. Electron Devices **ED-41** (1994) pp. 2093–2106
14. C. W. Gardiner: *Handbook of stochastic methods* (Springer-Verlag, New York, Berlin, Heidelberg, 1990) 2nd edition

15. K. Sobczyk: *Stochastic differential equations* (Kluwer Academic Publishers, London, 1991)
16. C. M. van Vliet: Macroscopic and microscopic methods for noise in devices, IEEE Trans. Electron Devices **ED-41** (1994) pp. 1902–1915
17. K. M. van Vliet: Markov approach to density fluctuations due to transport and scattering. I. Mathematical formalism, J. Math. Phys. **12** (1971) pp. 1981–1998
18. K. M. van Vliet: Markov approach to density fluctuations due to transport and scattering. II. Applications, J. Math. Phys. **12** (1971) pp. 1998–2012
19. K. M. van Vliet: Linear transport theory revisited. I. The many-body van Hove limit, J. Math. Phys. **19** (1978) pp. 1345–1370
20. K. M. van Vliet: Linear transport theory revisited. II. The master-equation approach, J. Math. Phys. **20** (1979) pp. 2573–2595
21. M. Charbonneau, K. M. van Vliet, P. Vasilopoulos: Linear transport theory revisited. III. One-body response formulas and generalized Boltzmann equations, J. Math. Phys. **23** (1982) pp. 318–336
22. C. M. van Vliet, P. Vasilopoulos: The quantum Boltzmann equation and some applications to transport phenomena, J. Phys. Chem. Solids **49** (1988) pp. 639–649
23. F. N. Hooge: $1/f$ noise sources, IEEE Trans. Electron Devices **ED-41** (1994) pp. 1926–1935
24. P. H. Handel: Fundamental quantum $1/f$ noise in semiconductor devices, IEEE Trans. Electron Devices **ED-41** (1994) pp. 2023–2033
25. G. E. Uhlenbeck, L. S. Ornstein: On the theory of the Brownian motion, Phys. Rev. **36** (1930) pp. 823–841
26. S. Chandrasekar: Stochastic problems in physics and astronomy, Rev. Modern Phys. **15** (1943) pp. 1–89
27. A. Papoulis: *Probability, random variables and stochastic processes* (McGraw-Hill, Kogakusha, 1965)
28. J. P. Nougier: Fluctuations and noise of hot carriers in semiconductor materials and devices, IEEE Trans. Electron Devices **ED-41** (1994) pp. 2034–2049
29. J. P. Nougier: Noise and diffusion of hot carriers, In *Physics of nonlinear transport in semiconductors*, ed. by D. K. Ferry, J. R. Barker, C. Jacoboni (Plenum Press, New York, 1980) p. 415
30. P. J. Price: Fluctuations of hot electrons, In *Fluctuations phenomena in solids*, ed. by R. E. Burgess (Academic Press, New York, 1965) pp. 355–380
31. J. P. Nougier: Identity between spreading and noise diffusion coefficients for hot carriers in semiconductors, Appl. Phys. Lett. **32** (1978) pp. 671–673
32. J. P. Nougier, M. Rolland: Differential relaxation times and diffusivities of hot carriers in isotropic semiconductors, J. Appl. Phys. **48** (1977) pp. 1683–1687
33. G. Ghione, F. Bonani, M. Pirola: High-field diffusivity and noise spectra in GaAs MESFETs, J. Phys. D: Appl. Phys. **27** (1994) pp. 365–375
34. K. M. van Vliet, J. R. Fassett: Fluctuations due to electronic transitions and transport in solids, In *Fluctuations phenomena in solids*, ed. by R. E. Burgess (Academic Press, New York, 1965) pp. 267–354
35. S. Selberherr: *Analysis and simulation of semiconductor devices* (Springer-Verlag, Wien, 1984)
36. M. R. Pinto: *Comprehensive semiconductor device simulation for silicon ULSI*, PhD thesis (Stanford University, 1990)
37. W. Shockley, W. T. Read: Statistics of the recombination of holes and electrons, Phys. Rev. **87** (1952) pp. 835–842
38. R. N. Hall: Electron-hole recombination in germanium, Phys. Rev. **87** (1952) p. 387

39. F. Bonani, G. Ghione: Generation-recombination noise modelling in semiconductor devices through population or approximate equivalent current density fluctuations, Solid-State Electron. **43** (1999) pp. 285–295
40. L. K. J. Vandamme, X. Li, D. Rigaud: $1/f$ noise in MOS devices, mobility or number fluctuations?, IEEE Trans. Electron Devices **ED-41** (1994) pp. 1936–1945
41. J. H. Scofield, N. Borland, D. M. Fleetwood: Reconciliation of different gate-voltage dependencies of $1/f$ noise in n-MOS and p-MOS transistors, IEEE Trans. Electron Devices **ED-41** (1994) pp. 1946–1952
42. D. M. Fleetwood, T. L. Meisenheimer, J. H. Scofield: $1/f$ noise and radiation effects in MOS devices, IEEE Trans. Electron Devices **ED-41** (1994) pp. 1953–1964
43. J. Chang, A. A. Abidi, C. R. Viswanathan: Flicker noise in CMOS transistors from subthreshold to strong inversion at various temperatures, IEEE Trans. Electron Devices **ED-41** (1994) pp. 1965–1971
44. T. G. M. Kleinpenning: Low-frequency noise in modern bipolar transistors: impact of intrinsic transistor and parasitic series resistances, IEEE Trans. Electron Devices **ED-41** (1994) pp. 1981–1991
45. J. C. Costa, D. Ngo, R. Jackson, N. Camilleri, J. Jaffee: Extracting $1/f$ noise coefficients for BJTs, IEEE Trans. Electron Devices **ED-41** (1994) pp. 1981–1991
46. C. Delseny, F. Pascal, S. Jarrix, G. Lecoy, J. Dangla, C. Dubon-Chevallier: Excess noise in AlGaAs/GaAs heterojunction bipolar transistors and associated TLM test structures, IEEE Trans. Electron Devices **ED-41** (1994) pp. 2000–2005
47. D. H. Held, A. R. Kerr: Conversion loss and noise of microwave and millimeter-wave mixers: Part I — Theory, IEEE Trans. Microwave Theory Tech. **MTT-26** (1978) pp. 49–55
48. H. Siweris, B. Schiek: Analysis of noise upconversion in microwave FET oscillators, IEEE Trans. Microwave Theory Tech. **MTT-85** (1985) pp. 233–242
49. V. Rizzoli, F. Mastri, D. Masotti: General noise analysis of nonlinear microwave circuits by the piecewise harmonic-balance technique, IEEE Trans. Microwave Theory Tech. **MTT-42** (1994) pp. 807–819
50. F. N. Hooge: $1/f$ noise is no surface effect, Phys. Lett. **29A** (1969) p. 139
51. Sh. Kogan: *Electronic noise and fluctuations in solids* (Cambridge University Press, Cambridge, 1996)
52. T. G. M. Kleinpenning, A. H. de Kuijper: Relation between variance and sample duration of $1/f$ noise signals, J. Appl. Phys. **63** (1988) p. 43
53. M. Keshner: $1/f$ noise, Proc. IEEE **70** (1982) pp. 212–218
54. L. K. J. Vandamme: Is the $1/f$ noise parameter α a constant?, In *Proc. Int. Conf. on Noise in Physical Systems and 1/f Noise*, ed. by M. Savelli, G. Lecoy, J. P. Nougier (North-Holland, Amsterdam, 1983) pp. 183–192
55. A. L. McWhorter: *$1/f$ noise and related surface effects in germanium*, Rep. 80 (Lincoln Lab, 1955)
56. M. Kogan, K. E. Nagaev: On the low-frequency current $1/f$ noise in metals, Solid State Commun. **49** (1984) p. 387
57. N. Giordano: Defect motion and low-frequency noise in disordered metals, Rev. Solid State Science **3** (1989) p. 27
58. J. Pelz, J. Clarke: Quantitative local interference model for $1/f$ noise in metal films, Phys. Rev. B **36** (1987) p. 4479
59. P. H. Handel: $1/f$ noise – an infrared phenomenon, Phys. Rev. Lett. **34** (1975) p. 1492

60. C. M. van Vliet: A survey of results and future prospects on quantum $1/f$ noise and $1/f$ noise in general, Solid-State Electron. **34** (1991) p. 1
61. R. Kubo: Statistical mechanical theory of irreversible processes – I. General theory and simple applications to magnetic conduction problems, J. Phys. Soc. Japan **12** (1957) p. 570
62. P. Vasilopoulos, C. M. van Vliet: Master hierarchy of kinetic equations for binary interaction in the adiabatic approximation, Can. J. Phys. **61** (1983) pp. 102–112
63. L. O. Chua, C. A. Desoer, E. S. Kuh: *Linear and nonlinear circuits* (McGraw-Hill International, New York, 1987)
64. P. Russer, S. Müller: Noise analysis of linear microwave circuits, Int. J. Num. Mod. **3** (1990) pp. 287–316
65. D. E. Meer: Noise figures, IEEE Trans. Education **32** (1989) pp. 66–72
66. H. Rothe, W. Dahlke: Theory of noisy fourpoles, Proc. IRE **44** (1956) pp. 811–818
67. C. Jacoboni, P. Lugli: *The Montecarlo method for semiconductor device simulation* (Springer-Verlag, Wien, 1990)
68. C. Moglestue: *Monte Carlo simulation of semiconductor devices* (Chapman and Hall, London, 1993)
69. L. Varani, L. Reggiani, T. Kuhn, T. González, D. Pardo: Microscopic simulation of electronic noise in semiconductor materials and devices, IEEE Trans. Electron Devices **ED-41** (1994) pp. 1916–1925
70. P. A. Markowich, C. A. Ringhofer, C. Schmeiser: *Semiconductor equations* (Springer-Verlag, Wien, 1990)
71. C.M. Snowden: *Semiconductor device modelling* (Peregrinus, London, 1988)
72. S. M. Sze: *Physics of semiconductor devices* (John Wiley & Sons, New York, 1981) 2nd edition
73. D. Schroeder: *Model of interface carrier transport for device simulation* (Springer-Verlag, Wien, 1994)
74. S. Maas: *Nonlinear microwave circuits* (Artech House, Norwood, 1988)
75. K. M. van Vliet, A. Friedman, R. J. J. Zijlstra, A. Gisolf, A. van der Ziel: Noise in single injection diodes. I: A survey of methods, J. Appl. Phys. **46** (1975) pp. 1804–1813
76. K. M. van Vliet: General transport theory of noise in *pn* junction-like devices – I: Three-dimensional Green's function formulation, Solid-State Electron. **15** (1972) pp. 1033–1053
77. K. K. Thornber, T. C. McGill, M.-A. Nicolet: Structure of the Langevin and Impedance Field methods of calculating noise in devices, Solid-State Electron. **17** (1974) pp. 587–590
78. V. Gružhinskis, E. Starikov, P. Shiktorov, L. Reggiani, M. Saraniti, L. Varani: Hydrodynamic analysis of dc and ac hot-carrier transport in semiconductors, Semiconductor Sci. Tech. **8** (1993) pp. 1283–1290
79. P. Shiktorov, E. Starikov, V. Gružhinskis, L. Reggiani, T. Gonzàlez, J. Mateos, D. Pardo, L. Varani: Acceleration fluctuation scheme for diffusion noise sources within a generalized impedance field method, Phys. Rev. B **57** (1998) pp. 11866–11869
80. P. Shiktorov, E. Starikov, V. Gružhinskis, T. Gonzàlez, J. Mateos, D. Pardo, L. Reggiani, L. Varani, J. C. Vaissiere, J. P. Nougier: Spatiotemporal correlation of conduction current fluctuations within a hydrodynamic-Langevin scheme, Appl. Phys. Lett. **74** (1999) pp. 723–725
81. V. Gružhinskis, E. Starikov, P. Shiktorov, L. Reggiani, L. Varani: A hydrodynamic approach to noise spectra in unipolar semiconductor structures, Appl. Phys. Lett. **64** (1994) pp. 1662–1664

82. A. Abou-Elnour, K. Schünemann: Two-dimensional noise calculations of submicrometer MESFETs, Microelectronic Engineering **19** (1992) pp. 43–46
83. A. van der Ziel: *Noise in solid-state devices and circuits* (Wiley-Interscience, New York, 1986)
84. H. A. Haus, H. Statz, R. A. Pucel: Noise characteristics of gallium arsenide field-effect transistors, IEEE Trans. Electron Devices **ED-21** (1974) pp. 549–562
85. R. A. Pucel, H. A. Haus, H. Statz: Signal and noise properties of gallium arsenide microwave field-effect transistors, Advances Electron. Electron Phys. **38** (1974) pp. 195–265
86. Y. Ando, T. Itoh: DC, small-signal and noise modelling for two-dimensional electron gas field-effect transistors based on accurate charge-control characteristics, IEEE Trans. Electron Devices **ED-37** (1990) pp. 67–78
87. K. M. van Vliet: The transfer-impedance method for noise in field-effect transistors, Solid-State Electron. **22** (1979) pp. 233–236
88. J. P. Nougier: Origine du bruit dans les dispositifs à semiconducteurs, Revue Phys. Appliquée **22** (1987) pp. 803–819
89. T. M. Brookes: The noise properties of high electron mobility transistors, IEEE Trans. Electron Devices **ED-33** (1986) pp. 52–57
90. Y. Tsividis: *Operation and modeling of the MOS transistor* (McGraw-Hill, New York, 1999) 2nd edition
91. A. J. Scholten, H. J. Tromp, L. F. Tiemeijer, R. van Langevelde, R. J. Havens, P. W. H. de Vreede, R. F. M. Roes, P. H. Woerlee, A. H. Montree, D. B. M. Klaassen: Accurate thermal noise model for deep-submicron CMOS, In *Proc. International Electron Device Meeting*, Washington, USA, 1999, pp. 155–158
92. F. Bonani, G. Ghione, C. U. Naldi, R. D. Schnell, H. J. Siweris: HEMT short-gate noise modelling and parametric analysis of NF performance limits, In *Proc. International Electron Device Meeting*, San Francisco, USA, 1992, pp. 581–584
93. A. van der Ziel: Unified presentation of $1/f$ noise in electronic devices – Fundamental $1/f$ noise sources, Proc. IEEE **76** (1988) pp. 233–258
94. A. Cappy: Noise modelling and measurement techniques, IEEE Trans. Microwave Theory Tech. **MTT-36** (1988) pp. 1–10
95. H. Fukui: Optimal noise figure of microwave GaAs MESFET's, IEEE Trans. Electron Devices **ED-26** (1979) pp. 1032–1037
96. M. W. Pospieszalski: Modelling the noise parameters of MESFETs and MODFETs and their frequency and temperature dependence, IEEE Trans. Microwave Theory Tech. **MTT-37** (1989) pp. 1340–1350
97. B. Hughes: A temperature noise model for extrinsic FETs, IEEE Trans. Microwave Theory Tech. **MTT-40** (1992) pp. 1821–1832
98. F. Danneville, H. Happy, G. Dambrine, J-M. Belquin, A. Cappy: Microscopic noise modeling and macroscopic noise models: how good a connection?, IEEE Trans. Electron Devices **ED-41** (1994) pp. 779–786
99. M. L. Tarng, K. M. van Vliet: General transport theory of noise in *pn* junction-like devices – II: Carrier correlations and fluctuations for high injection, Solid-State Electron. **15** (1972) pp. 1055–1069
100. H. S. Min, K. M. van Vliet, A. van der Ziel: General transport theory of noise in *pn* junction-like devices – III: Junction noise in p^+-n diodes at high injection, Phys. Stat. Sol. (a) **10** (1972) pp. 605–618
101. A. van der Ziel: *Noise: sources, characterization, measurement* (Prentice-Hall, Englewood Cliffs, 1970)

102. G. Ghione, A. Benvenuti: Discretization schemes for high-frequency semiconductor device models, IEEE Trans. Antennas Propagation **45** (1997) pp. 443–456
103. J. Sato-Iwanaga, K. Fujimoto, Y. Ota, K. Inoue, B. Troyanovsky, Z. Yu, R. W. Dutton: Distortion analysis of GaAs MESFETs based on physical model using PISCES-HB, In *Proc. International Electron Device Meeting*, San Francisco, USA, 1996, pp. 163–166
104. K. S. Kundert, A. Sangiovanni-Vincentelli, J. K. White: *Steady-state methods for simulating analog and microwave circuits* (Kluwer Academic Publishers, Boston, 1990)
105. S. Donati, F. Bonani, M. Pirola, G. Ghione: Sensitivity-based optimization and statistical analysis of microwave semiconductor devices through multidimensional physical simulation, Int. J. Microwave and Millimeter-Wave Comput.-Aided Eng. **7** (1997) pp. 129–143
106. F. Bonani, S. Donati, F. Filicori, G. Ghione, M. Pirola: Physics-based large-signal sensitivity analysis of microwave circuits using technological parametric sensitivity from multidimensional semiconductor device models, IEEE Trans. Microwave Theory Tech. **MTT-45** (1997) pp. 846–855
107. R. E. Bank, W. H. Coughran, Jr., W. Fichtner, E. H. Grosse, D. J. Rose, R. K. Smith: Transient simulation of silicon devices and circuits, IEEE Trans. Electron Devices **ED-32** (1985) pp. 1992–2006
108. C. F. Rafferty, M. R. Pinto, R. W. Dutton: Iterative methods in semiconductor device simulation, IEEE Trans. Electron Devices **ED-32** (1985) pp. 2018–2027
109. S. E. Laux: Techniques for small-signal analysis of semiconductor devices, IEEE Trans. Electron Devices **ED-32** (1985) pp. 2028–2037
110. A. Gnudi, P. Ciampolini, R. Guerrieri, M. Rudan, G. Baccarani: Small-signal analysis of semiconductor devices containing generation-recombination centers, In *Proc. Nasecode V*, ed. by J. J. H. Miller (Boole Press, Dublin, 1987) pp. 207–212
111. Y. Lin, L. Wang, M. Obrecht, T. Manku: Quasi-3D device simulation for microwave noise characterization of MOS devices, In *Proc. International Electron Device Meeting*, San Francisco, USA, 1998, pp. 77–80
112. S. W. Director, R. A. Roher: Automated network design – The frequency-domain case, IEEE Trans. Circuit Theory **CT–16** (1969) pp. 330–337
113. R. A. Roher, L. Nagel, R. Meyer, L. Weber: Computationally efficient electronic-circuit noise calculation, IEEE J. Solid State Circ. **SC–6** (1971) pp. 204–212
114. F. Bonani, G. Ghione, M. R. Pinto, R. K. Smith: A novel implementation of noise analysis in general-purpose PDE-based semiconductor device simulators, In *Proc. International Electron Device Meeting*, Washington, USA, 1995, pp. 777–780
115. F. H. Branin: Network sensitivity and noise analysis simplified, IEEE Trans. Circuit Theory **CT–20** (1973) pp. 285–288
116. A. van der Ziel, R. Jindal, S. K. Kim, H. Park, J. P. Nougier: Generation-recombination noise at 77 K in silicon bars and JFETs, Solid-State Electron. **22** (1979) pp. 177–179
117. M. N. Darwish, J. L. Lentz, M. R. Pinto, P. M. Zeitzoff, T. J. Krutsick, H. H. Vuong: An improved electron and hole mobility model for general purpose device simulation, IEEE Trans. Electron Devices **ED-44** (1997) pp. 1529–1537
118. S. Selberherr, W. Hänsch, M. Seavey, J. Slotboom: The evolution of the MINIMOS mobility model, Solid-State Electron. **33** (1990) p. 1425

119. C. Hu, W. Liu, X. Jin: *BSIM3v3.2.2 MOSFET model*, http://www-device.eecs.berkeley.edu/~bsim3 (Berkeley University, 1999)
120. H. C. de Graaf, F. M. Klaassen: *Compact transistor modelling for circuit design* (Springer-Verlag, Wien, 1990)
121. K. K. Hung, P. K. Ko, C. Hu, Y. C. Cheng: A physics-based MOSFET noise model for circuit simulators, IEEE Trans. Electron Devices **ED-37** (1990) pp. 1323–1333
122. B. Wang, J. R. Hellums, C. G. Sodini: MOSFET thermal noise modeling for analog integrated circuits, IEEE J. Solid State Circ. **29** (1994) pp. 833–835
123. D. P. Triantis, A. Birbas, D. Kondis: Thermal noise modeling for short-channel MOSFETs, IEEE Trans. Electron Devices **ED-43** (1996) pp. 1950–1955
124. P. A. Layman, S. G. Chamberlain: A compact thermal noise model for the investigation of soft error rates in MOS VLSI digital circuits, IEEE J. Solid State Circ. **24** (1989) pp. 79–89
125. A. Demir: Floquet theory and non-linear perturbation analysis for oscillators with differential-algebraic equations, Int. J. Circuit Theory and Applications **28** (2000) pp. 163–185
126. A. Demir, A. Mehrotra, J. Roychowdhury: Phase noise in oscillators: a unifying theory and numerical methods for characterization, IEEE Trans. Circuits Syst. I **47** (2000) pp. 655–674
127. B. Troyanovsky: *Frequency domain algorithms for simulating large signal distorsion in semiconductor devices*, PhD thesis (Stanford University, 1997)
128. A. Demir, A. Sangiovanni-Vincentelli: *Analysis and simulation of noise in nonlinear electronic circuits and systems* (Kluwer Academic Publishers, Boston, 1998)
129. C. Dragone: Analysis of thermal and shot noise in pumped resistive diodes, Bell Sys. Tech. J. **47** (1968) pp. 1883–1902
130. R. J. Gilmore, M. B. Steer: Nonlinear circuit analysis using the method of harmonic balance — A review of the art: Part II, Int. J. Microwave and Millimeter-Wave Comput.-Aided Eng. **1** (1991) pp. 159–180
131. V. Rizzoli, A. Lipparini, A. Costanzo, F. Mastri, C. Cecchetti, A. Neri, D. Masotti: State of the art harmonic balance simulation of forced nonlinear microwave circuit by the piecewise technique, IEEE Trans. Microwave Theory Tech. **MTT-40** (1992) pp. 12–28
132. J. Roychowdhury, D. Long, P. Feldmann: Cyclostationary noise analysis of large RF circuits with multitone excitations, IEEE J. Solid State Circ. **33** (1998) pp. 324–336
133. R. B. Bracewell: *The Fourier transform and its applications* (McGraw-Hill, New York, 1986) 2nd edition
134. A. Cappy, F. Danneville, G. Dambrine, B. Tamen: Noise analysis in devices under nonlinear operation, Solid-State Electron. **43** (1999) pp. 21–26
135. B. Troyanovsky, F. Rotella, Z. Yu, R. W. Dutton, J. Sato-Iwanaga: Large-signal analysis of RF/microwave devices with parasitics using harmonic balance device simulation, In *Proc. Sixth Workshop on Synthesis And System Integration of Mixed Technologies*, Fukuoka, Kyushu, Japan, 1996
136. R. W. Dutton, B. Troyanovsky, Z. Yu, T. Arnborg, F. Rotella, G. Ma, J. Sato-Iwanaga: Device simulation for RF applications, In *Proc. International Electron Device Meeting*, Washington, USA, 1997, pp. 301–304
137. B. Troyanovsky, Z. Yu, R. W. Dutton: Physics-based simulation of nonlinear distortion in semiconductor devices using the harmonic balance method, Comput. Methods Appl. Mech. Eng. **181** (2000) pp. 467–482

138. R. Sommet, E. Ngoya: Full implementation of an implicit nonlinear model with memory in a harmonic balance software, IEEE Microwave Guided Wave Lett. **7** (1997) pp. 153–155
139. E. Ngoya, R. Larcheveque: Envelope transient analysis: a new method for transient and steady-state analysis of microwave communication circuits and systems, In *Proc. Microwave Theory & Tech. Symposium*, San Francisco, USA, 1996, pp. 1365–1368
140. S. Donati Guerrieri: *Linear and nonlinear physics-based noise analysis of semiconductor devices through the impedance field method*, PhD thesis (University of Trento, 1999)
141. W. A. Gardner: *Introduction to random processes: with applications to signals & systems* (McGraw-Hill, New York, 1990) 2nd edition
142. W. A. Gardner: *Statistical spectral analysis – A nonprobabilistic theory* (Prentice-Hall, Englewood Cliffs, 1988)
143. W. A. Gardner (Editor): *Cyclostationarity in communications and signal processing* (IEEE Press, Hoes Lane, 1994)
144. N. G. van Kampen: *Stochastic processes in physics and chemistry* (North-Holland, Amsterdam, 1981)

Index[1]

[1] Boldface numbers indicate pages where the corresponding topic is mainly treated.

Printing (Computer to Film): Saladruck Berlin
Binding: Stürtz AG, Würzburg